Fisheries Science Series

This series presents cutting-edge studies and accumulated wisdom from Japan and Asia, with titles addressing many of the aqua-bioscience and aquatic environment fields that form the basis of fishery science. Most of our food originates on land, but we cannot overlook food from the sea. Fishery science provides us with substantial knowledge – knowledge of living organisms in water, their habitats and environment; the knowledge to utilize these organisms; the political and administrative knowledge to organize social activities and systems to distribute fishery products; and a technical and engineering knowledge of ships, fishing equipment, sea ports and harbors. The discipline covers a huge variety of subjects, and each subject includes basic and applicative aspects that are both related and essential to one another. For fishery science to prosper, none of them can be ignored. As a country with a long history of fish eating, Japan has created unique, world-class cultures and technologies in fisheries, aquaculture, aquatic environments, seafood science, and other fishery-related sciences. Through carefully selected state-of-the-art works on topics in the field of fishery science, this series aims to contribute to the development of fishery and the welfare of people around the globe. This is an official book series of the Japanese Society of Fisheries Science.

Katsumi Tsukamoto • Mari Kuroki •
Soichi Watanabe
Editors

Eel Science

Springer

Editors
Katsumi Tsukamoto
The University of Tokyo
Tokyo, Japan

Mari Kuroki
The University of Tokyo
Tokyo, Japan

Soichi Watanabe
The University of Tokyo
Tokyo, Japan

ISSN 2522-0470 ISSN 2522-0489 (electronic)
Fisheries Science Series
ISBN 978-981-99-5691-3 ISBN 978-981-99-5692-0 (eBook)
https://doi.org/10.1007/978-981-99-5692-0

This Springer imprint is published by the registered company Springer Nature Singapore Pte Ltd.
The registered company address is: 152 Beach Road, #21-01/04 Gateway East, Singapore 189721, Singapore

Preface

Eels have always fascinated scientists as interesting research subjects. Even the ancient Greek philosopher and naturalist Aristotle, who set the groundwork for the development of modern science, wondered how eels reproduced in the wild. His quest to understand the great mysteries of eels had inspired creative research and writing activities.

Twenty years have passed since the comprehensive 2003 publication, "Eel Biology" (Katsumi Aida, Katsumi Tsukamoto, Kohei Yamauchi eds.) by Springer. Since that time, eel science has greatly advanced. In particular, the collection of wild Japanese eel eggs along the West Mariana Ridge in the Pacific Ocean in 2009 was a landmark scientific event. This achievement represents the fruit of the tireless passion of Professor Katsumi Tsukamoto, who has led eel research for a long time, along with many other scientists. Other advances include the discovery of a new *Anguilla* species from the Philippines Luzon Island, progress in the phylogenetic origins of Anguilliformes and tropical eel research, and better understanding of adult spawning migrations using the latest telemetry technologies. In the field of applied eel science, artificially spawned Japanese eel larvae were successfully reared to the glass eel stage in 2002, and second-generation artificial hatchlings were born in 2010. At present, successive generations of artificially hatched and reared larvae can be produced, which has led to the emergence of new fields, such as breeding research.

Public concern for the conservation, management, and IUU (illegal, unreported, and unregulated) fishing of eel resources has progressively increased in recent years. Although international trade and fishing regulations have been implemented in various countries and regions, there is an urgent need for more effective measures. In terms of both wildlife and food resources, eels will become increasingly important as living organisms that symbolize the symbiosis between the global environment and human beings.

The background information in "Eel Science" originates from Professor Tsukamoto, who has led international eel research for years. He planned to publish this book as a condensed compilation of his research philosophy and the achievements in eel science over nearly 50 years. However, his decline in health made it difficult to complete the project. Therefore, in line with his wishes, his colleagues took charge of writing chapters in their respective fields of expertise to create this book. Thus, each chapter reflects the authors' views and personalities.

We hope that this book will contribute to a comprehensive understanding of eel science for a wide range of readers and will be used as a guidepost for future studies. As you read throughout this book, a complete understanding of eels has not yet been achieved. Based on the accumulation of research findings, we hope that eel science will have progressed in a few decades to such an extent that the information in this book will be outdated, and that eel populations will be stable and thriving on our planet.

Tokyo, Japan
May 2023

Mari Kuroki
Soichi Watanabe

Contents

Part I
Taxonomy

Chapter 1
Morphology and Taxonomy

Shun Watanabe

Contents

Freshwater eels (*Anguilla* spp.) have been shrouded in mystery in areas where they were present ever since the beginning of human thought because of their unusual bodies compared with other fishes and their mysterious life histories, which surprised Aristotle, one of the great thinkers of ancient Greece. With their slender, snake-like bodies, and barely noticeable scales and gills, it is easy to forget that they are fish. There are relatively few species and only two different sets of body coloration and fin positions, so the body forms of all anguillid eels are generally quite similar, with all northern temperate species of *Anguilla* essentially looking the same in their external appearance. The taxonomy and morphology of freshwater eels has attracted the attention of some ichthyologists though, going back to Kaup (1856), Günther (1870), and Ege (1939), and these previous works were more recently reexamined by Watanabe (2003). The genus was classified into 19 species and subspecies after several studies. Recently, genetic identification has been increasingly used to identify anguillid eels; however, as discussed in this chapter, morphology is still a valuable tool for distinguishing among anguillid eel species worldwide.

S. Watanabe (✉)
Department of Fisheries, Faculty of Agriculture, Kindai University, Nara, Japan
e-mail: swpc@nara.kindai.ac.jp

K. Tsukamoto et al. (eds.), *Eel Science*, Fisheries Science Series,
https://doi.org/10.1007/978-981-99-5692-0_1

1.1 Morphology

Anguilliform eels have a wide range of body forms, but freshwater eels have what is considered to be a typical eel-like body. After metamorphosing from the completely different, laterally compressed, and transparent bodies of their leptocephalus larvae into glass eels, young juveniles (elvers), yellow eels, and adults (silver eels) all have elongated round bodies. The bodies of anguillid eels are moderately elongated, and the anus lies slightly before the midbody (Fig. 1.1a). They do not have pelvic fins, but their pectoral, dorsal, anal, and caudal fins are well-developed. The dorsal and anal fins are confluent with the caudal fin and the tail is flexible and broadly rounded. The dorsal fin begins between the anus and gill openings, and this characteristic varies among species. The pectoral fins are well-developed and broadly rounded. The gill openings are crescentic and on the side of the head, just below the lateral midline. Small and embedded scales are present. The lateral line is complete on the

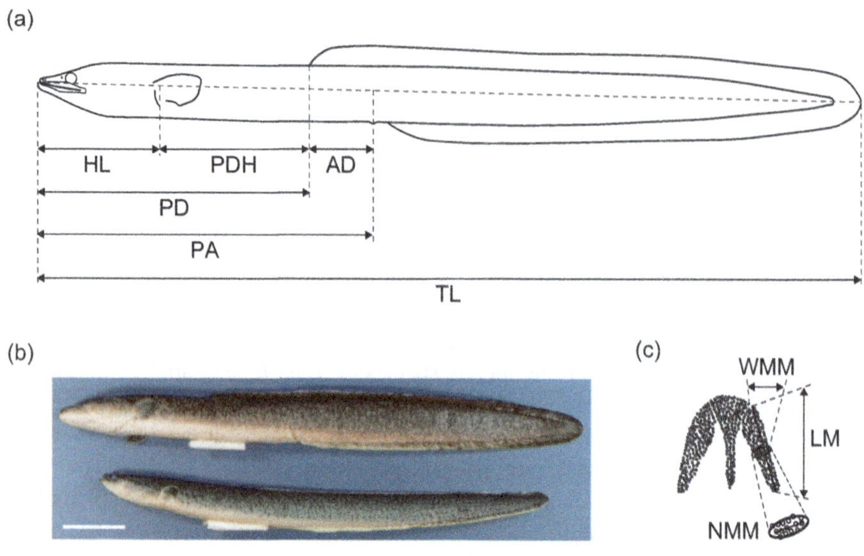

Fig. 1.1 The morphometric characters for body proportions of anguillid eels (**a**) (modified from Ege 1939, Fig. 1). TL: total length, measured as the distance from the tip of the lower jaw, mouth closed, to the end of the tail. HL: head length, measured as the distance from the tip of the lower jaw to lower point of the gill-opening. PD: the predorsal length, measured as the distance from the tip of the lower jaw to the vertical line through the origin of the dorsal fin. PA: preanal length, measured as the distance from the tip of the lower jaw to the vertical line through the anus. PDH: distance between the lower point of the gill-opening and origin of the dorsal fin, which is calculated as PD minus HL. AD: distance between the verticals through the anus and origin of the dorsal fin, which is calculated as PA minus PD. The difference in the marbled color pattern between *Anguilla marmorata* (upper specimen) and *A. reinhardtii* (lower specimen) (**b**). White scale bar shows 50 mm. New dentition characters defined by Watanabe et al. (2004) (modified from Watanabe et al. 2004, Fig. 1c) (**c**). LM: length of the left maxillary band. WMM: width of the mid part of the maxillary band. NMM: number of teeth of the mid part of the maxillary band

head and body, and the snout is relatively short and moderately acute to rounded in the lateral profile. The lower jaw projects beyond the upper jaw, and a well-developed flange is present on the upper and lower lips. The eyes are relatively large and become larger during the reproductive-migration silver eel stage. The anterior nostril tubulars are located near the tip of the snout and are oriented anterolaterally. The posterior nostrils are round and in front of the mid-eye. The teeth are small and granular in bands in the maxilla, vomer, and mandible.

The morphology of each species in the genus *Anguilla* is very similar, and it can be difficult to see differences among them upon first observing them. However, characteristics such as the presence or absence of variegated body markings, types of maxillary bands of the teeth, positions of the origin of the dorsal fin, and number of vertebrae are important morphological characteristics for identifying or distinguishing species (Table 1.1).

Several external body proportions have been used as morphological characters to distinguish species following the measurements used by Ege (1939), which are presently taken on the left side of specimens (Fig. 1.1a). The preanal length is measured as the distance from the tip of the lower jaw, with the mouth closed, to the vertical point through the anus; the predorsal length is measured from the tip of the lower jaw to the vertical point through the origin of the dorsal fin; the length of the head is the distance from the tip of the lower jaw to the lower point of the gill-opening. Important proportional morphological characteristics include the distance between the verticals through the anus and the origin of the dorsal fin (AD), and pre-dorsal length without the head (PDH) to the total length (TL). Ege (1939) referred to these characters as being related to the position of the dorsal fin and a long or short dorsal fin.

There are two main color patterns in the yellow eel stage of freshwater eels: plain (without variegated body markings) and marbled (with variegated body markings). The term yellow eel, which comes from the yellowish hue of their bodies, refers to the juvenile growth stage in their life history. The plain pattern is black, gray, or brown on the dorsal side and white on the ventral side, with a less distinct separation between the yellow-green coloring on the dorsal and ventral sides. The marbled color pattern and variegated body marking is referred to as a "mottled" skin coloration. Ege (1939) described that the dark colored areas in *A. reinhardtii* are more uniform in size, more isolated, round, or slightly oval without many spurs (Fig. 1.1b) compared with *A. marmorata*, which is a species in the same morphological group as *A. reinhardtii*. Tsutsui et al. (2019) determined the genomic DNA sequences of the mucosal galectin-encoding genes in all species and subspecies of the genus *Anguilla*. The nucleotide sequences of the galectin genes were ~2.3–2.5 kb long and the organization of their four exons and three introns was conserved in all species. An unusual sequence was found in the fourth exon of *A. reinhardtii*, resulting in a unique deduced amino acid sequence at the C-terminus. This indicates that the marbled color pattern and galectin genes in the skin of *A. reinhardtii* are unique compared with those in other anguillid eels.

These body colors change during development and are not present in eel larvae (leptocephali). Glass eels are transparent until they enter freshwater, which is when

Table 1.1 List of all anguillid eel species and their common names worldwide that are likely valid species and subspecies (following Tsukamoto et al. 2020), showing the eel species and subspecies recognized by the important revision of the genus *Anguilla* by Ege (1939) and more recent changes or new species with their important morphological characters and geographic regions

Species name		Common name	Morphological characters					Region
According to Ege (1939)	Names recognized in recent years		Variegated skin	Band of maxillary teeth	Type of dorsal fin	Range of AD/TL	Range of TV	
Anguilla celebesensis Kaup 1856	*A. celebesensis* Kaup 1856	Celebes eel	Presence	Broad	Longfin	6.3–12.8	101–106	Tropical
A. interioirs Whitley 1938	*A. interioirs* Whitley 1938	New Guinea eel	Presence	Broad	Longfin	7.0–14.4	101–106	Tropical
A. megastoma Kaup 1856	*A. megastoma* Kaup 1856	Polynesian longfin eel	Presence	Broad	Longfin	7.5–13.5	107–116	Tropical
	A. luzonensis Watanabe et al. 2009a	Luzon eel	Presence	Broad	Longfin	9.3–13.9	103–107	Tropical
A. ancestralis Ege 1939								
	A. bengalensis (Gray 1831)	Bengal eel	Presence	Narrow	Longfin	8.3–14.1	106–115	Tropical
A. nebulosa nebulosa McClelland, 1844	*A. bengalensis bengalensis* (Gray 1831)	Indian Bengal eel	Presence	Narrow	Longfin	8.3–13.1	106–113	Tropical
A. nebulosa labiata (Peters, 1852)	*A. bengalensis labiata* (Peters, 1852)	African Bengal eel	Presence	Narrow	Longfin	9.9–14.1	107–115	Tropical
A. marmorata Quoy and Gaimard, 1824	*A. marmorata* Quoy and Gaimard, 1824	Indo-Pacific eel	Presence	Narrow	Longfin	12.4–19.6	100–110	Tropical
A. reinhardtii Steindachner, 1867	*A. reinhardtii* Steindachner, 1867	Australian longfin eel	Presence	Narrow	Longfin	8.3–13.4	104–110	Tropical
A. borneensis Popta, 1924	*A. borneensis* Popta, 1924	Borneo eel	Absence	–	Longfin	9.7–13.4	103–108	Tropical
A. japonica Temminck and Schlegel, 1846	*A. japonica* Temminck and Schlegel, 1846	Japanese eel	Absence	–	Longfin	6.7–12.8	112–119	Temperate

A. rostrata (Lesueur, 1817)	*A. rostrata* (Lesueur, 1817)	American eel	Absence	–	Longfin	6.1–13.0	100–111	Temperate
A. anguilla (Linnaeus, 1758)	*A. anguilla* (Linnaeus, 1758)	European eel	Absence	–	Longfin	8.2–14.4	109–119	Temperate
A. dieffenbachii Gray, 1842	*A. dieffenbachii* Gray, 1842	New Zealand longfin eel	Absence	–	Longfin	9.4–14.4	109–116	Temperate
A. mossambica (Peters, 1852)	*A. mossambica* (Peters, 1852)	Mozambique eel	Absence	–	Longfin	12.3–16.9	100–106	Tropical
	A. bicolor McClelland, 1844	Bicolor eel	Absence	–	Shortfin	4.7–3.3	103–114	Tropical
A. bicolor bicolor McClelland, 1844	*A. bicolor bicolor* McClelland, 1844	Indian bicolor eel	Absence	–	Shortfin	2.4–3.3	103–111	Tropical
A. bicolor pacifica Schmidt, 1928	*A. bicolor pacifica* Schmidt, 1928	Pacific bicolor eel	Absence	–	Shortfin	−4.7–2.9	106–114	Tropical
A. obscura Günther, 1872	*A. obscura* Günther, 1872	Polynesian shortfin eel	Absence	–	Shortfin	2.0–7.6	101–107	Tropical
	A. australis Richardson, 1841	Australian eel	Absence	–	Shortfin	−1.7–5.3	108–116	Temperate
A. australis australis Richardson, 1841	*A. australis australis* Richardson, 1841	Australian shortfin eel	Absence	–	Shortfin	−1.7–3.6	109–116	Temperate
A. australis schmidtii Phillipps, 1925	*A. australis schmidtii* Phillipps, 1925	New Zealand shortfin eel	Absence	–	Shortfin	0.9–5.3	108–115	Temperate

The range of the distance between the verticals through the anus and the origin of the dorsal fin/total length (AD/TL) was taken from Watanabe et al. (2004). The total number of vertebrae (TV) were taken from Ege (1939) and Watanabe et al. (2005a), except *Anguilla luzonensis* (taken from Watanabe et al. 2009a) body pigmentation develops, and the glass eels turn into elvers. Once they grow to a sufficient size, yellow eels turn into silver eels, their eyes grow larger, and their bodies turn black or dark brown with a metallic sheen.

In addition to variations in color patterns, tooth patterns differ among anguillid eel species. Ege (1939) reported that the form and structure of the bands in the upper jaw vary in the presence or absence of a maxillary longitudinal furrow. Other more detailed features vary among species, such as the occurrence of small teeth, particularly in the maxillary bands, the relationship between the breadth of the intermaxillary vomerine and that of the maxillary band, and the relationship between the lengths of these bands. Watanabe et al. (2004) defined two new dentition characteristics (Fig. 1.1c): the width of the mid part of the maxillary band (WMM) to the length of the maxillary band (LM), and the number of teeth in the mid part of the maxillary band (NMM). Tooth bands are difficult to examine and, in some cases, may overlap in shape among species (Aoyama et al. 2001). Watanabe et al. (2004) did not use the characteristic of a toothless longitudinal groove in the maxillary and mandibular bands of teeth used by Ege (1939) because this characteristic was subjective and unclear.

Watanabe et al. (2004) proposed that two groups of freshwater eels could be clearly distinguished based on their color-marking patterns and then further separated by tooth patterns. The relationships between WMM/LM and NMM could be divided into two groups of variegated species (Fig. 1.2), and the relationships between PDH/TL and AD/TL could distinguish between two groups (long or short dorsal fins) of species without variegated body markings (Fig. 1.3). Based on these comparative analyses (Watanabe et al. 2004), the genus *Anguilla* can be divided into four morphological groups (Fig. 1.4):

(a) Variegated body marking with broad maxillary bands of teeth
(b) Variegated body marking with narrow maxillary bands of teeth
(c) Non-variegated body markings with a long dorsal fin
(d) Non-variegated body markings with a short dorsal fin

It was clear that only four groups of freshwater eels could be recognized based on these morphological characters. However, eel species within these four groups have very similar morphological features that make them difficult to distinguish (Watanabe et al. 2004). The total number of vertebrae (TV) is useful (Table 1.1) for separating some species within each group or for species with the same geographic distribution (Ege 1939; Watanabe et al. 2005a), but there are many cases of overlap in this characteristic, and the TV can only be counted using difficult procedures such as X-rays or computed tomography scanning. Furthermore, Aoyama et al. (2001) suggested that color characteristics are adaptive and do not reflect phylogenetic relationships.

Tesch (2003) divided freshwater eels into tropical and temperate species according to the geographic distribution of their growth habitats. Tropical and temperate eels consist of 11 and 5 species, respectively. Seven of the 11 tropical species have variegated body markings, live in tropical areas, and do not live in

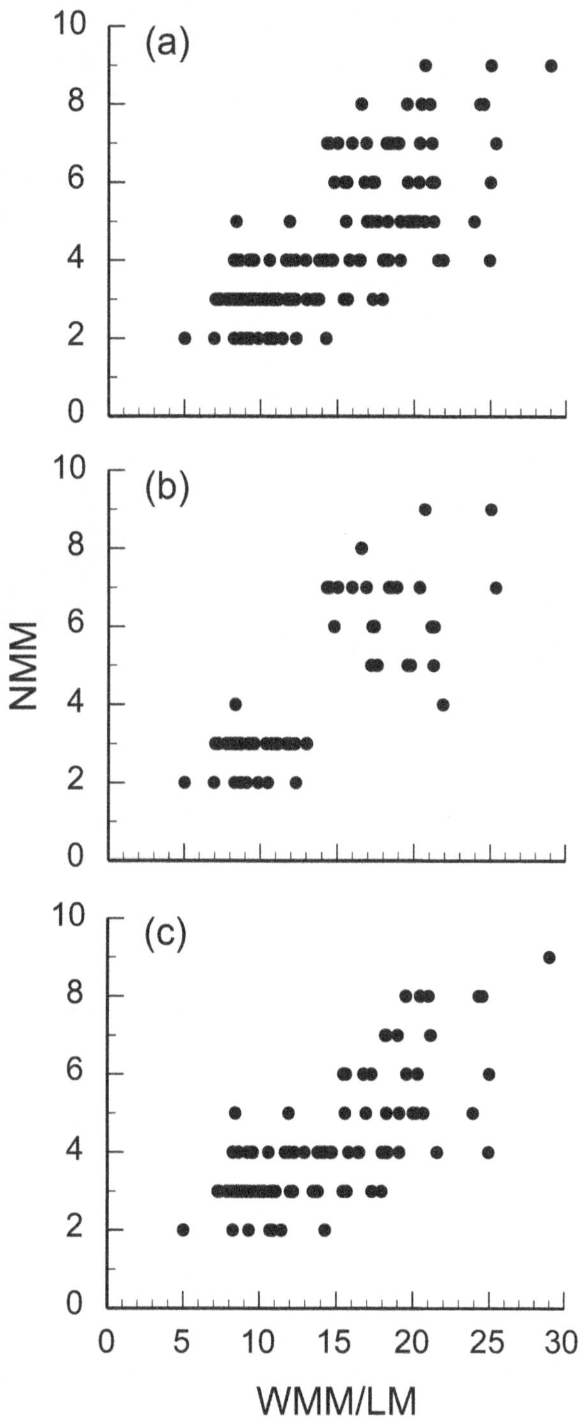

Fig. 1.2 Relationship between the dentition characters: width of the mid part of the maxillary band/ length of the maxillary band (WMM/LM) and number of teeth of the mid part of the maxillary band (NMM). All groups (N = 168) (**a**); specimens with variegated body marking groups (N = 74) (**b**); specimens without variegated body marking groups (N = 94) (**c**) (modified from Watanabe et al. 2004, Fig. 6)

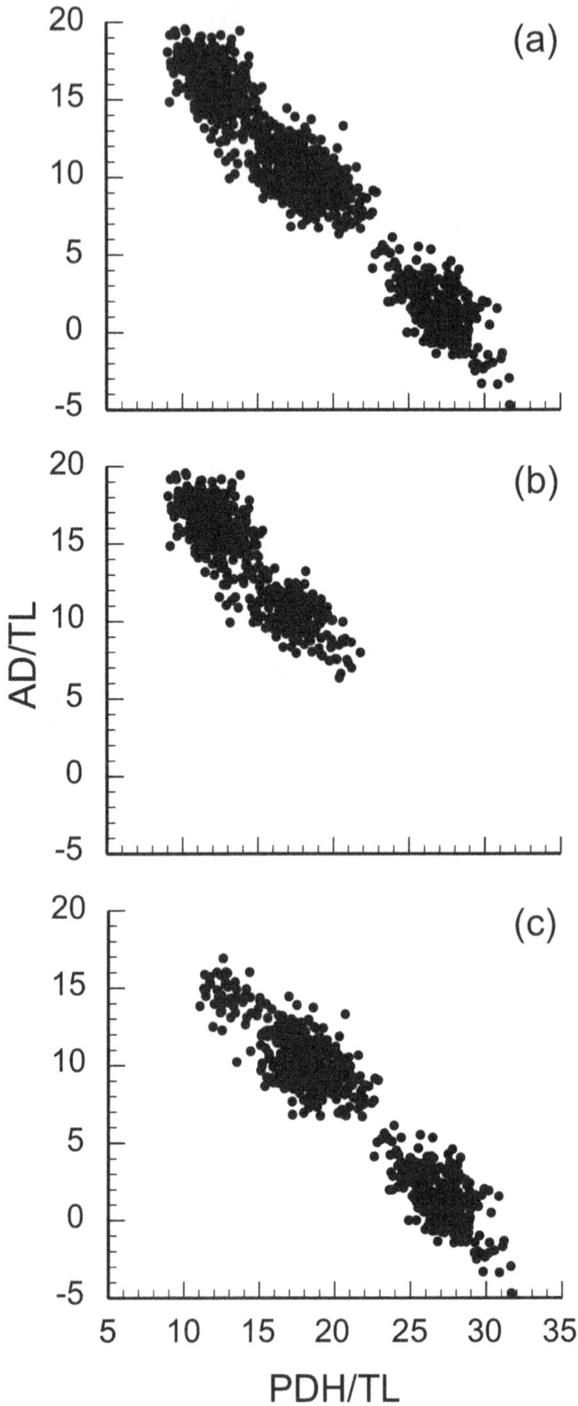

Fig. 1.3 The relationship between the dorsal fin characters: distance between the lower point of the gill-opening and origin of the dorsal fin/total length (PDH/TL) and the distance between the verticals through the anus and origin of the dorsal fin (AD)/TL. All groups (N = 1614) (**a**); specimens with variegated body marking groups (N = 763) (**b**); specimens without variegated body marking groups (N = 851) (**c**) (modified from Watanabe et al. 2004, Fig. 5)

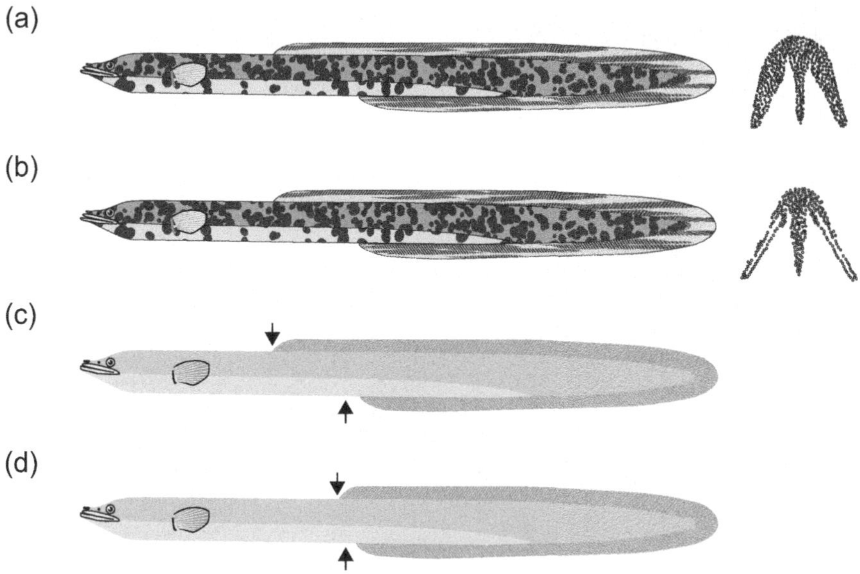

Fig. 1.4 Morphological characteristics of the four groups of anguillid eels (modified from Watanabe et al. 2004, Fig. 5). Arrows show the position of the origin of the dorsal fin and anus

temperate areas, except for a few *A. marmorata* that are present in parts of southern mainland Japan. Variegated body markings may be a morphological adaptation to clear streams in tropical islands that serve as a form of camouflage.

1.2 Taxonomy

Freshwater eels of the family Anguillidae are members of the order Anguilliformes, which includes 16 families, 156 genera, and more than 1000 species worldwide, with new species still being described (Fricke et al. 2022). The only comprehensive revisions of the genus *Anguilla* Schrank, 1798 were made by Kaup (1856), Günther (1870), and Ege (1939) using morphological analyses. In the last revision by Ege (1939), the genus was classified into 16 species, three of which were divided into two subspecies. However, 35 years later and based on morphological analyses, Castle and Williamson (1974) reported that *A. ancestralis* Ege 1939, which was described by Ege (1939) using only glass eels, was synonymous with *A. celebesensis* Kaup 1856.

Two new species, *A. luzonensis* Watanabe, Aoyama and Tsukamoto, 2009 (Watanabe et al. 2009a) and *A. huangi* Teng, Lin and Tzeng, 2009 (Teng et al. 2009) were described from the Cagayan River system on northern Luzon Island in the Philippines and from glass eels collected from the Cagayan River estuary in the Philippines, respectively, the latter of which were reared in a culture pond in Taiwan.

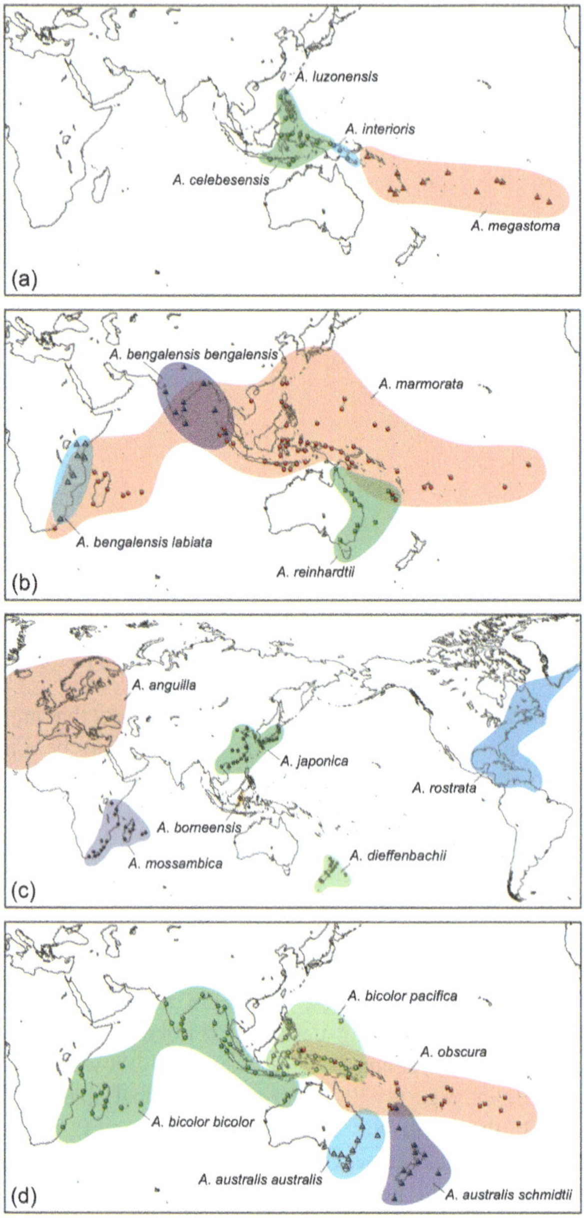

Fig. 1.5 Geographical distribution of each species and subspecies from four morphologically divided groups of the genus *Anguilla*, indicating the distribution of the North Atlantic eels (*A. anguilla* and *A. rostrata*) by Schmidt (1909), 13 species reported by Ege (1939), and *Anguilla luzonensis* reported by Watanabe et al. (2009a) (modified from Tsukamoto et al. 2020, Fig. 1)

Comparisons of their morphological and molecular genetic characteristics clearly showed that *A. luzonensis* and *A. huangi* are the same species. *Anguilla luzonensis* is a valid species name and *A. huangi* is a junior synonym of *A. luzonensis* (Watanabe et al. 2013). The genus *Anguilla* comprises 16 species, three of which are further divided into two subspecies (Table 1.1; Fig. 1.5).

The Borneo eel *A. borneensis* Popta, 1924, which is endemic to eastern Borneo Island in Malaysia and Indonesia, has been widely recognized as a valid species name since Ege (1939). However, Bauchot et al. (1993) revived the name *A. malgumora* Kaup, 1856 as a senior synonym of *A. borneensis*, according to the principle of priority, but without morphological examination. Smith (1999) also listed *A. malgumora* Schlegel in Kaup (1856) as the Borneo eel in his species identification guide, and the name *A. malgumora* Kaup was treated as valid in other studies (e.g., Lin et al. 2005; Teng et al. 2009). Ege (1939) proposed that *A. malgumora* Schlegel should be treated as a synonym of the European eel *A. anguilla* (Linnaeus 1758). Silfvergrip (2009) argued that the Borneo eel had been erroneously cited as *A. malgumora* Kaup.

To resolve this inconsistency and to determine the valid name of the Borneo eel that has plain body coloration and is endemic to Borneo Island, the morphological characteristics of three type specimens of *A. borneensis* Popta, 1924, *A. malgumora* Kaup 1856, and *Muraena malgumora* Bleeker, 1864 were examined and compared with 42 longfin eels from Borneo, *A. anguilla* (N = 44), *A. japonica* (N = 37), and *A. dieffenbachii* (N = 14), all of which have a long dorsal fin and plain body color with the exception of *M. malgumora*, which has a short dorsal fin (Watanabe et al. 2014a). The discriminant analysis by Watanabe et al. (2014a) used 13 proportional characters and the relationship among three vertebral characters to clearly show that the type specimen of *A. malgumora* Kaup 1856, upon which Bauchot et al. (1993) proposed the revival of the name to replace *A. borneensis* Popta, 1924, based on priority without examining the specimen, was clearly not a Borneo eel, as was previously concluded by Ege (1939). Therefore, *A. borneensis* Popta, 1924 is a valid name for the Borneo eel, *A. malgumora* Kaup 1856 is considered a junior synonym of *A. anguilla* (Linnaeus 1758), and *M. malgumora* Bleeker is a junior synonym of *A. bicolor* McClelland, 1844.

Ege (1939) used the name *A. nebulosa* McClelland, 1844 to describe the two subspecies. However, the species name *A. bengalensis* (Gray 1831) was recently used by several authors (Watanabe and Miller 2012; Arai 2016) because, according to the principle of priority in zoological nomenclature, *A. bengalensis* (Gray 1831) is the valid species name, while *A. nebulosa* McClelland, 1844 is a junior synonym of this. Gray (1831) presented a sketch of the origin of the dorsal fin in *A. bengalensis*, which is further backward, distinguishing this species from *A. marmorata*.

These taxonomic issues are not isolated occurrences for anguillid eels; however, 152 species of anguillid eels have been reported to date (Fricke et al. 2022). Classification of the synonyms of these species names will also be necessary in the future.

1.3 Common Names

Although the scientific names of anguillid eels have been established, the common names of some species have varied over time. Scientific naming rules for animals are strictly defined by the International Code of Zoological Nomenclature. However, the procedure for selecting common names is vague and not well defined. Therefore, the use of common names has caused confusion among countries and regions. Anguillid eels are an important food and cultural resource worldwide (Kuroki et al. 2014a). The endangered status of some *Anguilla* species has been frequently broadcast in the mass media; hence, common names have also been used frequently. In addition, the shortage of temperate eel resources has led to the exploitation of tropical eels in temperate regions. Such eel resources were previously consumed at the local level only, and far less scientific research has been conducted on tropical eels than on temperate eels. These factors have resulted in an urgent need standardizing common names, which is important for ensuring the recognition of all *Anguilla* species among citizens and consumers. To standardize common names for both scientists and society in general, Tsukamoto et al. (2020) proposed the application of defined rules for determining the common names of animals using the common names of freshwater eels in the genus *Anguilla* as an example (Table 1.1), many of which have become confused in the scientific literature, pictorial books, and online resources in recent years. The system uses a combination of five factors to determine the most appropriate common names: the geographic distribution or location, distinguishing shortfin or longfin species, using a name helps recall the scientific name, valuing longstanding names if they adhere to the previous three factors, and using the shortest possible name. The only common name that was consistently used in the same way in the past was the "giant mottled eel" for *A. marmorata*, which was changed to the "Indo-Pacific eel" because it is the only eel species that is present across the entire Indo-Pacific.

1.4 Migration Loops and the Species Concept

A unique aspect of tropical and temperate species of anguillid eels is that they use distinct locations as spawning areas, which were recognized by Tsukamoto et al. (2002) as part of the establishment of these species. The adults migrate to the spawning areas, while the larvae move back to the recruitment areas or juvenile growth habitats of the species; this is referred to as the migration loop. Therefore, speciation of the two Atlantic eels occurred when the larvae of the American eel *A. rostrata* and European eel *A. anguilla* diverged by creating different migration loops, with the larvae of *A. rostrata* moving west to recruit to North America and further south, and *A. anguilla* larvae moving to the east side of the North Atlantic. This larval separation seems to require late-stage swimming by leptocephali to complete the two different migration loops (Miller and Tsukamoto 2017), which is essential for the creation of the two closely related species.

The concept can be applied to other species, particularly subspecies such as *A. bicolor*, which spawns in different ocean basins (Minegishi et al. 2012), and *A. australis*, which recruits to different areas (Australia and New Zealand) (Jellyman 1987). It is also important for widespread species, such as *A. marmorata*, which have different populations. The Northern Population of *A. marmorata* has been found to spawn in an overlapping area with the Japanese eel *A. japonica*, but based on genetic data, there is another population in the North Pacific that lives further east in the Pacific Islands region (Minegishi et al. 2008). The spawning locations of this species in the Indian Ocean are unknown; however, the eels there form a separate population. In addition, the distance between the western landmasses in the western South Pacific (WSP) and the many small islands of French Polynesia in the east suggests that *A. marmorata* likely has different spawning populations with different migration loops in both the east and west, which is likely true for other similarly widespread species of *A. megastoma*, as suggested by vertebral counts (Watanabe et al. 2011). Genetic studies have mostly used mtDNA, which, due to maternal inheritance, is not suitable for population studies on anguillid eels that have a long larval duration, which can cause occasional mixing of larvae. Therefore, nuclear DNA studies are required to understand the migration loops and potential taxonomic implications of the population structure of widespread anguillid species.

Another factor is that the spawning locations of tropical anguillids are known to include short-distance migrations for local spawning in the Indonesia region (Miller and Tsukamoto 2017), so the types of migration loops used by tropical WSP species are not yet known (Schabetsberger et al. 2021). Even if understanding the types of migration loops and the presence of subpopulations within existing species is difficult to define taxonomically in relation to the species concept because of probable gene exchange among populations over time, identifying these spawning populations with distinct migration loops is important for taxonomy, conservation, and management. Therefore, further research is needed.

1.5 Taxonomic Problems and Geographic Factors

After finding the spawning areas of Atlantic eels, Johannes Schmidt conducted a round-the-world voyage in search of other anguillid spawning areas by collecting leptocephali from the seas worldwide (Schmidt 1935). The expedition of Dana II in 1923–1930 spurred dramatic advances in eel biology. Schmidt and his colleagues gathered freshwater eels at each port they visited, which made it possible for their colleagues to make the most important advancements in the morphology and taxonomy of both the larval and yellow eel stages of freshwater eels. After Schmidt's death, two of his younger colleagues, Vilhelm Ege and Poul Jespersen, wrote scholarly works on the taxonomy of freshwater eels and the distribution of Indo-Pacific leptocephali of the genus *Anguilla* (Ege 1939; Jespersen 1942). Ege's taxonomy was a monumental achievement and likely considered eel life history, including part of the migration loop concept, from the perspective of the importance of the catadromy of anguillid species because the validity of Ege's taxonomy to separate species depended somewhat on where each species was distributed.

The use of geographic distribution as an essential characteristic of each species in Ege's taxonomy of the genus *Anguilla* is a primary problem because the location of origin is unknown for some specimens, which then cannot be strictly identified (Watanabe et al. 2004). In addition, there are many cases of partial sympatric distribution between two species, for which all morphological characters overlap (Table 1.1). In these cases, the specimens collected from the place of sympatric distribution cannot be confidently identified as a particular species. Another potential problem is that knowledge of the geographic distribution of species has increased dramatically since the study by Ege (1939). Passive transport of leptocephali for long periods of 3 months or more can easily change the recruitment range of glass eels, which means that the geographic distribution of species can change year by year. There have been many new findings regarding the geographic distribution of freshwater eels since the Ege (1939) monograph, such as the well-known cases of *A. rostrata* in Iceland, *A. reinhardtii* in New Zealand (Jellyman et al. 1996), and *A. marmorata* in the Galápagos Islands (McCosker et al. 2003). Furthermore, various reports have documented the introduction of non-native species of freshwater eels in several areas of the world as a result of international trade in glass eels for rearing yellow eels in aquaculture (see Okamura et al. 2008).

Watanabe et al. (2005a) examined molecular genetic data in combination with morphological information to evaluate the taxonomy of Ege. These results confirmed that 15 taxa belonged to the genus *Anguilla* and were the same species proposed by Ege (1939). Molecular genetic characteristics are useful for understanding the taxonomy of this genus and can replace geographical distribution as a taxonomic character in Ege's (1939) taxonomy. The only complete morphological key for species identification of the genus *Anguilla* was provided by Ege (1939), but this has been suggested to be unsatisfactory because of critical underestimates of intraspecific variation among characters (Aoyama et al. 2000b). Aoyama et al. (2000b) suggested that 16S rRNA sequences were appropriate for identifying eels at the species level. It is possible to identify any eel specimen using genetic characteristics, regardless of the stage of growth (eggs, leptocephali, glass eels, elvers, yellow eels, and silver eels). This DNA identification method is now routinely used to study anguillid leptocephali in the Atlantic and Indo-Pacific regions.

The advantage of morphological identifications is that it is easy, while the disadvantages are that it involves problems of overlapping characters among different species and growth stages in their life history, and for samples with unknown origins, as described above. The advantage of molecular identification is that it is accurate regardless of growth stage and geographical distribution; however, it requires considerable labor in the laboratory.

Another important issue that has been carefully considered in recent years is the distinctiveness and levels of genetic divergence of the three subspecies and multiple regional populations of *A. marmorata*. At least 6 subspecies appear to be morphologically and genetically distinct (e.g., Watanabe 2003; Watanabe et al. 2005b, 2006, 2008b, 2014b; Shen and Tzeng 2007; Minegishi et al. 2012), but considering the multiple populations that have now been found within *A. marmorata* (see Ishikawa et al. 2004; Minegishi et al. 2008; Watanabe et al. 2008a, 2009b) and *A. megastoma* (Watanabe et al. 2011), the taxonomy of these subspecies and populations may require further evaluation, as mentioned above. There have been questions about how to approach the use of species or subspecies as taxonomic categories and populations as biological or ecological units for the taxonomy of the genus *Anguilla*, which has unique catadromy and migration loops. The two subspecies of *A. bicolor* appear to have clearly separated distribution areas and likely spawning areas; however, the two subspecies of *A. australis* may spawn in a similar region but would require different larval behaviors to reach either Australia or New Zealand. The two subspecies of *A. bengalensis* are separated on the east and west sides of the northern Indian Ocean, but little is known about their distribution and life history.

Another factor that must be considered when morphologically or genetically identifying anguillid eels in certain regions of the world is the possibility of hybridization between two closely related species. The two North Atlantic eel species, *A. anguilla* and *A. rostrata*, spawn in a partially overlapping area of the Sargasso Sea, providing ample opportunities for interbreeding. Both eels are known to hybridize; however, hybrids have been observed almost exclusively in Iceland (see Pujolar et al. 2014). In the Indo-Pacific, Barth et al. (2020) recently analyzed the possible levels of historical hybridization/introgression of 7 species (*A. marmorata*, *A. megastoma*, *A. obscura*, *A. luzonensis*, *A. bicolor*, *A. interioris*, and *A. mossambica*). They used genome-wide sequence data from more than 450 individuals sampled across the tropical Indo-Pacific, morphological information, and three newly assembled draft genomes to compare contemporary hybridization patterns with the signatures of past introgression across a time-calibrated phylogeny. The analyses showed that the seven species appeared to have remained distinct for up to ten million years and current hybridization frequencies across species pairs differed from the genomic signatures of past introgressions. These results show that the evolution of eels involves not only divergence but also convergence through hybridization. Information about hybrids provide interesting insights into how present-day species and subspecies of the genus *Anguilla* came into existence through genetic intermixing over time.

Presently, in eel science, subspecies are treated as a taxonomic category, and populations as biological or ecological units are treated almost equally. If more ecological and genetic information indicates that each migration loop is clearly distinguished from other subspecies or populations, they might be considered as species in the taxonomy of anguillid eels.

1.6 Perspectives on the Morphology and Taxonomy of Anguillid Eels

In the years after studies on the morphological characteristics of eel species, genetic identification tools have increased, and other types of morphological studies have been conducted. After the discovery of the new species, *A. luzonensis*, it seems possible that other cryptic species may exist in poorly studied regions. For example, while *A. interioris* is endemic to Papua New Guinea (Ege 1939; Aoyama et al. 2000a), Kuroki et al. (2014b) reported that leptocephali of this species were collected from both the Indonesian Seas and along the Indonesian coast of the eastern Indian Ocean. However, there is limited ecological information or data on the geographical distribution of this species. Other studies have examined morphological characteristics, such as the detailed sequential development and pigmentation of glass eels (e.g., Tesch 2003; Fukuda et al. 2013), but many species have yet to be studied carefully for these characteristics. The development of artificial spawning and larval rearing of anguillid eels has also enabled morphological studies, such as observing ontogenetic patterns of ossification during metamorphosis into the larval stage (Masuda et al. 2019). Another development is that some species, such as *A. japonica* and *A. anguilla*, seem to have two morphologically distinct head shapes (Kaifu et al. 2013) that also appear in glass eels (De Meyer et al. 2018), suggesting these differences may represent two different genotypes or a form of adaptive phenotypic plasticity for different feeding modes that are likely adapted for different types of prey. These recent studies and the recent finding of a new species in the genus *Anguilla* indicate that there is more to be learned about the morphology and taxonomy of these unique eels that live in freshwater and estuaries in many parts of the world.

References

Aoyama J, Watanabe S, Ishikawa S, Nishida M, Tsukamoto K (2000a) Are morphological characters distinctive enough to discriminate between two species of freshwater eels, *Anguilla celebesensis* and *A. interioris*? Ichthyol Res 47:157–161. https://doi.org/10.1007/BF02684236

Aoyama J, Watanabe S, Nishida M, Tsukamoto K (2000b) Discrimination of catadromous eel species, genus *Anguilla*, using PCR-RFLP analysis of the mitochondrial 16SrRNA domain. Trans Am Fish Soc 129:873–878. https://doi.org/10.1577/1548-8659(2000)129%3C0873:DOCEOG%3E2.3.CO;2

Aoyama J, Nishida M, Tsukamoto K (2001) Molecular phylogeny and evolution of the freshwater eel, genus *Anguilla*. Mol Phylogenet Evol 20:450–459. https://doi.org/10.1006/mpev.2001.0959

Arai T (2016) Taxonomy and distribution. In: Arai T (ed) Biology and ecology of anguillid eels. CRC Press, pp 1–20

Barth JMI, Gubili C, Matschiner M, Trresen OK, Watanabe S, Egger B, Han Y-S, Feunteun E, Sommaruga R, Jehle R, Schabetsberger R (2020) Stable species boundaries despite ten million years of hybridization in tropical eels. Nat Commun 11:1433. https://doi.org/10.1038/s41467-020-15099-x

Bauchot ML, Desoutter M, Castle PHJ (1993) Catalogue of type specimens of fishes in the Muse'um national d'Histoire naturelle, Paris. Anguilliformes and Saccopharyngiformes. Cybium 17:91–151. https://sfi-cybium.fr/fr/catalogue-critique-des-types-de-poissons-du-mus%C3%A9um-national-dhistoire-naturelle-suite-ordres-des. Accessed 17 May 2023

Castle PHJ, Williamson GR (1974) On the validity of the freshwater eel species *Anguilla ancestralis* Ege from Celebes. Copeia 1974:569–570. https://doi.org/10.2307/1442564

De Meyer J, Wassenbergh SV, Bouilliart M, Dhaene J, Adriaens D (2018) Built to bite? Differences in cranial morphology and bite performance between narrow- and broad-headed European glass eels. J Morphol 279:349–360. https://doi.org/10.1002/jmor.20776

Ege V (1939) A revision of the genus *Anguilla* Shaw, a systematic, phylogenetic and geographical study. Dana Rep 16:1–256

Fricke R, Eschmeyer WN, van der Laan R (eds) (2022) Eschmeyer's catalog of fishes: genera, species, references. http://researcharchive.calacademy.org/research/ichthyology/catalog/fishcatmain.asp. Electronic version accessed 26 June 2022

Fukuda N, Miller MJ, Aoyama J, Shinoda A, Tsukamoto K (2013) Evaluation of the pigmentation stages and body proportions from the glass eel to yellow eel in *Anguilla japonica*. Fish Sci 79: 425–438. https://doi.org/10.1007/s12562-013-0621-x

Gray JE (1831) Illustrations of Indian zoology, chiefly selected from the collection of Major-General Hardwicke, F. R. S, vol 2. Adolphus Richter, London. 202 pls

Günther A (1870) Catalogue of the fishes in the British Museum, vol VIII. The Trustees, London

Ishikawa S, Tsukamoto K, Nishida M (2004) Genetic evidence for multiple geographic populations of the giant mottled eel *Anguilla marmorata* in the Pacific and Indian oceans. Ichthyol Res 51: 343–353. https://doi.org/10.1007/s10228-004-0241-7

Jellyman DJ (1987) Review of the marine life history of Australasian temperate species of *Anguilla*. Am Fish Soc Symp 1:276–285

Jellyman DJ, Chisnall BL, Dijkstra LH, Boubee JAT (1996) First record of the Australian longfinned eel, *Anguilla reinhardtii*, in New Zealand. Mar Freshw Res 47:1037–1040. https://doi.org/10.1071/MF9961037

Jespersen P (1942) Indo-Pacific leptocephali of the genus *Anguilla*. Dana Report 22:1–128

Kaifu K, Yokouchi K, Miller MJ, Aoyama J, Tsukamoto K (2013) Head-shape polymorphism in Japanese eels *Anguilla japonica* in relation to differences of somatic growth in freshwater and brackish habitats. J Fish Biol 82:1308–1320. https://doi.org/10.1111/jfb.12070

Kaup JJ (1856) Catalogue of apodal fish in the collection of the British Museum, London. British Museum, London. 163 pp, 19 pls

Kuroki M, Righton D, Walker AM (2014a) The importance of Anguillids: a cultural and historical perspective introducing papers from the World Fisheries Congress. Ecol Freshw Fish 23:2–6. https://doi.org/10.1111/eff.12089

Kuroki M, Miller MJ, Tsukamoto K (2014b) Diversity of early life-history traits in freshwater eels and the evolution of their oceanic migrations. Can J Zool 92:749–770. https://doi.org/10.1139/cjz-2013-0303

Lin YS, Tzeng CS, Hwang JK (2005) Reassessment of morphological characteristics in freshwater eels (genus *Anguilla*, Anguillidae) shows congruence with molecular phylogeny estimates. Zool Scr 34:225–234. https://doi.org/10.1111/j.1463-6409.2005.00192.x

Masuda Y, Shima Y, Tamaru O, Takahashi Y, Ohmura Y, Iwasaki T, Kamoshida M, Arimoto M, Yamano K, Yatabe T (2019) Japanese eel jaw and vertebra ossification occurring respectively during the larval stage and metamorphosis. Fish Sci 85:1045–1054. https://doi.org/10.1007/s12562-019-01352-w

McCosker JE, Bustamante RH, Wellington GM (2003) The freshwater eel, *Anguilla marmorata*, discovered at Galápagos. Noticias de Galápagos 62:2–6

Miller MJ, Tsukamoto K (2017) The ecology of oceanic dispersal and survival of anguillid leptocephali. Can J Fish Aquat Sci 74:958–971. https://doi.org/10.1139/cjfas-2016-0281

Minegishi Y, Aoyama J, Tsukamoto K (2008) Multiple population structure of the giant mottled eel, *Anguilla marmorata*. Mol Ecol 17:3109–3122. https://doi.org/10.1111/j.1365-294X.2008.03822.x

Minegishi Y, Gagnaire P-A, Aoyama J, Bosc P, Feunteun E, Tsukamoto K, Berrebi P (2012) Present and past genetic connectivity of the Indo-Pacific tropical eel *Anguilla bicolor*. J Biogeogr 39:408–420. https://doi.org/10.1111/j.1365-2699.2011.02603.x

Okamura A, Zhang H, Mikawa N, Kotake A, Yamada Y, Utoh T, Horie N, Tanaka S, Oka HP, Tsukamoto K (2008) Decline in non-native freshwater eels in Japan: ecology and future perspectives. Environ Biol Fish 81:347–358. https://doi.org/10.1007/s10641-007-9205-x

Pujolar JM, Jacobsen MW, Als TD, Frydenberg J, Magnussen E, Jónsson B, Jiang X, Cheng L, Bekkevold D, Maes GE, Bernatchez L, Hansen MM (2014) Assessing patterns of hybridization between North Atlantic eels using diagnostic single-nucleotide polymorphisms. Heredity 112: 627–637. https://doi.org/10.1038/hdy.2013.145

Schabetsberger R, Chang Y-L, Miller MJ (2021) Spawning migration and larval dispersal of tropical Pacific eels in the centre of their distribution ranges. Mar Ecol Prog Ser 670:167–184. https://doi.org/10.3354/meps13745

Schmidt J (1909) On the distribution of the freshwater eels (*Anguilla*) throughout the world. I. Atlantic Ocean and adjacent region. Meddr Kommn Havunders Ser Fisk 3:1–45

Schmidt J (1935) Danish eel investigations during 25 years (1905–1930). The Carlsberg Foundation, Copenhagen, pp 1–16

Shen KN, Tzeng WN (2007) Genetic differentiation among populations of the shortfinned eel *Anguilla australis* from East Australia and New Zealand. J Fish Biol 70:177–190. https://doi.org/10.1111/j.1095-8649.2007.01399.x

Silfvergrip AMC (2009) CITES identification guide to freshwater eels (Anguillidae), with focus on the European eel *Anguilla anguilla*. Swedish Environmental Protection Agency, Stockholm

Smith DG (1999) Anguillidae: freshwater eels. In: Carpenter KE, Niem VH (eds) FAO species identification guide for fishery purposes: the living marine resources of the western central Pacific, vol 3. FAO, Rome, pp 1630–1636

Teng H-Y, Lin Y-S, Tzeng C-S (2009) A new *Anguilla* species and reanalysis of the phylogeny of freshwater eels. Zool Stud 48:808–822. https://zoolstud.sinica.edu.tw/Journals/48.6/808.pdf

Tesch F-W (2003) The eel. Blackwell, Oxford, p vii. + 408

Tsukamoto K, Aoyama J, Miller MJ (2002) Migration, speciation and the evolution of diadromy in anguillid eels. Can J Fish Aquat Sci 59:1989–1998. https://doi.org/10.1139/f02-165

Tsukamoto K, Kuroki M, Watanabe S (2020) Common names for all species and subspecies of the genus *Anguilla*. Environ Biol Fish 103:985–991. https://doi.org/10.1007/s10641-020-00988-3

Tsutsui S, Yoshinaga T, Watanabe S, Aoyama J, Tsukamoto K, Nakamura O (2019) Mucosal galectin genes in all freshwater eels of the genus *Anguilla*. J Fish Biol 94:660–670. https://doi.org/10.1111/jfb.13936

Watanabe S (2003) Taxonomy of the freshwater eels, genus *Anguilla* Schrank, 1798. In: Aida K, Tsukamoto K, Yamauchi K (eds) Eel Biology. Springer, Tokyo, pp 3–18

Watanabe S, Miller MJ (2012) Species, geographic distribution, habitat and conservation of freshwater eels. In: Nakashima S, Fujimoto M (eds) Eels: physiology, habitat and conservation. Nova Science Publishers Inc, New York, pp 1–44

Watanabe S, Aoyama J, Tsukamoto K (2004) Reexamination of Ege's (1939) use of taxonomic characters of the genus *Anguilla*. Bull Mar Sci 74:337–351. https://www.ingentaconnect.com/contentone/umrsmas/bullmar/2004/00000074/00000002/art00006. Accessed 1 June 2023

Watanabe S, Aoyama J, Nishida M, Tsukamoto K (2005a) A molecular genetic evaluation of the taxonomy of eels of the genus *Anguilla* (Pisces: Anguilliformes). Bull Mar Sci 76:675–690. https://www.ingentaconnect.com/contentone/umrsmas/bullmar/2005/00000076/00000003/art00006. Accessed 1 June 2023

Watanabe S, Aoyama J, Nishida M, Tsukamoto K (2005b) Evaluation of the population structure of *Anguilla bicolor bicolor* using total number of vertebrae and the mtDNA control region. Coast Mar Sci 29:165–169. https://doi.org/10.15083/00040815

Watanabe S, Aoyama J, Tsukamoto K (2006) Reconfirmation of morphological differences between *A. australis australis* Richardson and *A. australis schmidtii* Phillipps. NZ J Mar Freshw Res 40:325–331. https://doi.org/10.1080/00288330.2006.9517424

Watanabe S, Aoyama J, Miller MJ, Ishikawa S, Feunteun E, Tsukamoto K (2008a) Evidence of population structure in the giant mottled eel, *Anguilla marmorata*, using total number of vertebrae. Copeia 2008:680–688. https://doi.org/10.1643/CG-07-080

Watanabe S, Aoyama J, Tsukamoto K (2008b) The use of morphological and molecular genetic variations to evaluate subspecies issues in the genus *Anguilla*. Coastal Mar Sci 32:19–29. https://doi.org/10.15083/00040709

Watanabe S, Aoyama J, Tsukamoto K (2009a) A new species of freshwater eel *Anguilla luzonensis* (Teleostei: Anguillidae) from Luzon Island of the Philippines. Fish Sci 75:387–392. https://doi.org/10.1007/s12562-009-0087-z

Watanabe S, Miller MJ, Aoyama J, Tsukamoto K (2009b) Morphological and meristic evaluation of the population structure of *Anguilla marmorata* across its range. J Fish Biol 74:2069–2093. https://doi.org/10.1111/j.1095-8649.2009.02297.x

Watanabe S, Miller MJ, Aoyama J, Tsukamoto K (2011) Analysis of vertebral counts of the tropical anguillids, *Anguilla megastoma*, *A. obscura*, and *A. reinhardtii*, in the western South Pacific in relation to their possible population structure and phylogeny. Environ Biol Fish 91:353–360. https://doi.org/10.1007/s10641-011-9791-5

Watanabe S, Aoyama J, Hagihara S, Ai B, Azanza RV, Tsukamoto K (2013) *Anguilla huangi* Teng, Lin, and Tzeng, 2009, is a junior synonym of *Anguilla luzonensis* Watanabe, Aoyama, and Tsukamoto, 2009. Fish Sci 79:375–383. https://doi.org/10.1007/s12562-013-0620-y

Watanabe S, Aoyama J, Tsukamoto K (2014a) On the identities of *Anguilla borneensis*, *A. malgumora*, and *Muraena malgumora*. Copeia 2014:568–576. https://doi.org/10.1643/CI-14-059

Watanabe S, Miller MJ, Aoyama J, Tsukamoto K (2014b) Evaluation of the population structure of *Anguilla bicolor* and *A. bengalensis* using total number of vertebrae and consideration of the subspecies concept for the genus *Anguilla*. Ecol Freshw Fish 23:77–85. https://doi.org/10.1111/eff.12076

Chapter 2
Population Structure and Speciation

Jun Aoyama

Contents

Population genetic analysis enables us to understand genetic differences within and across populations or species, and provides a basis for understanding how these groups evolved over time and space. An accurate understanding of evolutionarily significant units is critical not only to progress ecology but also for effective stock management. Owing to the morphological similarities between species of the genus *Anguilla* (see Chap. 1), genetic markers have been applied to anguillid eels to define their taxonomic status since the early 1960s to aid in differentiation of species in this genus (van Ginneken and Maes 2005). Recent developments in molecular techniques and algorithms have allowed the analysis of various genome-wide genetic data and the detection of detailed signatures of population structure, gene flow, past species demography, and adaptive genetic variation. In the history of population genetic research for anguillid species, subtle but statistically significant spatial or temporal genetic differences within a species have been repeatedly shown for northern temperate species, suggesting various genetic population structures. However, these were later challenged by updated techniques that provided evidence for the panmixia of these species. Sexually mature silver eels from a wide range of species migrate to a single spawning site, where hatched larvae, called leptocephali, are randomly and passively transported by ocean currents back to their growth habitats. Based on such migratory traits, despite there being subtle genetic differences due to variable environmental conditions across large-scale geographic ranges,

J. Aoyama (✉)
Otsuchi Coastal Research Center, Atmosphere and Ocean Research Institute, The University of Tokyo, Iwate, Japan
e-mail: jaoyama@aori.u-tokyo.ac.jp

K. Tsukamoto et al. (eds.), *Eel Science*, Fisheries Science Series,
https://doi.org/10.1007/978-981-99-5692-0_2

large-population Northern Hemisphere anguillid eels form single panmictic populations within a growth habitat connected by the same oceanic current system. This is not the case for the widespread tropical Indo-Pacific species *A. marmorata*, which has multiple spawning populations distributed across various ocean basins. Other widespread tropical eels in the South Pacific also appear to have more than one spawning area, but their population structures have not yet been clearly defined using nuclear DNA markers.

2.1 Population Structure of Temperate Species

Detailed genetic analyses conducted in the early 2000s brought panmixia in the European eel *A. anguilla* into debate. Wirth and Bernatchez (2001) were the first to use highly polymorphic microsatellite DNA markers for population analysis of the European eel and found a weak but highly significant genetic difference among specimens collected from the North Atlantic, Baltic and North Seas, and Iceland, suggesting restricted gene flow between the geographic groups or maritime units. Genetic differences in the European eel population were later repeatedly suggested as being a "genetic mosaic," based on the evidence of isolation by distance (IDB) or isolation by time (IDT) (van Ginneken and Maes 2005). Als et al. (2011) conducted population genetic analyses using 21 microsatellite loci for leptocephali and glass eels and provided strong evidence for panmixia in the European eel. Recently, Enbody et al. (2021) concluded that European eels should be conclusively considered as a single panmictic population based on whole-genome sequencing. In comparison with the European eel, significantly fewer studies have been conducted to analyze the population genetic structure of the American eel *A. rostrata*. However, Côté et al. (2013) conducted an extensive analysis of the genetic structure of the American eel using 18 microsatellite loci for more than 2000 individuals from 32 localities and showed strong evidence to support panmixia of the American eel.

For the Japanese eel *A. japonica* in the western North Pacific, the understanding of their population structure has progressed similarly to that of European eels in the Atlantic. Panmixia in Japanese eels has been suggested by studies using various genetic markers (Aoyama 2009). Chan et al. (1997) found a geographic cline in the allele frequency of isozyme loci, suggesting a geographic population structure in Japanese eels, however, differential selection was also proposed as a possible explanation. Tseng et al. (2006) reported 2 genetically different groups of Japanese eels that were defined as low- (southern China and Taiwan) and high-latitude groups (Japan, Korea, and northeast China) by analyzing 8 polymorphic microsatellite loci. Chang et al. (2007) found that glass eels of consecutive monthly cohorts (estimated by the otolith microstructure) that recruited the same habitat exhibited subtle genetic patchiness, although they did not find significant temporal genetic variations. Studies later conducted using microsatellite loci also showed weak but significant levels of genetic differentiation at both temporal and spatial scales.

Japanese eels spawn at a very restricted location in synchrony with new moon periods, and their larvae are passively transported by the North Equatorial Current and Kuroshio (Aoyama 2009), which shows considerable changes in speed, eddy structure, and route at daily, monthly, or even decadal scales (see Chap. 5). These facts strongly suggest that the oriented migration of leptocephali or glass eels to the same specific growth habitat as their parents is highly unlikely, making geographic population structure unlikely to occur. Han et al. (2010) analyzed 8 microsatellite loci using specimens collected exclusively across the species range and confirmed panmixia in the Japanese eel. A more recent RAD-seq approach by Gong et al. (2019) for the Japanese eel population structure revealed a small amount of genetic heterogeneity within specimens from Japan and China. However, the authors concluded that this genetic difference was likely to reset each generation by the panmixia of the Japanese eel. Igarashi et al. (2018), who also used the RAD-seq approach for the Japanese eel population structure, showed that eels in a river estuary (Kuma River, Japan) are genetically different from other eels inhabiting the Kuroshio region, including silver eels captured around the spawning site (near the West Mariana Ridge), however, no explanation was offered for how the genetically varied eels might only be recruited to that area.

In the South Pacific, Dijkstra and Jellyman (1999) questioned the status of the Australian eel *A. australis* classically being divided into 2 subspecies, *A. australis australis* in Australia and *A. australis schmidtti* in New Zealand. Dijkstra and Jellyman (1999) analyzed the control region of the mitochondrial DNA (mtDNA) of *A. australis* collected from Australia and New Zealand. However, the number of specimens and genetic markers used in that study were limited (importantly, only mtDNA was used), and thus, the authors found no geographic genetic structure in *A. australis*. Therefore, they suggested that this species shares a common gene pool and that their subspecies designation is invalid. Smith et al. (2001) revisited this issue, including the endemic New Zealand longfin eel *A. diffenbachii* by analyzing allozyme loci of glass eels and adults of both species collected from rivers in the North and South Islands of New Zealand. The results suggested panmixia in the South Pacific temperate anguillid eels, *A. australis* and *A. dieffenbachii,* despite heterogeneity among adult samples found within a species, which suggests sweepstake events or different selective pressures in the adult environments. However, Shen and Tzeng (2007a, b) again analyzed both species using a more appropriate method of microsatellite loci and found that Australian and New Zealand populations of *A. australis* showed significant genetic differentiation, likely reflecting their subspecies designation. In contrast, the Australian longfin eel *A. reinhardtii* populations were not significantly different among 799 glass eels collected from 5 estuaries across the range of the species in East Australia in different years. Watanabe et al. (2006) also found a statistical difference in the overlapping ranges of the total vertebrae between the 2 *A. australis* subspecies. This example illustrates that only using mtDNA for population structure analysis of subspecies of different anguillid-eel populations is not appropriate because of the long larval duration of the leptocephalus larvae, for which only a few larval exchanges over extensive periods could result in the false appearance of no population structure.

Most studies on anguillid eels use various nuclear DNA markers, and recent developments in genetic analyses have made it possible to detect very slight genetic differences. However, considering the ecological/migratory traits of anguillid eels, which migrate back to their growth habitats within their large-scale species range by global oceanic currents, such as western boundary currents, it seems reasonable to assume that temperate species of anguillid eels are generally panmictic.

2.2 Population Structure of Tropical Species

In contrast to their temperate counterparts, tropical anguillid species have been studied considerably less, and their population structures are not well known. The first detailed population genetic study of a tropical species was conducted by Ishikawa et al. (2004) on the widely distributed species *A. marmorata*, which was unlikely to be panmictic because it is distributed throughout most of the Indo-Pacific region, ranging from the eastern side of Africa to French Polynesia. Furthermore, this study used 449 eels collected from 10 localities throughout the species range and found 5 apparently different populations using mtDNA and AFLP markers. These populations include the North Pacific population that spawns in the North Equatorial Current region near the Japanese eel spawning site (Kuroki et al. 2009), and other populations in the western Indian Ocean, eastern Indian Ocean, Fiji region, and French Polynesia (Tahiti). A subsequent, more extensive study by Minegishi et al. (2008) used longer sequences of mtDNA and 8 microsatellite loci from eels collected from 13 sites, including 5 sites that were not analyzed by Ishikawa et al. (2004). The data shown by Minegishi et al. (2008) generally supported the results of a previous study but revealed different genetic structures for each population. The authors concluded that *A. marmorata* has 4 genetically different populations (North Pacific, South Pacific, Indian Ocean, and Guam region; Fig. 2.1a). The North Pacific population is fully panmictic, whereas the South Pacific and Indian Ocean populations possess metapopulation structures. Minegishi et al. (2008) found a significantly different population in Guam in the Mariana Islands of the western North Pacific (Fig. 2.1a). Watanabe et al. (2008) also found strong morphological evidence for a separate Micronesia population in Guam and other Micronesian islands in an analysis of the total number of vertebrae. The eels from Guam were found to have a significantly higher number of vertebrae than in any other region, however, the geographic distribution of the eels in this population is not yet known, as illustrated by Donovan et al. (2012).

Another interesting finding was that individuals from at least 2 populations (North Pacific and South Pacific) appeared to be mixed in Ambon, which is a small island to the east of Sulawesi Island, Indonesia (Minegishi et al. 2008; Fig. 2.1b). The authors concluded that Ambon is geographically located in the border region between the North and South Pacific oceans and is affected by waters from the North and South Pacific through the Molucca Sea and the Halmahera Sea, respectively (Minegishi et al. 2008, Gordon 2005; Fig. 2.1b). Therefore, the eel

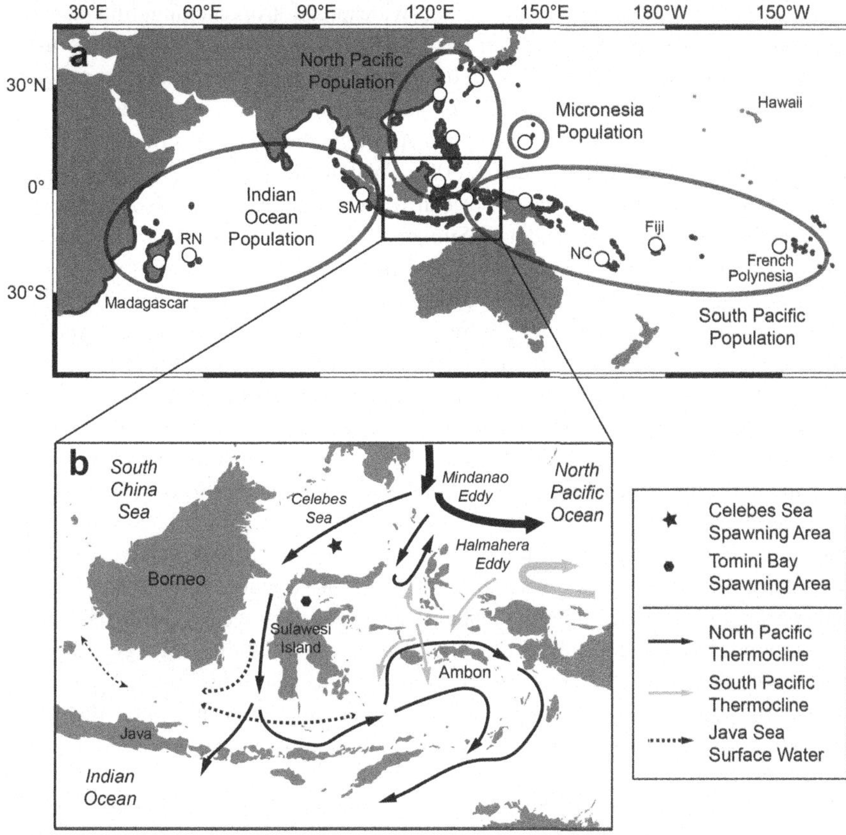

Fig. 2.1 (**a**) Genetic population structure of *Anguilla marmorata* that was indicated by analysis of both mtDNA and microsatellite DNA, which was modified from Minegishi et al. (2008). White circles indicate sampling locations of the specimens used for the analysis, which include labeled locations, Sumatra (SM), Réunion Island (RN), New Caledonia (NC), and unlabeled locations, New Guinea, Ambon, Sulawesi Island, Philippines, Taiwan and Southern Japan. Bold coastlines show the regions where that species occurs. The South Pacific and Indian Ocean eels may form metapopulations with more than one spawning location. (**b**) Map showing Indonesian throughflow pathways around Ambon and the Indonesian Seas region modified from Gordon (2015); the 2 locations were collections of small leptocephali that indicated where spawning by *Anguilla celebesensis* occurs as indicated by Aoyama et al. (2003, 2018)

populations in Ambon are likely composed of *A. marmorata* individuals transported from both the North and South Pacific populations. Gagnaire et al. (2009) focused on the detailed gene flow of *A. marmorata* within the Indian Ocean basin, using 10 microsatellite loci and 444 eels from a wide range of locations. Their results suggested that there was evidence of east-west differentiation, as found in previous studies, and unidirectional gene flow from Sumatra to the Madagascar region was indicated.

The population structure of the other widespread tropical anguillid species, *A. bicolor*, which is distributed from the eastern coast of Africa through the Indonesian Seas to New Guinea adjacent to the Pacific Ocean, has been studied morphologically, along with consideration of its subspecies designation of being divided into *A. bicolor bicolor* (Indian Ocean) and *A. bicolor pacifica* (Pacific and Indonesian Seas) by Watanabe et al. (2005). Watanabe et al. (2014) examined *A. bengalensis* inhabiting the Indian Ocean, which was designated as 2 subspecies, *A. bengalensis bengalensis* (India to Sumatra) and *A. bengalensis labiata* (southeastern Africa). Significant morphological differences in the total number of vertebrae exist between individuals within each geographic group, corresponding to their historical subspecies designation, however, no further subpopulation signatures within each subspecies were found. Recently, Arai and Taha (2021) genetically analyzed *A. b. bengalensis* for the first time using a portion of the mitochondrial COI gene sequences and provided molecular evidence to support the results of Watanabe et al. (2014). Minegishi et al. (2012) examined the population structure of *A. bicolor* using the entire mitochondrial control region sequence and 6 microsatellite loci for 234 specimens collected from localities where the 2 subspecies have historically been recognized. This analysis showed genetic divergence between specimens from the Indian and Pacific oceans, whereas no significant variation was observed within each ocean, as suggested by Watanabe et al. (2005). However, Minegishi et al. (2012) found 2 mitochondrial sub-lineages that coincided with neither geographical regions nor microsatellite markers were found, as was also found by Arai and Taha (2021). The analysis of gene flow and demographic history indicated that these 2 mitochondrial sublineages probably represented haplotype groups of relict ancestral populations.

The tropical eel *A. reinhardtii,* which has a limited species range in the northwest margin of the western South Pacific and along the east coast of Australia, has been genetically studied by Shen and Tzeng (2007b). To understand both spatial and temporal genetic differentiation, 6 microsatellite loci were examined for 799 glass eels collected at various months at localities across the species range. The results showed no significant genetic differences among specimens from different months or localities, suggesting pamixia in this species.

The tropical anguillids *A. megastoma* and *A. obscura* have much larger ranges that extend from New Guinea to French Polynesia. However, no population structure was found in these species using both mitochondrial and nuclear markers from eels in Samoa, New Caledonia, or the Solomon Islands in the western region (Gubili et al. 2019). Further research using nuclear DNA is needed to examine whether the eels of these species are genetically different farther east in French Polynesia.

A species inhabiting the western Indian Ocean, *A. mossambica*, was recently subjected to genetic analysis of its population structure. Frankowski et al. (2020) collected 151 specimens from southeast Africa and Madagascar to infer their population structures and demographic parameters. A segment of the mitochondrial cytochrome b gene sequence showed the absence of genetic structuring in this species, thereby supporting the panmixia of the endemic *A. mossambica*, nuclear DNA has not been examined.

2.3 Speciation

The current information available for both temperate and tropical species provides insights into the evolutionary history of anguillid eels. A conceptual circular route connecting the spawning area and growth habitat of anguillid species has been defined as a "migration loop" (Tsukamoto et al. 2002). In this concept, each species or distinct genetic population within a species spatially and temporally shares a migration loop, which is peculiar to their geographic distribution and ecological traits, such as spawning season and location, larval duration, and recruitment timing. Spatial or temporal shifts of the specific migration loop would potentially cause reproductive isolation and speciation (Fig. 2.2). Accumulating evidence indicates that the long-distance spawning migrations of anguillid eels by temperate species have evolved from short-scale migrations of their tropical ancestral congeners. Tsukamoto et al. (2002) proposed a model to explain the evolution of long-distance spawning migration in anguillid eels, as discussed more recently by Aoyama (2009) and Kuroki et al. (2014) in relation to early life history data of anguillid eels. Larva of anguillid species inhabiting the tropics appear to have faster growth rates, smaller maximum larval sizes, and shorter larval durations than temperate species (Kuroki et al. 2006, 2014). This suggests that ancestral eels that originated in the tropics had small-scale migration loops between their freshwater or estuarine growth habitats and spawning sites in the ocean. However, their leptocephali would have dispersed not only to their original species range in the tropics but could also accidentally disperse to higher latitudes over an evolutionary time scale (Fig. 2.2). If these eels that colonized higher latitudes successfully migrated back to the original spawning

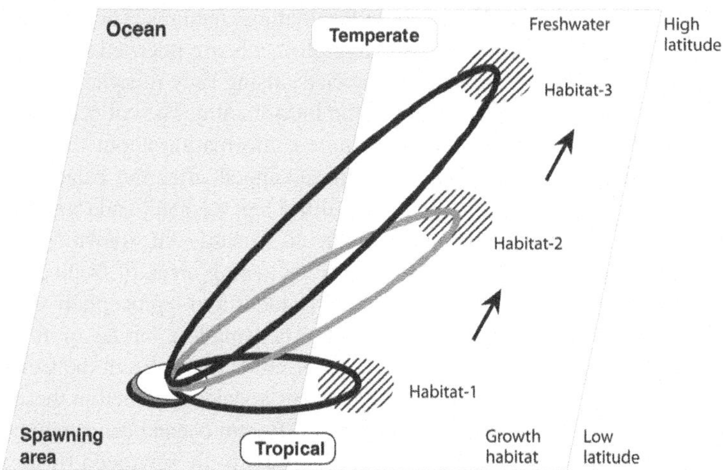

Fig. 2.2 A diagram showing shifts in the migration loop of an anguillid species that could lead to speciation, which was modified from Aoyama (2009). The leptocephali are dispersed and recruit to higher latitudes causing an eventual change in growth habitats, whereas the adults still return to spawn at tropical latitudes

area, it could eventually result in the creation of a new large-scale migration loop between their tropical spawning area and higher latitude growth habitat (Fig. 2.2). Kuroki et al. (2006) found that even individuals of the same eel species recruited to the same location showed considerably large variations in their larval periods, indicating an intrinsically wide variation in their larval periods. This variation may be related to a mechanism such as delayed metamorphosis, which has been reported for coral reef fish (McCormick 1999) and carapid fish (Parmentier et al. 2004). Kuroki et al. (2006) suggested that the key to determining the migration scale of anguillid species is their larval growth and maximum larval size, which directly influence their planktonic period and species range.

In the Atlantic Ocean, genetic backgrounds related to migratory traits, and thus speciation of anguillid species, have rapidly accumulated. A recent transcriptome analysis of leptocephali in both European and American eels collected in the Sargasso Sea suggested genetic control of their specific larval duration (Bernatchez et al. 2011). In addition, a genome-wide approach for the 2 Atlantic species found a high divergence for genes related to developmental processes and phosphorylation, suggesting genetic control of larval-phase variations and subsequent migration distance (Jacobsen et al. 2014). Molecular approaches will provide detailed information on the speciation mechanisms of anguillid species, possibly with the aid of more whole-genome research that is underway.

2.3.1 A Question About A. celebesensis *on Sulawesi Island*

Opening the door to the mysterious life of anguillid eels provides a hint for the next step to comprehensively understand these fascinating creatures. The spawning areas and oceanic larval migration of tropical anguillid eels are not well known because their leptocephali could not be identified exactly using only morphological characters, such as during the Danish surveys of the Indo-Pacific. The collection of data on tropical anguillid leptocephali provided limited information about their spawning areas and migration ecology. However, this changed after the establishment of genetic tools for species identification of anguillid leptocephali, and research cruises using this identification method were conducted to study the spawning areas and larval distributions of tropical anguillids. The spawning areas of tropical anguillid species based on the distribution of genetically identified leptocephali were determined in Indonesian waters (see Chap. 4). The smallest larvae of the species *A. celebesensis* (12–20 mm in total length) ever collected, were in the Celebes Sea and Tomini Bay of Indonesia, suggesting that this species spawned in these regions (Aoyama et al. 2003; Fig 2.1b). Spawning in 2 different ocean basins may be owing to which side of Sulawesi Island they were living on (only separated by the narrow northern arm of Sulawesi Island). Different spawning areas of a species located just a few hundred kilometers apart raise the question of whether individuals from these spawning sites are genetically different, since some larvae could possibly drift out of each area. Genetic data from the nuclear DNA of *A. celebesensis* inhabiting the northern part of Sulawesi Island will provide new insights into the migration and spawning ecology of anguillid species.

2.4 Summary and Future Perspectives

From the brief review of population genetic studies on anguillid eels, it is obvious that their genetic population structures, or their taxonomic status in some cases, have yet to be fully elucidated, particularly for species inhabiting the tropics. Many questions remain, such as how some tropical species, such as *A. bornensis* (Borneo area) and *A. luzonensis* (Philippines area), can maintain populations strictly in very specific areas. In the South Pacific, it is still unclear whether each species travels to their specific spawning areas or if they spawn locally (Schabetsberger et al. 2021), which would significantly influence their population structures. Future studies on more species of tropical anguillid eels that inhabit areas with unique patterns of growth habitats and spawning sites, such as *A. celebesensis* in Sulawesi Island, *A. megastoma* and *A. obscura* in the South Pacific, will surely add valuable new perspectives on the ecology and evolution of this remarkable group of fish found in many parts of the world. These studies should focus on using nuclear DNA, which reflects recent genetic exchange, rather than the historical signatures of mitochondrial DNA.

References

Als TD, Hansen MM, Maes GE, Castonguay M, Riemann L, Aarestrup K, Munk P, Sparholt H, Hanel R, Bernatchez L (2011) All roads lead to home: panmixia of European eel in the Sargasso Sea. Mol Ecol 20:1333–1346. https://doi.org/10.1111/j.1365-294X.2011.05011.x

Aoyama J (2009) Life history and evolution of migration in catadromous eels (genus *Anguilla*). Aqua-BioSci Monogr 2:1–42

Aoyama J, Wouthuyzen S, Miller MJ, Inagaki T, Tsukamoto K (2003) Short-distance spawning migration of tropical freshwater eels. Biol Bull 204:104–108. https://doi.org/10.2307/1543500

Aoyama J, Wouthuyzen S, Miller MJ, Sugeha HY, Kuroki M, Watanabe S, Syahailatua A, Tantu FY, Hagihara S, Triyanto OT, Tsukamoto K (2018) Reproductive ecology and biodiversity of freshwater eels around Sulawesi Island Indonesia. Zool Stud 57:e30. https://doi.org/10.6620/ZS.2018.57-30

Arai T, Taha H (2021) Contrasting patterns of genetic population structure in tropical freshwater eels of genus *Anguilla* in the Indo-Pacific. Heiyon 7:e07097. https://doi.org/10.1016/j.heliyon.2021.e07097

Bernatchez L, Jérôme SC, Normandeau E, Maes GE, Als TD, Kalujnaia S, Cramb G, Castonguay M, Hansen MM (2011) Differential timing of gene expression regulation between leptocephali of the two *Anguilla* eel species in the Sargasso Sea. Ecol Evol 1:459–467. https://doi.org/10.1002/ece3.27

Chan IKK, Chan DKO, Lee SC, Tsukamoto K (1997) Genetic variability of the Japanese eel *Anguilla japonica* (Temminck & Schlegel) related to latitude. Ecol Fresh Fish 6:4549–4549. https://doi.org/10.1111/j.1600-0633.1997.tb00141.x

Chang KC, Han YS, Tzeng WN (2007) Population genetic structure among intra-annual arrival waves of the Japanese eel *Anguilla japonica* in Northern Taiwan. Zool Stud 46:583–590. http://ntur.lib.ntu.edu.tw/bitstream/246246/64352/1/no5.950912.pdf. Accessed 1 June 2023

Côté CL, Gagnaire PA, Bourret V, Verrault G, Castonguay M, Bernatchez L (2013) Population genetics of the American eel (*Anguilla rostrata*). F_{ST}=0 and North Atlantic oscillation effects on demographic fluctuations of a panmictic species. Mol Ecol 22:1763–1776. https://doi.org/10.1111/mec.12142

Dijkstra LH, Jellyman DJ (1999) Is the subspecies classification of the freshwater eels *Anguilla australis australis* Richardson and *A. a. schmidtii* Phillipps still valid? Mar Freshw Res 50:261–263. https://doi.org/10.1071/MF98144

Donovan S, Pezold F, Chen Y, Lynch B (2012) Phylogeography of *Anguilla marmo*rata (Teleostei: Anguilliformes) from the eastern Caroline Islands. Ichthyol Res 59:70–76. https://doi.org/10.1007/s10228-011-0245-z

Enbody ED, Pettersson ME, Sprehn CG, Palm S, Wickström H, Andersson L (2021) Ecological adaptation in European eels is based on phenotypic plasticity. Proc Natl Acad Sci U S A 118: e2022620118. https://doi.org/10.1073/pnas.2022620118

Frankowski J, Lübke K, Coke M, Weyl OLF (2020) Genetic variability and demographic history of *Anguilla mossambica* (Peters, 1852) from continental Africa and Madagascar. J Fish Biol 96: 1251–1259. https://doi.org/10.1111/jfb.14220

Gagnaire P-A, Minegishi Y, Aoyama J, Reveillac E, Robinet T, Bosc P, Tsukamoto K, Feunteun E, Berrebi P (2009) Ocean currents drive secondary contact between *Anguilla marmorata* populations in the Indian Ocean. Mar Ecol Prog Ser 379:267–278. https://doi.org/10.3354/meps07895

Gong X, Davenport ER, Wang D, Clark AG (2019) Lack of spatial and temporal genetic structure of Japanese eel (*Anguilla japonica*) populations. Conserv Genet 20:467–475. https://doi.org/10.1007/s10592-019-01146-8

Gordon AL (2005) Oceanography of the Indonesian seas and their throughflow. Oceanography 18: 14–27. https://doi.org/10.5670/oceanog.2005.01

Gubili C, Schabetsberger R, Poellabauer C, Bates B, Wagstaff RM, Woodward LM, Sichrowsky U, Scheck A, Boseto DT, Feunteun E, Acou A, Jehle R (2019) High genetic diversity and lack of pronounced population structure in five species of sympatric Pacific eels. Fish Manag Ecol 26: 31–41. https://doi.org/10.1111/fme.12287

Han Y-S, Hung C-L, Liao Y-F, Tzeng W-N (2010) Population genetic structure of the Japanese eel *Anguilla japonica*: panmixia at spatial and temporal scales. Mar Ecol Prog Ser 401:221–232. https://doi.org/10.3354/meps08422

Igarashi Y, Zhang H, Tan E, Sekino M, Yoshitake K, Kinoshita S, Mitsuyama S, Yoshinaga T, Chow S, Kurogi H, Shinoda A, Han YS, Wakiya R, Mochioka N, Yamamoto T, Kuwada H, Kaji Y, Suzuki Y, Gojobori T, Kobayashi T, Saitoh K, Watabe S, Asakawa S (2018) Whole-genome sequencing of 84 Japanese eels reveals evidence against panmixia and support for sympatric speciation. Genes 9:474

Ishikawa S, Tsukamoto K, Nishida M (2004) Genetic evidence for multiple geographic populations of the giant mottled eel *Anguilla marmorata* in the Pacific and Indian oceans. Ichthyol Res 51: 343–353. https://doi.org/10.1007/s10228-004-0241-7

Jacobsen M, Pujolar J, Gilbert M, Moreno-Mayar JV, Bernatchez L, Als TD, Lobon-Cervia J, Hansen MM (2014) Speciation and demographic history of Atlantic eels (Anguilla anguilla and A. rostrata) revealed by mitogenome sequencing. Heredity 113:432–442. https://doi.org/10.1038/hdy.2014.44

Kuroki M, Aoyama J, Miller MJ, Wouthuyzen S, Arai T, Tsukamoto K (2006) Contrasting patterns of growth and migration of tropical anguillid leptocephali in the western Pacific and Indonesian seas. Mar Ecol Prog Ser 309:233–246. https://doi.org/10.3354/meps309233

Kuroki M, Aoyama J, Miller MJ, Yoshinaga T, Shinoda A, Hagihara S, Tsukamoto K (2009) Sympatric spawning of *Anguilla marmorata* and *Anguilla japonica* in the western North Pacific Ocean. J Fish Biol 74:1853–1865. https://doi.org/10.1111/j.1095-8649.2009.02299.x

Kuroki M, Miller MJ, Tsukamoto K (2014) Diversity of early life-history traits in freshwater eels and the evolution of their oceanic migrations. Can J Zool 92:749–770. https://doi.org/10.1139/cjz-2013-0303

McCormick MI (1999) Delayed metamorphosis of a tropical reef fish (*Acanthurus triostegus*): a field experiment. Mar Ecol Prog Ser 176:25–38. https://doi.org/10.3354/meps176025

Minegishi Y, Aoyama J, Tsukamoto K (2008) Multiple population structure of the giant mottled eel, *Anguilla marmorata*. Mol Ecol 17:3109–3122. https://doi.org/10.1111/j.1365-294X.2008.03822.x

Minegishi Y, Gagnaire P-A, Aoyama J, Feunteun E, Tsukamoto K, Berrebi P (2012) Present and past genetic connectivity of the Indo-Pacific tropical eel *Anguilla bicolor*. J Biogeogr 39:408–420. https://doi.org/10.1111/j.1365-2699.2011.02603.x

Parmentier E, Lecchini D, Lagardere F, Vandewalle P (2004) Ontogenic and ecological control of metamorphosis onset in a carapid fish, *Carapus homei*: experimental evidence from vertebra and otolith comparisons. J Exp Zool 301:617–628. https://doi.org/10.1002/jez.a.50

Schabetsberger R, Chang Y-L, Miller MJ (2021) Spawning migration and larval dispersal of tropical Pacific eels in the centre of their distribution ranges. Mar Ecol Prog Ser 670:167–184. https://doi.org/10.3354/meps13745

Shen KN, Tzeng WN (2007a) Genetic differentiation among populations of the shortfinned eel *Anguilla australis* from East Australia and New Zealand. J Fish Biol 70:177–190. https://doi.org/10.1111/j.1095-8649.2007.01399.x

Shen KN, Tzeng WN (2007b) Population genetic structure of the year-round spawning tropical eel, *Anguilla reinhardtii*, in Australia. Zool Stud 46:441–453

Smith PJ, Benson PG, Stanger C, Chisnall BL, Jellyman DJ (2001) Genetic structure of New Zealand eels *Anguilla dieffenbachii* and *A. australis* with allozyme markers. Ecol Freshw Fish 10:132–137. https://doi.org/10.1034/j.1600-0633.2001.100302.x

Tseng MC, Tzeng WN, Lee SC (2006) Population genetic structure of the Japanese eel *Anguilla japonica* in the Northwest Pacific Ocean: evidence of non-panmictic populations. Mar Ecol Prog Ser 308:221–230. https://doi.org/10.3354/meps308221

Tsukamoto K, Aoyama J, Miller MJ (2002) Migration, speciation and the evolution of diadromy in anguillid eels. Can J Fish Aquat Sci 59:1989–1998. https://doi.org/10.1139/f02-165

van Ginneken VJT, Maes GE (2005) The European eel (*Anguilla anguilla*, Linnaeus), its lifecycle, evolution and reproduction: a literature review. Rev Fish Biol Fisheries 15:367–398. https://doi.org/10.1007/s11160-006-0005-8

Watanabe S, Aoyama J, Nishida M, Tsukamoto K (2005) Evaluation of the population structure of *Anguilla bicolor bicolor* using total number of vertebrae and the mtDNA control region. Coast Mar Sci 29:165–169. https://doi.org/10.15083/00040815

Watanabe S, Aoyama J, Tsukamoto K (2006) Confirmation of morphological differences between Anguilla australis australis and A. australis schmidtii. N Z J Mar Freshw Res 40:325–331. https://doi.org/10.1080/00288330.2006.9517424

Watanabe S, Aoyama J, Miller MJ, Ishikawa S, Feunteun E, Tsukamoto K (2008) Evidence of population structure in the giant mottled eel, *Anguilla marmorata*, using total number of vertebrae. Copeia 2008:680–688. https://doi.org/10.1643/CG-07-080

Watanabe S, Miller MJ, Aoyama J, Tsukamoto K (2014) Evaluation of the population structure of *Anguilla bicolor* and *A. bengalensis* using total number of vertebrae and consideration of the subspecies concept for the genus *Anguilla*. Ecol Freshw Fish 23:77–85. https://doi.org/10.15083/00040815

Wirth T, Bernatchez L (2001) Genetic evidence against panmixia in the European eel. Nature 409: 1037–1040. https://doi.org/10.1038/35059079

Part II
Ecology

Chapter 3
Life History

Mari Kuroki

Contents

All animals have a "life history", which defines where they live and how they survive and reproduce. The long life history of anguillid eels begins with spawning areas in the ocean and subsequently leads to various juvenile growth habitats in the coastal sea, estuaries, freshwater rivers, ponds, and lakes. They migrate between different aquatic habitats, marine and freshwater, at least twice during their life-history transitions. The first migration of anguillid eels is a larval migration. The larvae hatch in the ocean where they feed and grow while drifting in ocean currents until they are near their nursery habitats. This is followed by the upstream migration of glass eels and elvers that reach the estuaries, and eventually become young yellow eels that seek out growth habitats in coastal and freshwater areas. The second is spawning migration, in which maturing adults return to their spawning areas in the ocean for reproduction. These dynamic migrations make them typical model species of catadromous fish owing to their separated spawning and growth habitats. To make that possible, anguillid eels have evolved unique life history stages, which include their oceanic larvae, the post-metamorphosis juvenile stage that swims from the ocean near the continental shelf to coastal waters and estuaries, the early upstream migration and growth stages, and finally, the maturing reproduction stage. The morphological, ecological, and physiological characteristics of eels vary according to their life history stages, making them a unique creature that might be difficult to identify as a single species. This chapter provides an overview of the unique life history of anguillid eels and their various life history stages.

M. Kuroki (✉)
Graduate School of Agricultural and Life Sciences, The University of Tokyo, Tokyo, Japan
e-mail: mari@g.ecc.u-tokyo.ac.jp

K. Tsukamoto et al. (eds.), *Eel Science*, Fisheries Science Series,
https://doi.org/10.1007/978-981-99-5692-0_3

3.1 Migration as a Key of Life History

3.1.1 Diadromy

Migration occurs in species of fish that have spatial separations in their seasonally/ontogenetically specific habitats. The diadromous fish species migrate between marine and freshwater environments during certain stages of their life history for either spawning (breeding) or growth (feeding).

Diadromous migrations are generally categorized into three forms: anadromy, in which the adults migrate from the ocean to natal freshwater to spawn (e.g., salmon and lampreys); catadromy, in which the adults migrate from freshwater to the natal ocean/estuary to spawn (e.g., eels and sculpins); and amphidromy, in which the individuals migrate between ocean and freshwater habitats unrelated to spawning (e.g., ayu *Plecoglossus altivelis*, gobies, and galaxiids) (McDowall 1988).

The adaptive nature of these migratory forms has generally been considered to be attributed to global patterns of marine and freshwater productivity. McDowall (1987) and Gross et al. (1988) proposed a conceptual model of the productivity hypothesis that catadromous species should evolve in tropical latitudes where freshwater productivity exceeds that of the ocean, whereas anadromous species should evolve in temperate latitudes where ocean productivity exceeds that of freshwater. The diadromous life history could theoretically have evolved only when migration across the ocean-freshwater boundary enhanced individual growth and, thus, fitness under the constraints of natal site selection.

The anguillid eels originated in the tropics from a group of mesopelagic marine eels of the Anguilliformes, containing only pelagic families that live in the deep sea away from land, and based on the phylogenetic relationships of the superorder Elopomorpha (Inoue et al. 2010). This means that anguillid eels have retained their ancestral marine spawning areas, but were able to switch from an oceanic mesopelagic lifestyle to benthic behavior after invading freshwater as advantageous habitats for growth (Tsukamoto et al. 2002, 2009a). Thus, their catadromous migrations between the ocean and freshwater were established (Fig. 3.1).

It has been hypothesized that the origin of fish migrations, including those of eels, might have initially been to escape from an unfavorable environment (Tsukamoto et al. 2009a). This was referred to as the random escapement hypothesis, which proposes that escape is caused by environmental stresses, such as poor feeding, increasing population density, or threats of predation. It was then suggested that migratory behavior might have become established when conditions were more favorable to individuals that engaged in the migratory behavior. The anguillid eels may have gained a significant advantage when they acquired the ability to enter freshwater in the tropics where there are no other congeneric competitors and fewer large-sized predators compared to the ocean.

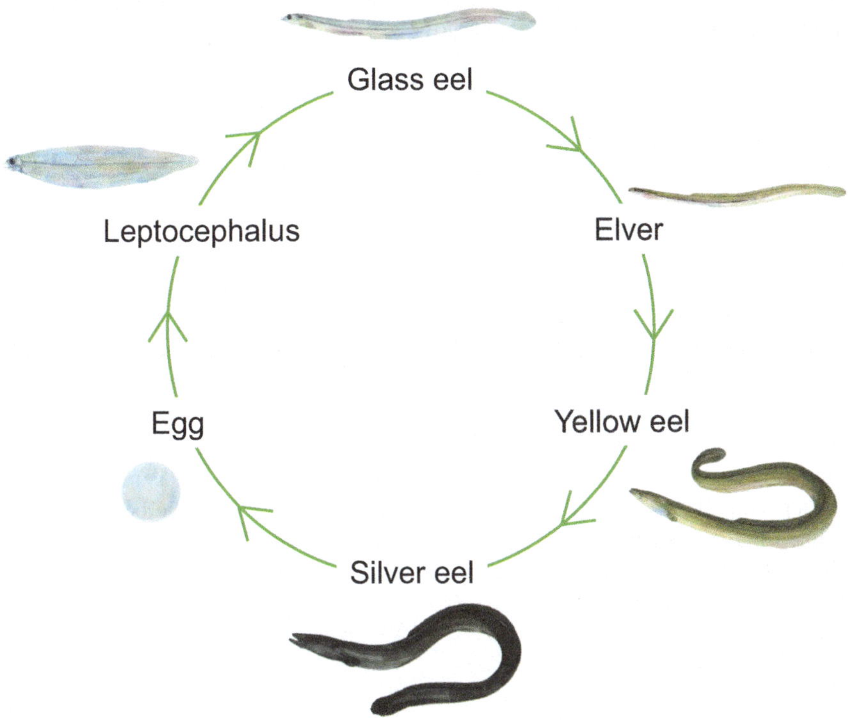

Fig. 3.1 The life history of anguillid eels. All illustrations except the egg were drawn by Yusei Nagashima for the Web Dictionary of Eels and Salmon (https://salmoneel.com) with permission

3.1.2 Migration Scales

All anguillid eel spawning areas are located in the ocean at low latitudes in the tropics and subtropics. In contrast, their growth (feeding) habitats range widely from equatorial tropics to temperate regions at high latitudes in the northern and southern hemispheres. Therefore, temperate eels, mainly in temperate regions, have extensive migrations over thousands of kilometers (Schmidt 1922; Tsukamoto 1992; Béguer-Pon et al. 2015). In contrast, tropical eels, which mainly use tropical and subtropical regions as their growth habitats, mostly appear to have short migrations connecting their spawning and growth habitats and show localized small-scale migrations of tens to hundreds of kilometers (Aoyama et al. 2003; Kuroki et al. 2006; Robinet et al. 2008).

Considering the phylogenetic evolutionary process and biogeography of eels, it is likely that the local-scale migrations of tropical anguillid eels distributed at low latitudes were the ancestral life-history patterns of freshwater eels. Furthermore, the migration scales expanded owing to the larvae being carried farther away by ocean currents and strong western boundary currents, which led to the establishment of large-scale migrations of eels living in temperate regions at high latitudes (Tsukamoto et al. 2002; Kuroki et al. 2014).

3.2 Life-History Stages

3.2.1 Egg

Similar to many other offshore spawning fish species, anguillid eels produce transparent pelagic eggs. After oceanic spawning, fertilized eggs of the Japanese eel *Anguilla japonica* (Fig. 3.2a) measure ~1.6 mm in diameter (Tsukamoto et al. 2011). This egg size is approximately average and slightly larger than the median size of 1.1 mm for marine teleost fish eggs (Chambers and Leggett 1996). The yolk-sac larvae (Fig. 3.2b) hatched ~1.5–2 days after fertilization based on artificially induced spawning experiments for some eel species (Yamamoto and Yamauchi 1974; Bezdenezhnykh et al. 1983; Lokman and Young 2000). Eggs have been estimated to have hatched in the upper layer of the thermocline at approximate depths of 150–300 m, based on the depth ranges where small larvae, preleptocephali, and eggs

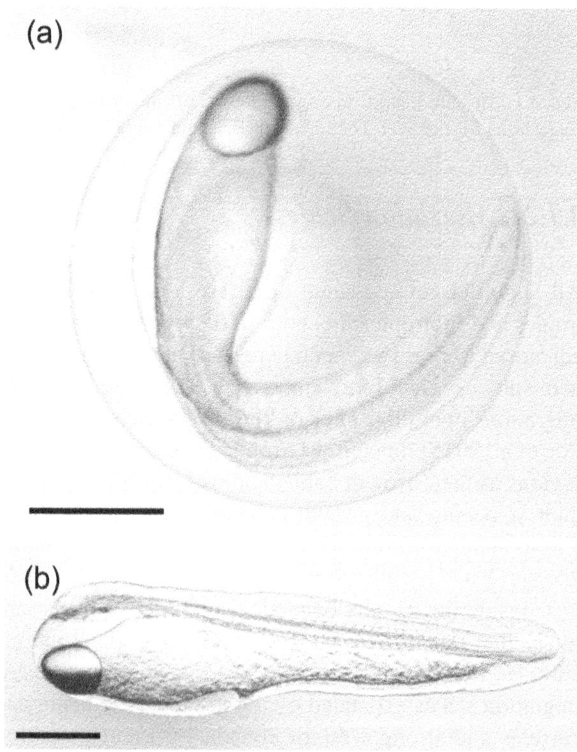

Fig. 3.2 Photographs of an egg (embryo) and hatched larva of the Japanese eel *Anguilla japonica* collected in the western North Pacific during the KH-12-2 cruise. Scale bars are 0.5 mm

were collected (Castonguay and McCleave 1987; Tsukamoto et al. 2011; Aoyama et al. 2014).

The spawning season for temperate eels is well defined, with seaward migration of adult eels generally occurring during the autumn season. In the case of the two Atlantic eels, European eel *A. anguilla* and American eel *A. rostrata*, whose growth habitats are separated between the east and west Atlantic, their spawning areas partly overlap in the Sargasso Sea, but peak spawning for the European eel is estimated to be slightly later than that of the American eel (Miller et al. 2015). On the other hand, the spawning season of tropical eels in a constant high-temperature environment may be almost year-round for some species, although seaward migrations may occur during the rainy season in each local region (Kuroki et al. 2006).

3.2.2 *Leptocephalus*

The ~5 mm offspring, referred to as preleptocephali, consumes the yolk, their eyes become pigmented, and thin teeth appear within several days. The body shape of anguillid eel larvae develop a flattened olive-leaf-like appearance, which is called a leptocephalus (Fig. 3.3). All fish of the superorder Elopomorpha, which includes the marine and freshwater eels of the Anguilliformes and their relatives (tarpons, bonefishes, and notacanths), have leptocephalus larvae (Miller and Tsukamoto 2004). The laterally compressed, transparent bodies of leptocephali have large surface areas and contain materials such as polysaccharide glycosaminoglycans (GAG). The extracellular matrix of hydrophilic GAG in leptocephali contains a high percentage of water, ranging from ~90% to 95% of the total body mass (Pfeiler 1999). This unique morphology provides a buoyant body that allows leptocephali to migrate long distances within currents without using large amounts of energy (Tsukamoto et al. 2009b). When leptocephali reach sizes of approximately >10 mm in total length, they appear to start diel vertical migrations from deeper depths during the day to shallower depths at night (Castonguay and McCleave 1987; Otake et al. 1998). Leptocephali have been found to obtain nutrition by feeding on marine snow-associated materials that include visible objects, such as discarded appendicularian houses, zooplankton fecal pellets, small organisms, and carbohydrates that aggregate into marine snow (Tsukamoto and Miller 2021).

Fig. 3.3 Photograph of a leptocephalus larva of the Indo-Pacific eel *Anguilla marmorata* collected in the western South Pacific during the KH-13-2 cruise. Scale bar is 2 mm

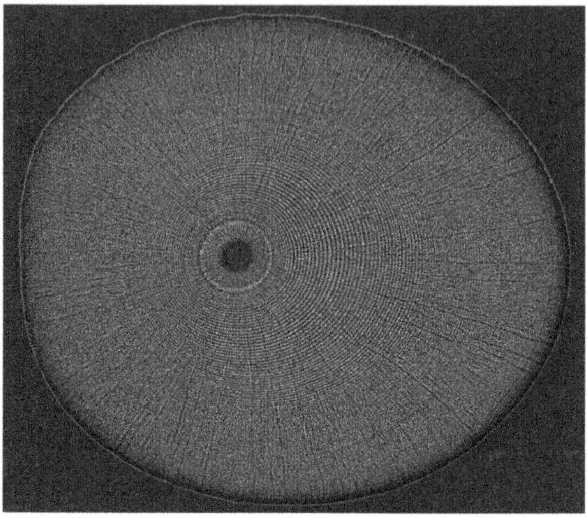

Fig. 3.4 SEM image of an otolith of the Borneo eel *Anguilla borneensis* leptocephalus. Otolith daily increments are used to estimate the hatching dates and subsequent growth rates

The larval duration of anguillid leptocephali can be estimated using their otolith daily rings (Fig. 3.4). The larval duration of each species has been estimated to be several months, with a maximum size of ~50 mm in total length for tropical eels with short-distance larval migration, whereas the temperate European eel that migrates long-distance reaching up to 8000 km have longer larval durations with a maximum size of ~90 mm in total length (Kuroki et al. 2014). It is difficult to estimate the larval duration of the European eel, which migrates the longest distance across the Atlantic Ocean. Despite numerous studies, larval durations remain unclear, ranging from 7 months to more than 2 years (Bonhommeau et al. 2010).

When approaching the outer continental shelf areas, the anguillid eel body shape completely transforms during metamorphosis from leptocephali to glass eels from compressed to rounded with increased muscle tissue; the gut moves forward, the pointed snout becomes rounded, and the teeth disappear. The skeletal system is then formed from both the caudal fin and head, before completing the vertebrae. Leptocephali have no red blood cells (erythrocytes) and very few white blood cells (leukocytes), and erythrocytes appear in the blood when their metamorphosis is complete (Miller 2009). The central nervous system, sensory organs, and internal organs are also transformed, significantly affecting their various functions. They do not feed during metamorphosis. Metamorphosis from the leptocephalus stage to the glass eel stage represents a remarkable transition from planktonic life to the actively moving juvenile stage as a subsequent life cycle.

3.2.3 Glass Eel and Elver

Glass eels that have completed metamorphosis are almost transparent, with no pigmentation on their entire body, except for the caudal fin of some anguillid eel species, and caudal fin pigmentation has different expression patterns in each species (Leander et al. 2012). The body shape of the glass eel is elongated and almost identical to those of the elvers and yellow eels. In tropical regions where multiple anguillid species are sympatrically distributed, pigmentation patterns are useful morphological traits for species identification. The elongated cylindrical body shape of glass eels allows for swimming motion with the entire body, which is the characteristic eel-like (anguilliform) mode of locomotion. The skin of glass eels develops through several stages of pigmentation, beginning from the head and tail tips of the elongated body, and eventually covering the entire body. Once they reach estuaries, they either begin upstream migration into freshwater, such as rivers and lakes, or remain in the estuaries (see Chap. 6); during this period, body length shrinks slightly.

Eels shift from the glass eel to the elver (called *kuroko* in Japanese) when the deposition of body pigmentation is morphologically completed. The developmental stages have been identified in detail in the European eel (Strubberg 1913) and Japanese eel (Fukuda et al. 2013), and are classified according to the degree of progression of pigment expression. The glass eels of other anguillid species are similarly pigmented according to their ontogenetic stages (Jellyman 1977; Haro and Krueger 1988).

Feeding begins at the time of transition from glass eels reaching the estuary into elvers, with opportunistic feeding on aquatic and terrestrial animals. The diversity of prey expands during the yellow eel stage.

3.2.4 Yellow Eel

During the yellow eel stage, a yellowish coloration develops on the ventral side of the body. This immature eel growth stage is the longest period in their life history. Some tropical eel species have distinct mottled patterns on their skin, whereas others have plain-colored skin (see Chap. 1). When eels grow to ~300 mm in total length, histologically identifiable testes or ovaries become differentiated (see Chap. 12). The sexual differentiation of eels occurs later than that in other typical fish taxa. This process appears to be environmentally determined, with factors such as density and stress response being crucial (Geffroy and Bardonnet 2016). Males typically occur in the lower reaches of river systems where there are higher densities, whereas females often occur in the upper reaches where eel density is low and their growth conditions are better (Oliveira and McCleave 2000; Walsh et al. 2004; Yokouchi et al. 2012); however, these processes may have been disrupted by anthropogenic factors such as channelization and damming.

The anguillid eels are known as euryhaline fish, and the aquatic habitats they use vary among individuals. This has been referred to as a form of facultative catadromy, wherein not all individuals migrate into freshwater habitats. It has been found that a portion of Japanese eel individuals remain in seawater rather than moving to freshwater, based on the analysis of the otolith microchemistry profiles of Sr:Ca ratios (Tsukamoto et al. 1998). Such variable migratory tactics within a species may be referred to as partial migration. Otolith microchemistry analysis used to reconstruct the habitat profiles of individuals has now been used to examine the types of habitats used by anguillid eels, which has provided valuable data for various teleost fish species. It was found that individual eels could be generally classified as river eels, that inhabit freshwater rivers and lakes; estuarine eels, that inhabit brackish water; and sea eels, that inhabit saltwater tidal flats and inner bays. In addition, eels that change habitats or move between different habitats are referred to as habitat shifters. The abundance of eels in each habitat depends on the eel species and environmental conditions, which have been heavily influenced by human activities, such as dams and various types of revetments. A common pattern is observed, wherein young eels that initially moved upstream for a few years have later moved downstream into the brackish estuary (Daverat et al. 2005; Yokouchi et al. 2012). The American eel at high latitudes have been reported to migrate to estuaries in summer and return to freshwater rivers in winter (Thibault et al. 2007). In temperate eels distributed across broad latitudes, a greater proportion of sea resident tends to be present at higher latitudes because of the higher productivity in the ocean than that in freshwater rivers (Tsukamoto et al. 2009a). This phenomenon mirrors the phenomenon in which freshwater resident forms tend to be more common at low latitudes in anadromous salmonids.

Various degrees of plasticity occur in the life history of eels, such as their growth rate and age at maturation. Age in years can be estimated based on the number of otolith annuli. Environmental factors, such as water temperature, salinity, feeding availability, density, and predation risk, likely affect the onset of the maturation–migration process. Thus, the growth rates of yellow eels differ greatly even within the same species. In general, individuals distributed at low latitudes with higher temperatures grow faster (Jessop 2010). The body size at the end of the yellow eel stage differs between females and males, with females becoming larger and older ages than males in all anguillid eel species. This pattern of sexual size dimorphism reflects the importance of selection for fecundity in anguillid eels, in which the fecundity of females increases isometrically with body weight, but the reproductive success of males is less size-dependent (Stearns 1992). In addition, the maturation ages of temperate eel species inhabiting high latitudes are generally older than those of tropical eels. Meanwhile, the maximum body size of each species is not related to their distribution locations because, for example, both tropical female eels of the Indo-Pacific eel *A. marmorata* and temperate eels of the New Zealand longfin eel *A. dieffenbachii* reach the largest sizes of nearly 2 m in total length.

Reproductive tactics are different for males, with the time until sexual maturity and downstream migration to the ocean for reproduction being generally shorter, resulting in different life spans between the sexes, which can range widely among

Table 3.1 Total length and estimated age of silver eels for anguillid eels

Species	Collection location	Collection year	Sex	Total length (cm)	Age (years)	References[a]
A. anguilla	Ireland	1987, 1988	Female	40.5–92.9	8–57	1
			Male	28.9–46.0	10–33	1
A. rostrata	Canada	1998	Female	37.8–74.0	11–29	2
			Male	32.6–41.2	10–25	2
A. japonica	Japan	2003–2008	Female	47.6–85.6	4–22	3,4
			Male	41.3–63.4	3–14	3,4
A. dieffenbachii	New Zealand	1972–1980	Female	73.7–156.0	25–60	5
			Male	48.2–73.5	12–35	5
A. australis	New Zealand	1972–1980	Female	48.3–102.4	10–35	5
			Male	37.0–59.8	6–24	5
A. reinhardtii	Australia	1999–2001	Female	74–142.3	10–30	6,7
			Male	44.6–62	7–19	6,7
A. marmorata	Japan	2009, 2010	Female	80.0–163.0	7–23	8
			Male	68.3[b]	7	8
A. marmorata	Réunion Island	2001	Male	70.8[b]	10	9
A. bicolor bicolor	Réunion Island	2001	Female	74.4[b]	8	9
A. celebesensis	Indonesia	2009, 2010	Female	58.5–108.3	5–11	8

[a](1) Poole and Reynolds (1996), (2) Jessop et al. (2004); (3) Yokouchi et al. (2009); (4) Sudo et al. (2013), (5) Todd (1980), (6) Walsh et al. (2003), (7) Walsh et al. (2004), (8) Hagihara et al. (2018), (9) Robinet and Feunteun (2002)
[b]Data from one specimen

anguillid eel species and locations. Sexual selection that favors larger male size in polygynous species would not be adaptive in anguillid eel species. There also appear to be sex-dependent threshold sizes that eels must attain before starting to mature and become silver eels. The maturity sizes and ages of females were generally larger and older than those of males in all anguillid eel species (Table 3.1).

Eels have one of the greatest longevity among teleosts. Long-term monitoring of individuals in closed waters, such as lakes and wells, has revealed incredibly old individuals. At Lake Rotoiti, surrounded by mountains within the Nelson Lakes National Park in New Zealand, the oldest known age has been reported to be 106 in a New Zealand longfin eel (Jellyman 1995).

3.2.5 Silver Eel

Silvering of anguillid eels is accompanied by drastic changes in morphological features and physiological processes that occur to prepare the body for adaptation to their long migration to the natal ocean site for reproduction. As they begin to mature, the dorsal body skin becomes darker, and the ventral skin becomes metallic silver colored with guanine pigmentation, called silver eels (Okamura et al. 2007). The eye size increases and the pectoral fins of silver eels become longer and black colored. The wall of the swim bladder thickens, and its function in regulating the internal gas pressure that controls buoyancy is enhanced (Yamada et al. 2001). It is estimated that during oceanic spawning migration the eels do not feed, due to their teeth dissolving and their digestive tracts constricting (Chow et al. 2010).

Seaward downstream migration of silver eels occurs in the autumn in temperate eels. These downstream movements are considered to be triggered by a drop in water temperature, rainfall, high river flow, or possibly atmospheric pressure (see Chap. 13). In the ocean, silver eels swim toward their spawning areas while conducting vertical migrations to avoid the photic zone (see Chap. 7). Swimming speed is estimated to be ~0.5 of body length per second (van Ginneken et al. 2005a). In a laboratory swim trial, European silver eels continued to swim the equivalent of 5500 km to their spawning area in the Sargasso Sea over a period of 6 months without feeding (van Ginneken et al. 2005a). The experiment revealed that eels can swim at a remarkably low energy expenditure, which is 4–6 times more efficient than non-anguilliform fish such as salmonids. Eels store large amounts of fat before starting their migration, and consume it efficiently as they migrate through the ocean.

Fully mature silver eels of the Japanese eel and the Indo-Pacific eel were captured from their spawning areas at the West Mariana Ridge in the Pacific Ocean (see Chap. 7). However, the reproductive behavior of male and female anguillid eels has never been observed in the ocean. In aquarium experiments, the spawning behavior of artificially matured eels, showed that males swim around a female and stimulate her to induce spawning, leading to sperm release (Boëtius and Boëtius 1980; Dou et al. 2007). These laboratory observations and genetic analyses suggest that anguillid eels spawn by polyandry or group mating (van Ginneken et al. 2005b; Takeuchi et al. 2022). Anguillid eels are categorized as semelparous fish, but it has been reported that some evidence from the developmental stages of oocytes in their ovaries indicates that the Japanese eel is capable of multiple spawning at least 2 or 3 times over several months within a single year (Shimizu et al. 2021). When the reproduction is complete and fertilized eggs are hatched in the ocean, the life history of semelparous eel species is complete.

3.3 The Unique Life History of Anguillid Eels

A basic overview highlights how unique freshwater eels are compared to all other fish and marine eels. Much of this information is also covered in other chapters on ecology and physiology. The spawning area and leptocephalus larval stage of the diadromous life history of anguillid eels are not flexible, and are constrained to specific oceanic areas (see Chap. 4). Larvae are passively transported from these areas by ocean currents and eddies to their growth habitats. In contrast, growth-stage yellow eels appear to be highly plastic to adapt to a wide range of geographic regions and continental aquatic environments.

Anguillid eels are not found in all regions of the world (absent in the eastern Pacific and South Atlantic Oceans), possibly owing to the necessity of warm ocean currents for larval migration and growth. However, where they are present, they appear to be highly adaptable to environmental variations. They use only deep-ocean oligotrophic waters for reproduction, but almost all types of nearshore marine or freshwater aquatic environments can be used as growth habitats. The types of habitats needed for reproduction and larval nursery areas are present in areas such as the Indonesian Seas. Therefore, some tropical eel species can spawn locally and utilize larval retention within the region. Other species have evolved to use far offshore areas with westward-flowing ocean currents to transport their larvae over long distances toward their growth habitats. Once they reach coastal waters, the high degree of plasticity of the yellow eel stage results in some individuals using various available habitats in river drainages, from the estuary to the small streams far upstream. Few species of fish or other animals seem to use such a wide range of aquatic habitats, ranging from the far offshore open ocean to the smallest freshwater streams far from the ocean.

As a creature with such a unique life cycle, the anguillid eel has been studied by many scientists, both biologically and anthropologically, in many parts of the world. However, there are still many aspects of its life history that remain unresolved, particularly during the period from its spawning migration after entering the ocean as maturing silver eels to the next generation. Thus, those parts of the oceanic life history would be useful subjects for future research.

References

Aoyama J, Wouthuyzen S, Miller MJ, Inagaki T, Tsukamoto K (2003) Short-distance spawning migration of tropical freshwater eels. Biol Bull 204:104–108. https://doi.org/10.2307/1543500

Aoyama J, Watanabe S, Miller MJ, Mochioka N, Otake T, Yoshinaga T, Tsukamoto K (2014) Spawning sites of the Japanese eel in relation to oceanographic structure and the West Mariana Ridge. PLoS One 9:e88759. https://doi.org/10.1371/journal.pone.0088759

Béguer-Pon M, Castonguay M, Shan S, Benchetrit J, Dodson JJ (2015) Direct observations of American eels migrating across the continental shelf to the Sargasso Sea. Nat Commun 6:8705. https://doi.org/10.1038/ncomms9705

Bezdenezhnykh VA, Prokhorchik GA, Petrikov AM, Petukhov VB, Plyuta MV (1983) Obtaining the larvae of European eel *Anguilla anguilla* L. (Pisces, Anguillidae) under experimental conditions. Dokl Akad Nauk SSSR 268:1264–1266

Boëtius I, Boëtius J (1980) Experimental maturation of female silver eels, *Anguilla anguilla*. Estimates of fecundity and energy reserves for migration and spawning. Dana 1:1–28. https://www.aqua.dtu.dk/-/media/institutter/aqua/publikationer/dana/dana_vol_1_pp_1_28.pdf. Accessed 8 June 2023

Bonhommeau S, Castonguay M, Rivot E, Sabatié R, Le Pape O (2010) The duration of migration of Atlantic *Anguilla* larvae. Fish Fish 11:289–306. https://doi.org/10.1111/j.1467-2979.2010.00362.x

Castonguay M, McCleave JD (1987) Vertical distributions, diel and ontogenetic vertical migrations and net avoidance of leptocephali of *Anguilla* and other common species in the Sargasso Sea. J Plankton Res 9:195–214. https://doi.org/10.1093/plankt/9.1.195

Chambers RC, Leggett WC (1996) Maternal influences on variation in egg sizes in temperate marine fishes. Am Zool 36:180–196. https://doi.org/10.1093/icb/36.2.180

Chow S, Kurogi H, Katayama S, Ambe D, Okazaki M, Watanabe T, Ichikawa T, Kodama M, Aoyama J, Shinoda A, Watanabe S, Tsukamoto K, Miyazaki S, Kimura S, Yamada Y, Nomura K, Tanaka H, Kazeto Y, Hata K, Handa T, Tawa A, Mochioka N (2010) Japanese eel *Anguilla japonica* do not assimilate nutrition during the oceanic spawning migration: evidence from stable isotope analysis. Mar Ecol Prog Ser 402:233–238. https://doi.org/10.3354/meps08448

Daverat F, Tomas J, Lahaye M, Palmer M, Elie P (2005) Tracking continental habitat shifts of eels using otolith Sr/ca ratios: validation and application to the coastal, estuarine and riverine eels of the Gironde–Garonne–Dordogne watershed. Mar Freshw Res 56:619–627. https://doi.org/10.1071/MF04175

Dou SZ, Yamada Y, Okamura A, Tanaka S, Shinoda A, Tsukamoto K (2007) Observations on the spawning behavior of artificially matured Japanese eels *Anguilla japonica* in captivity. Aquaculture 266:117–129. https://doi.org/10.1016/j.aquaculture.2007.02.032

Fukuda N, Miller MJ, Aoyama J, Shinoda A, Tsukamoto K (2013) Evaluation of the pigmentation stages and body proportions from the glass eel to yellow eel in *Anguilla japonica*. Fish Sci 79: 425–438. https://doi.org/10.1007/s12562-013-0621-x

Geffroy B, Bardonnet A (2016) Sex differentiation and sex determination in eels: consequences for management. Fish Fish 17:375–398. https://doi.org/10.1111/faf.12113

Gross MR, Coleman RM, McDowall RM (1988) Aquatic productivity and the evolution of diadromous fish migration. Science 239:1291–1293. https://doi.org/10.1126/science.239.4845.1291

Hagihara S, Aoyama J, Limbong D, Tsukamoto K (2018) Age and growth of migrating tropical eels, *Anguilla celebesensis* and *Anguilla marmorata*. J Fish Biol 92:1526–1544. https://doi.org/10.1111/jfb.13608

Haro AJ, Krueger WH (1988) Pigmentation, size, and migration of elvers (*Anguilla rostrata* (Lesueur)) in a coastal Rhode-Island stream. Can J Zool 66:2528–2533. https://doi.org/10.1139/z88-375

Inoue JG, Miya M, Miller MJ, Sado T, Hanel R, Hatooka K, Aoyama J, Minegishi Y, Nishida M, Tsukamoto K (2010) Deep-ocean origin of the freshwater eels. Biol Lett 6:363–366. https://doi.org/10.1098/rsbl.2009.0989

Jellyman DJ (1977) Invasion of a New Zealand freshwater stream by glass-eels of two *Anguilla* spp. NZ J Mar Freshw Res 11:193–209. https://doi.org/10.1080/00288330.1977.9515673

Jellyman DJ (1995) Longevity of longfinned eels *Anguilla dieffenbachii* in a New Zealand high country lake. Ecol of Freshw Fish 4:106–112. https://doi.org/10.1111/j.1600-0633.1995.tb00123.x

Jessop BM (2010) Geographic effects on American eel (*Anguilla rostrata*) life history characteristics and strategies. Can J Fish Aquat Sci 67:326–346. https://doi.org/10.1139/F09-189

Jessop BM, Shiao JC, Iizuka Y, Tzeng WN (2004) Variation in the annual growth, by sex and migration history, of silver American eels *Anguilla rostrata*. Mar Ecol Prog Ser 272:231–244. https://doi.org/10.3354/meps272231

Kuroki M, Aoyama J, Miller MJ, Wouthuyzen S, Arai T, Tsukamoto K (2006) Contrasting patterns of growth and migration of tropical anguillid leptocephali in the western Pacific and Indonesian seas. Mar Ecol Prog Ser 309:233–246. https://doi.org/10.3354/meps309233

Kuroki M, Miller MJ, Tsukamoto K (2014) Diversity of early life-history traits in freshwater eels and the evolution of their oceanic migrations. Can J Zool 92:749–770. https://doi.org/10.1139/cjz-2013-0303

Leander NJ, Shen KN, Chen RT, Tzeng WN (2012) Species composition and seasonal occurrence of recruiting glass eels (*Anguilla* spp.) in the Hsiukuluan River, eastern Taiwan. Zool Stud 51: 59–71

Lokman PM, Young G (2000) Induced spawning and early ontogeny of New Zealand freshwater eels (*Anguilla dieffenbachii* and *A. australis*). NZ J Mar Freshw Res 34:135–145. https://doi.org/10.1080/00288330.2000.9516921

McDowall RM (1987) Evolution and importance of diadromy: the occurrence and distribution of diadromy among fishes. In: Dadswell MJ, Klauda RJ, Moffitt CM, Saunders RL, Rulifson RA, Cooper JE (eds) Common strategies of anadromous and catadromous fishes, vol 1. American Fisheries Society, Bethesda, MD, pp 1–13

McDowall RM (1988) Diadromy in fishes: migrations between freshwater and marine environments. Croom Helm, London, p 308

Miller MJ (2009) Ecology of anguilliform leptocephali: remarkable transparent fish larvae of the ocean surface layer. Aqua-BioSci Monogr 2(4):1–94

Miller MJ, Tsukamoto K (2004) An introduction to leptocepahali: biology and identification. Ocean Research Institute, the University of Tokyo, Tokyo, 96 pp

Miller MJ, Bonhommeau S, Munk P, Castonguay M, Hanel R, McCleave JD (2015) A century of research on the larval distributions of the Atlantic eels: a reexamination of the data. Biol Rev 90: 1035–1064. https://doi.org/10.1111/brv.12144

Okamura A, Yamada Y, Yokouchi K, Horie N, Mikawa N, Utoh T, Tanaka S, Tsukamoto K (2007) A silvering index for the Japanese eel *Anguilla japonica*. Environ Biol Fish 80:77–89. https://doi.org/10.1007/s10641-006-9121-5

Oliveira K, McCleave JD (2000) Variation in population and life history traits of the American eel, *Anguilla rostrata*, in four rivers in Maine. Environ Biol Fish 59:141–151. https://doi.org/10.1023/A:1007631108201

Otake T, Inagaki T, Hasumoto H, Mochioka N, Tsukamoto K (1998) Diel vertical distribution of *Anguilla japonica* leptocephali. Ichthyol Res 45:208–211. https://doi.org/10.1007/BF02678565

Pfeiler E (1999) Developmental physiology of elopomorph leptocephali. Comp Biochem Physiol A 123:113–128. https://doi.org/10.1016/S1095-6433%2899%2900028-8

Poole WR, Reynolds JD (1996) Growth rate and age at migration of *Anguilla anguilla*. J Fish Biol 48:633–642. https://doi.org/10.1111/j.1095-8649.1996.tb01458.x

Robinet T, Feunteun E (2002) First observations of shortfinned *Anguilla bicolor bicolor* and longfinned *Anguilla marmorata* silver eels in the Réunion Island. Bull Fr Pêche Piscic 364: 87–95. https://doi.org/10.1051/kmae:2002004

Robinet T, Réveillac E, Kuroki M, Aoyama J, Tsukamoto K, Rabenevanana MW, Valade P, Gagnaire P-A, Berrebi P, Feunteun E (2008) New clues for freshwater eels (*Anguilla* spp.) migration routes to eastern Madagascar and surrounding islands. Mar Biol 154:453–463. https://doi.org/10.1007/s00227-008-0938-7

Schmidt J (1922) The breeding places of the eel. Philos Trans R Soc Lond Ser B 211:179–208

Shimizu A, Ijiri S, Izumi H, Gen K, Kurogi H, Hashimoto H, Tanaka H, Jinbo T, Saito H, Chow S (2021) Histological evidence of multiple spawning in wild female Japanese eel *Anguilla japonica*. Zool Stud 60:61. https://doi.org/10.6620/ZS.2021.60-61

Stearns SC (1992) The evolution of life histories. Oxford University, London

Strubberg AC (1913) The metamorphosis of elvers as influenced by outward conditions. Medd Komm Havunders ser Fisk 4:1–11

Sudo R, Fukuda N, Aoyama J, Tsukamoto K (2013) Age and body size of Japanese eels, *Anguilla japonica*, at the silver-stage in the Hamana Lake system, Japan. Coast Mar Sci 36:13–18. https://doi.org/10.15083/00040623

Takeuchi A, Sawayama E, Kuroki M, Miller MJ, Watanabe S, Tsukamoto K (2022) Preliminary insight into parental contributions to Japanese eel (*Anguilla japonica*) preleptocephali spawned on different nights. J Fish Biol 101:1601–1605. https://doi.org/10.1111/jfb.15215

Thibault I, Dodson JJ, Caron F (2007) Yellow-stage American eel movements determined by microtagging and acoustic telemetry in the St Jean river watershed, Gaspé, Quebec, Canada. J Fish Biol 71:1095–1112. https://doi.org/10.1111/j.1095-8649.2007.01584.x

Todd PR (1980) Size and age of migrating New Zealand freshwater eels (*Anguilla* spp.). NZ J Mar Freshw Res 14:283–293. https://doi.org/10.1080/00288330.1980.9515871

Tsukamoto K (1992) Discovery of the spawning area for Japanese eel. Nature 356:789–791. https://doi.org/10.1038/356789a0

Tsukamoto K, Miller MJ (2021) The mysterious feeding ecology of leptocephali: a unique strategy of consuming marine snow materials. Fish Sci 87:11–29. https://doi.org/10.1007/s12562-020-01477-3

Tsukamoto K, Nakai I, Tesch W (1998) Do all freshwater eels migrate? Nature 396:635–636. https://doi.org/10.1038/25264

Tsukamoto K, Aoyama J, Miller MJ (2002) Migration, speciation and the evolution of diadromy in anguillid eels. Can J Fish Aquat Sci 59:1989–1998. https://doi.org/10.1139/f02-165

Tsukamoto K, Miller MJ, Kotake A, Aoyama J, Uchida K (2009a) The origin of fish migration: the random escapement hypothesis. In: Haro AJ, Smith KL, Rulifson RA, Moffitt CM, Klauda RJ, Dadswell MJ, Cunjak RA, Cooper JE, Beal KL, Avery TS (eds) Challenges for diadromous fishes in a dynamic global environment, vol 69. American Fisheries Society, Bethesda, MD, pp 45–61

Tsukamoto K, Yamada Y, Okamura A, Kaneko T, Tanaka H, Miller MJ, Horie N, Mikawa N, Utoh T, Tanaka S (2009b) Positive buoyancy in eel leptocephali: an adaptation for life in the ocean surface layer. Mar Biol 156:835–846. https://doi.org/10.1007/s00227-008-1123-8

Tsukamoto K, Chow S, Otake T, Kurogi H, Mochioka N, Miller MJ, Aoyama J, Kimura S, Watanabe S, Yoshinaga T, Shinoda A, Kuroki M, Oya M, Watanabe T, Hata K, Ijiri S, Kazeto Y, Nomura K, Tanaka H (2011) Oceanic spawning ecology of freshwater eels in the western North Pacific. Nat Commun 2:179. https://doi.org/10.1038/ncomms1174

van Ginneken V, Antonissen E, Müller UK, Booms R, Eding E, Verreth J, van den Thillart G (2005a) Eel migration to the Sargasso: remarkably high swimming efficiency and low energy costs. J Exp Biol 208:1329–1335. https://doi.org/10.1242/jeb.01524

van Ginneken V, Vianen G, Muusze B, Palstra A, Verschoor L, Lugten O, Onderwater M, Schievan S, Niemantsverdriet P, van Heeswijk R, Eding E, van den Thillart G (2005b) Gonad development and spawning behavior of artificially matured European eel (*Anguilla anguilla* L.). Anim Biol 55:203–218. https://doi.org/10.1163/1570756054472791

Walsh CT, Pease BC, Booth DJ (2003) Sexual dimorphism and gonadal development of the Australian longfinned river eel. J Fish Biol 63:137–152. https://doi.org/10.1046/j.1095-8649.2003.00136.x

Walsh CT, Pease BC, Booth DJ (2004) Variation in the sex ratio, size and age of longfinned eels within and among coastal catchments of South-Eastern Australia. J Fish Biol 64:1297–1312. https://doi.org/10.1111/j.0022-1112.2004.00392.x

Yamada Y, Zhang H, Okamura A, Tanaka S, Horie N, Mikawa N, Utoh T, Oka HP (2001) Morphological and histological changes in the swim bladder during maturation of the Japanese eel. J Fish Biol 58:804–814. https://doi.org/10.1111/j.1095-8649.2001.tb00532.x

Yamamoto K, Yamauchi K (1974) Sexual maturation of Japanese eel and production of eel larvae in the aquarium. Nature 251:220–222. https://doi.org/10.1038/251220a0

Yokouchi K, Sudo R, Kaifu K, Aoyama J, Tsukamoto K (2009) Biological characteristics of silver-phase Japanese eels, *Anguilla japonica*, collected from Hamana Lake, Japan. Coast Mar Sci 33: 54–63. https://doi.org/10.15083/00040704

Yokouchi K, Fukuda N, Miller MJ, Aoyama J, Daverat F, Tsukamoto K (2012) Influences of early habitat use on the migratory plasticity and demography of Japanese eels in Central Japan. Estuar Coast Shelf Sci 107:132–140. https://doi.org/10.1016/j.ecss.2012.05.009

Chapter 4
Spawning Areas

Michael J. Miller

Contents

Where anguillid eels go to reproduce was historically considered to be one of the major biological mysteries of these species, and when Johannes Schmidt discovered the large overlapping Sargasso Sea spawning areas of the Atlantic eels of the genus *Anguilla*, it became a famous revelation about fish migration. This was followed by Katsumi Tsukamoto's discovery of the Japanese eel spawning area in the North Equatorial Current (NEC) of the western North Pacific (WNP), which was also a widely reported scientific discovery. The spawning area of the northern population of *A. marmorata* was found to overlap with the Japanese eel spawning area; however, *A. marmorata* has spawning populations in all three Indo-Pacific subtropical gyres. Continued research found that Japanese eels spawn during new moon periods at sites that vary latitudinally according to the salinity structure. Adult, egg, and pre-feeding larvae (preleptocephali) were eventually caught in the Japanese eel spawning area, and then underwater camera systems and environmental DNA (eDNA) techniques were then used to search for spawning eels. Internationally-collaborative larval sampling surveys were conducted to search for anguillid spawning areas in the Indonesian Seas and across the Indo-Pacific, and pop-up satellite archival transmitter (PSAT) tracking studies were conducted in the western South Pacific (WSP), both of which indicated that the spawning areas of anguillid species can occur both near or far from freshwater habitats. Eels from Australia and

M. J. Miller (✉)
Graduate School of Agricultural and Life Sciences, The University of Tokyo, Tokyo, Japan

K. Tsukamoto et al. (eds.), *Eel Science*, Fisheries Science Series,
https://doi.org/10.1007/978-981-99-5692-0_4

New Zealand migrate far offshore to South Equatorial Current (SEC) spawning areas, but WSP tropical anguillid spawning areas likely vary depending on the species range and ocean-current geography. No direct evidence has been obtained about some other spawning areas (eastern/western Indian Ocean and French Polynesia); however, larval sampling surveys have made great progress in the last century in understanding the locations of anguillid spawning areas in both the Atlantic and Indo-Pacific. This chapter provides an overview of the present state of understanding of anguillid eel spawning areas, with an emphasis on recent literature.

4.1 Atlantic Eel Spawning Areas

After concluding that the European eel does not appear to spawn in the Mediterranean Sea, Schmidt (1922) used ships of opportunity (i.e., cargo or navy vessels) and then research vessels to make plankton tows for eel larvae (leptocephali) across the North Atlantic, and the size data of the collected leptocephali enabled him to make a remarkably accurate map (Fig. 4.1) of the locations of the estimated 2 Atlantic eel spawning areas (Miller et al. 2015). Over 50 years later (1979–1989), German (F.-W. Tesch) and American (J.D. McCleave) research teams returned to the Sargasso Sea to further understand the 2 spawning areas, which Schmidt had previously only generally outlined with his now well-known overlapping spawning area ovals. The accuracy of his findings was assessed using a database constructed from all the available catch data of the leptocephali of both species, including Schmidt's data, German and American team surveys, all available incidental catches, and a recent

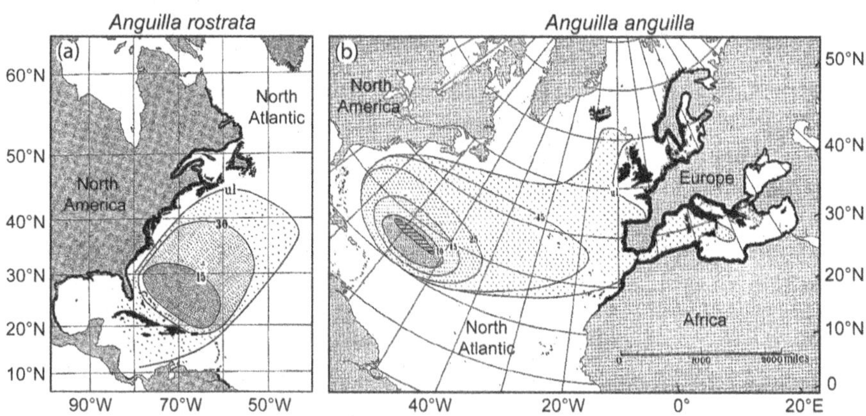

Fig. 4.1 Maps showing the estimated distributions of several size categories (ovals: size mm, ul: unmetamorphosed larvae) of American eel, *Anguilla rostrata* (**a**) and European eel, *A. anguilla* (**b**) leptocephali in the North Atlantic from the catch data of Johannes Schmidt that was modified from Schmidt (1935). Black lines on coastlines show the recruitment ranges estimated at that time (inland ranges not shown)

2007 Danish survey, which totaled more than 30,000 larvae of both species (Miller et al. 2015). The spatial and temporal plots of the larval sizes of each species showed remarkably different patterns because both species recruit to opposite sides of the North Atlantic at different ages; however, the distributions of their smallest larvae appear to outline their spawning areas fairly clearly.

The European eel spawns in a narrow band of latitude (Fig. 4.2a), which is determined by a northern temperature front that forms at 22 °C, with spawning extending south to, and a little past, a second front at 24 °C (Kleckner and McCleave 1988; Miller et al. 2015, 2019). The smallest larvae were collected across a wide range of longitudes (70°–50°W). A German survey in 1979 suggested that eels may spawn across the whole region within one spawning season (Schoth and Tesch 1982); however, in 2014, this was shown more clearly because both German and Danish research vessels concurrently conducted separately scheduled larval sampling surveys and alternated station transects across the entire spawning area (Miller et al. 2019). The results showed that small European eel larvae were present across 2000 km of the Sargasso Sea during a single period of the spawning season of 1 year (Fig. 4.2a). German surveys to evaluate the relative abundance of European eel leptocephali from a stock abundance perspective were conducted in 2011, 2014, and 2017, with a more widely spaced non-fisheries survey conducted in 2015, all of which found larval distribution patterns that were consistent with a well-defined spawning area in relation to the frontal positions (Hanel et al. 2014, unpublished data). Larval abundances due to fewer spawners or lower larval survival have been observed (Hanel et al. 2014; Westerberg et al. 2018), and global warming may cause spawning to occur at slightly higher latitudes if frontal positions are moving north.

In contrast, the spawning area of the American eel has an entirely different shape based on the catches of small larvae, but may also be affected by changes in frontal positions. Small larvae were previously caught in an oval-shaped region of the southwestern Sargasso Sea (Fig. 4.2a), but there has been insufficient sampling in recent decades to define the western or southern boundaries of the spawning area (Miller et al. 2015). This more western and southern spawning area is consistent with where the larvae eventually recruit, which includes areas along the margins of the Caribbean Sea, Gulf of Mexico, and Atlantic coast of North America from Florida to Atlantic Canada. Spawning in the southwestern Sargasso Sea would result in larvae moving south through the passages between the Greater Antilles where American eels live and into the westward flow of the Caribbean that eventually enters the Gulf of Mexico, where some eels recruit. Larvae would also move northwest with the flow of the Antilles Current and eventually enter the Florida Current and Gulf Stream for recruitment into Atlantic coast drainages.

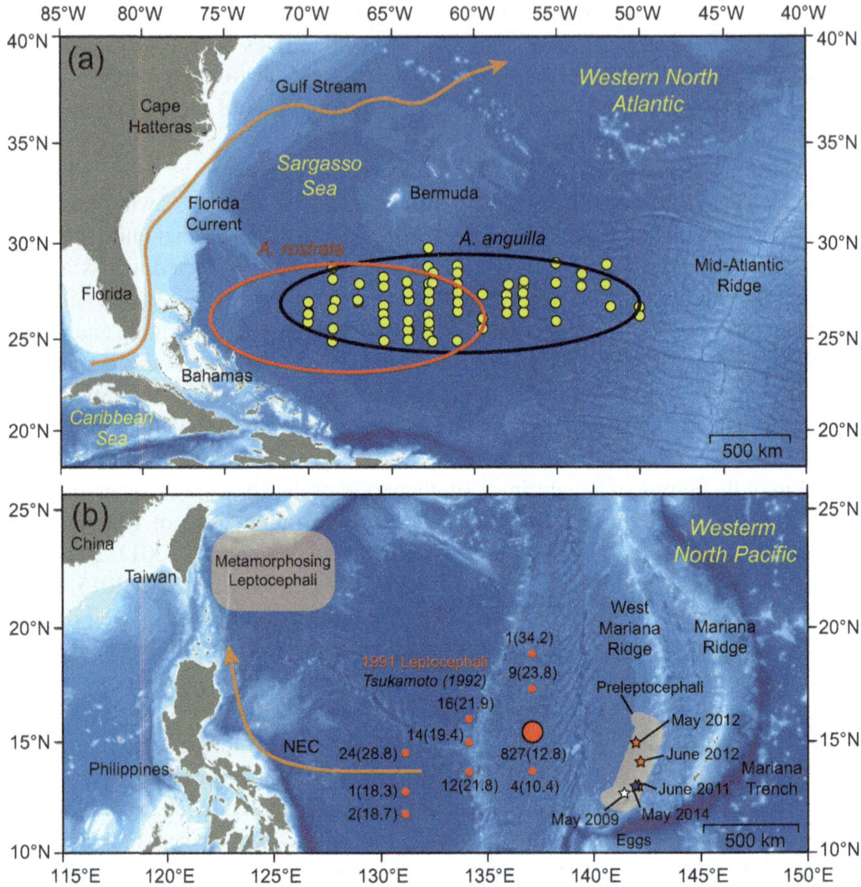

Fig. 4.2 Maps of Atlantic eel and *Anguilla japonica* spawning areas, showing (**a**) the general outlines (ovals) of where most small *A. rostrata* and *A. anguilla* leptocephali (≤12 mm) were collected (Miller et al. 2015). This roughly corresponds to typical areas of spawning, and stations where small ≤12 mm *A. anguilla* leptocephali (yellow circles) were collected by German and Danish research ships in Mar-Apr of 2014 show that spawning had occurred across a ~ 2000 km wide area in one season (Miller et al. 2019). Panel (**b**) shows a map of the *A. japonica* spawning and larval distribution region, the general collection areas of preleptocephali (shaded area), and where eggs (stars) have been collected along the WMR during 5 different new moon periods of 4 different years (see text for references). To the west, red circles show stations where leptocephali were collected (number caught and average size in parentheses) in the large-scale 1991 survey that discovered the spawning area (Tsukamoto 1992). The large red circle shows where 827 leptocephali were collected in multiple IKMT tows during one night, south of a salinity front. The region where metamorphosing leptocephali have been collected near Taiwan is shown (Otake et al. 2006; Fukuda et al. 2018). Blue (deeper) and white (shallow) color shades show bathymetric depths, with oceanic trenches in purple

4.2 Japanese Eel Spawning Area

Decades after the Atlantic eel spawning area was discovered, Japanese scientists began searching for the spawning area of the Japanese eel, *A. japonica*, along the western margin of the WNP (Shinoda et al. 2011). In 1991, Tsukamoto and his colleagues designed and sampled a large grid of Issacs-Kidd Midwater Trawl (IKMT) stations arranged in 6 long latitudinal transects spread out on both sides of the Mariana Ridge (including Guam). This resulted in the first small (~10 mm) *A. japonica* leptocephali (Fig. 4.2b) being collected that were found in the NEC (Tsukamoto 1992), which began a new era of Japanese eel spawning-area research (Tsukamoto et al. 2003).

After it was determined in 1991 that the Japanese eel spawns far offshore in the NEC (Tsukamoto 1992), efforts were made to determine the size and shape of the spawning area, considering that the spawning areas of the Atlantic eels are quite large. No small larvae were collected east of the ridge, so the catches of larvae seemed to point to the seamount chain of the West Mariana Ridge (WMR) (Tsukamoto et al. 2003). Therefore, survey cruises to the spawning area from 1998 to 2017 focused on the region along the seamount chain of the WMR. The first preleptocephali were collected in 2005 downstream of one of the seamounts (Tsukamoto 2006), and newly hatched larvae were collected during 2007 and 2008 at various latitudes along the west side of the WMR, until the first eggs were collected in 2009, along with more preleptocephali (Tsukamoto et al. 2011). The timing and locations of these collections (Fig. 4.2b) and otolith back-calculated hatching dates of the leptocephali indicated that the Japanese eel spawns along the west side of the WMR during the new moon periods of each month during the April–August spawning season (Shinoda et al. 2011; Tsukamoto et al. 2011).

Starting from the collection of the first small leptocephali (Tsukamoto 1992), the salinity front that forms at the NEC latitudes appeared to be an important landmark of the spawning sites. Later surveys also found spawning to occur south of where a distinct salinity front was present (Kimura and Tsukamoto 2006). However, the formation, strength, and presence of these near-surface fronts (upper 100 m) are influenced by tropical rainfall and other factors; therefore, distinct fronts are not always present. In 2011, 2012, and later surveys, a clear set of hypotheses (New Moon, Seamount Ridge, and Salinity Front) and sampling strategies were used to try to find individual spawning sites (by catching eggs) and to determine the latitudinal range of spawning during each new moon period of a sampling survey (Aoyama et al. 2014). This consisted of first determining the salinity structure along the WMR before a new moon period and then estimating a likely site for spawning based on if there was a clear salinity front or not. After sampling a grid of stations for eggs in each of these surveys, sampling was then conducted along the entire WMR for preleptocephali to approximate the range of latitudes at which spawning occurred during that new moon. This strategy was used in 2011, 2012, and 2014, which resulted in both eggs and preleptocephali being collected at several locations

(Fig. 4.2b) south of fronts or across a wider range of latitudes (Aoyama et al. 2014; Takeuchi et al. 2021).

The surveys also found that in some months, there was no distinct salinity front, and low salinity extended north across the entire spawning area latitude. In these cases, spawning occurred at several latitudes (based on the presence of preleptocephali) (Aoyama et al. 2014). Whenever a distinct salinity front was present, spawning appeared to occur only south of the fronts, but in several locations along the southern part of the WMR (Tsukamoto et al. 2011; Aoyama et al. 2014; Takeuchi et al. 2021). However, how far west of WMR spawning might occur still remains unclear.

The set of hypotheses regarding when and where spawning sites form was also used to deploy underwater camera systems to search for spawning Japanese eel adults, which marked the first attempt to visualize anguillid eels in their spawning area. These efforts included the use of the YOKOSUKA Deep-Tow underwater camera system (2001, 2012, and 2017), Hyper-Dolphin ROV, UnaCam drifting camera systems, and Shinkai 6500 submersible (Tsukamoto et al. 2013; Fukuba et al. 2015). A relatively new eDNA technique was also used to detect the presence of Japanese eels at 250 and 400 m in likely spawning locations (Takeuchi et al. 2019). Various marine animals were recorded, but no clear anguillid eels were observed until one eel consistent with being a Japanese eel was video-recorded in May 2017 at approximately 220 m in the same area where a high concentration of Japanese eel eDNA was found at 400 m at a time and place consistent with it originating from a new moon spawning event (Takeuchi et al. 2022).

Therefore, the long-term research survey efforts of Tsukamoto and his colleagues and students were able to eventually predict spawning sites and times precisely enough to detect the resulting eDNA from apparent spawning. This also resulted in a video recording being obtained of an apparent Japanese eel in the same area after achieving other remarkable milestones of collecting eggs and spawning-condition adults, which were collected in separate trawling surveys targeting the spawning adults (Chow et al. 2009; Tsukamoto et al. 2011). These achievements make the spawning ecology of the Japanese eel the most well-known of any species of anguillid eel in the world, which makes it a model species for further research on the behavioral ecology of anguillid eel spawning.

4.3 Central Indo-Pacific Eel Spawning Areas

From 1928 to 1930, Schmidt led the Danish Round the World Expedition that sailed west across the Atlantic and eventually through French Polynesia, over to New Caledonia, down to New Zealand, and then over to Australia (Schmidt 1935); however, they did not have sampling stations in the areas where anguillids were later found to spawn in the WSP. They then went south of New Guinea, through the Indonesian Seas, and over to China without collecting any small leptocephali (Wouthuyzen et al. 2009), until returning to the south where a spawning area of

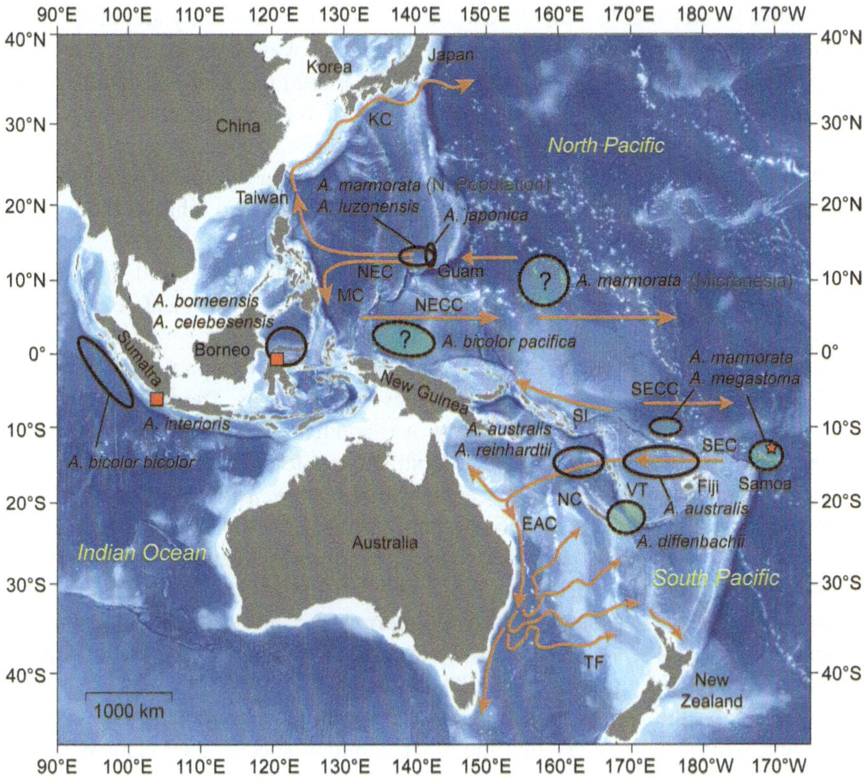

Fig. 4.3 Map of East and Southeast Asia and central Indo-Pacific regions showing known (solid-line ovals) and estimated (shaded, dashed-line ovals) spawning areas, with question marks showing spawning areas that have the least information about their exact locations. Red squares show where a 12.4 mm *A. interioris* larva was collected off southeastern Sumatra and where a few larvae were caught in Tomini Bay of Sulawesi Island, which indicates at least 2 spawning areas of that species. The red star near Samoa shows were a 7.8 mm *A. marmorata* larva was caught, suggesting local spawning there. Orange arrows shown the general locations of major current systems such as Kuroshio Current (KC), North Equatorial Current (NEC), Mindanao Current (MC), North Equatorial Countercurrent (NECC), South Equatorial Counter Current (SECC), South Equatorial Current (SEC), East Australia Current (EAC), and Tasman Front (TF), and the geographic locations of Solomon Islands (SI), New Caledonia (NC), Vanuatu (VN) are labeled

A. bicolor bicolor was found through collections of many small leptocephali off West Sumatra (Fig. 4.3; Aoyama et al. 2007). The expedition then passed through the northern Indian Ocean and down the west side of Madagascar before entering the South Atlantic, but it did not enter any likely areas of spawning in the western Indian Ocean. Therefore, the locations of all Indo-Pacific anguillid spawning areas were unknown after the Round the World Expedition, except for the spawning area that was found off West Sumatra in the eastern Indian Ocean.

The Indonesian Seas form a high-biodiversity central region of the Indo-Pacific, and progress has been made in understanding the spawning areas of most of the

6 anguillid species that live there. Aoyama et al. (2007) genetically identified larger *A. bicolor biolor* larvae from off West Sumatra, which was consistent with the small leptocephali that were collected there during the Round the World Expedition also being of that species. This suggested that the eels of West Sumatra make very short migrations to spawn not far offshore from their juvenile habitats. It remains unclear whether eels of that species living to the north in areas adjacent to the Bay of Bengal also spawn in the same area off Sumatra. This subspecies also inhabits the western side of the Indian Ocean, where no small leptocephali have been found.

The Danish Expedition did not succeed to collect small anguillid leptocephali within the Indonesian Seas, and the next efforts to search for tropical eel spawning areas began in 2000 with the "millennium eel cruise" on the research vessel (R/V) Hakuho Maru, which first had sampling stations in the western part of the WSP and then returned to the north, through the Celebes and Sulu seas. The following year in 2001, an international collaborative cruise was conducted between Japan and Indonesia on the R/V Baruna Jaya VII, and both surveys found that local spawning of 2 anguillid species occurred in the Celebes Sea (*A. bornensis* and *A. celebesensis*) (Aoyama et al. 2003). It was also found that 2 species (*A. celebesensis* and probably *A. interioris*) spawned in Tomini Bay of northern Sulawesi Island (Fig. 4.3; Aoyama et al. 2003). Subsequent surveys in 2002 and 2010 provided further evidence that some of these species spawn locally (*A. borneensis* and *A. celebesensis* in the Celebes Sea), and *A. celebesensis* also spawns in the semi-enclosed Tomini Bay of Sulawesi Island (Wouthuyzen et al. 2009; Aoyama et al. 2018).

Interestingly, the results of all of those surveys and other genetic research, are consistent with the discovery that the other common species on Sulawesi Island, *A. marmorata*, spawns in the NEC (Fig. 4.3), where *A. japonica* spawns after relatively long offshore migrations (Kuroki et al. 2009). Similarly, there is no evidence of *A. bicolor pacifica* spawning locally in the Indonesian Seas, but relatively small leptocephali of that subspecies have been collected in the region just north of New Guinea (Fig. 4.3), suggesting that their spawning area is somewhere in that region (Kuroki et al. 2020); as was indirectly indicated previously (Miller and Tsukamoto 2017). The recently discovered species, *A. luzonensis*, which is mostly endemic to the Philippines, has also been suggested to spawn offshore in the NEC (Kuroki et al. 2012). Small and large leptocephali of the other species in this region, *A. interioris*, have been collected in both Tomini Bay and off West Sumatra, indicating that it spawns locally (Aoyama et al. 2007, 2018).

4.4 South Pacific Eel Spawning Areas

The South Pacific is an interesting but complicated region for trying to understand the spawning areas of the 7 species or subspecies of anguillid eels that live there, because various species are present all the way from eastern Australia and New Zealand to Tahiti and the other islands of French Polynesia. Several Hakuho Maru surveys found evidence that eels from Australia and New Zealand spawned in the SEC that flows into current systems that reach both land masses (Kuroki et al. 2020 and references within). However, the Danish Expedition and the more recent surveys did not provide evidence of where the 3 species of eels inhabiting both the western and eastern regions of the WSP spawn in French Polynesia (*A. marmorata, A. megastoma*, and *A. obscura*).

Two surveys by Hakuho Maru in the WSP confirmed the presence of small *A. australis* leptocephali in the westward flow of the SEC (Fig. 4.3) on both sides of the Vanuatu island chain (Kuroki et al. 2020). This suggests the possibility that the Australian subspecies (*A. australis australis*) spawns on the west side of Vanuatu (would have a shorter adult migration and larval transport distance) and the New Zealand subspecies (*A. australis schmidtii*) spawns on the east side (direct northward adult migration and longer transport distance). There is no evidence of that occurring, but spawning on either side of Vanuatu might be an effective way to prevent interbreeding, even though both subspecies presumably drift westward in the SEC into the southward flow of the East Australian Current, before the larvae of the New Zealand subspecies must be transported to the east across the Tasman Front. Small larvae of the more tropical anguillid species, *A. reinhardtii* (overlapping latitudinal range of eastern Australia with *A. australis australis*), were only collected on the west side of Vanuatu, indicating that they spawn there (Kuroki et al. 2020). The details of the spawning areas in the SEC cannot yet be determined because the small larvae were collected in 2 widely separated transects of stations along the same longitudes (160°E and 175°E) during research cruises in both 1995 and 2016, and there are countercurrents at those latitudes that can transport larvae eastward.

Even less is known about the spawning area of the larger-sized New Zealand longfin eel, *A. dieffenbachii*, which is endemic to New Zealand, because their leptocephali have never been collected and identified. Previous studies that collected small numbers of larger leptocephali were not able to make genetic identifications, and the species was also not detected in recent studies using that definitive method (Kuroki et al. 2020). However, the PSAT tagging method was used to tag large female silver eels, which eventually resulted in clear evidence that the 3 migrating eels were swimming towards the area between New Caledonia and Fiji, while another tag from a previous study was found just northeast of New Caledonia (Jellyman and Tsukamoto 2010). This information indicates that *A. dieffenbachii* migrates to spawn at higher latitudes like other anguillids, but the location of its spawning area and larval migration route remain a mystery.

PSAT tagging studies were also used as a method to attempt to determine the western spawning locations of the widely distributed WSP tropical anguillids, *A. marmorata* and *A. megastoma*. All PSAT-tagged eels that appear to behave normally show clear patterns of diel vertical migrations (DVM), and some individuals that retain their tags swim long distances towards possible spawning areas; this occurred for a few *A. marmorata* and *A. megastoma* eels released from Vanuatu that swam northeast towards the area far northwest of Fiji (Schabetsberger et al. 2021 and references within). This area also corresponds to the location where relatively small *A. marmorata* (16–18 mm) leptocephali were collected during a 2013 Hakuho Maru survey (Miller et al. 2022). In contrast, no PSAT-tagged *A. marmorata* or

A. megastoma swam long distances away from Samoa after being released, even though they had showed DVM behavior for up to 25 days (Schabetsberger et al. 2021). Their lack of swimming far away from Samoa could indicate that they used a local spawning strategy, which would be consistent with the collection of a 7.8 mm *A. marmorata* larva just to the northeast of Samoa (star in Fig. 4.3) in August 2016 (Kuroki et al. 2020; Schabetsberger et al. 2021). Although there are various types of information about glass eel recruitment of other species for which no small larvae have been collected in either the western (New Caledonia, Fiji, etc.) or eastern regions (French Polynesia), for *A. megastoma* and *A. obscura*, it seems likely that both of those species, along with *A. marmorata,* have separate spawning areas in both regions because of the long distance (and few recruitment habitats) between them (Schabetsberger et al. 2021). The only sampling for leptocephali in the French Polynesia region was the 2016 Hakuho Maru survey, which collected no anguillid leptocephali from the 2 transects on either side of French Polynesia.

4.5 Unknown Spawning Areas

Although great progress has been made in narrowing down where anguillid eels spawn, there is still a lack of information about the spawning areas of some species. This is likely related to the fact that most anguillid eels do not spawn throughout the year, even though some tropical locations have some glass eel recruitment during much of the year. Therefore, each individual sampling survey for their small leptocephali has a risk of being conducted in the wrong season or location. This factor (or a simple lack of sampling surveys in some areas) has resulted in little or no information being obtained for several anguillid species or subspecies (Miller and Tsukamoto 2017). This has occurred in the northeastern and western Indian Ocean, where both subspecies *of A. bengalensis bengalensis* (western region) and *A. bengalensis labiata* (northeastern region), along with *A. marmorata* and *A. bicolor bicolor* (present on both sides of the basin) are distributed. There has likely also not been sufficient appropriate sampling along the southern margin of the WNP where *A. bicolor pacifica* might spawn, or where the Micronesian population of *A. marmorata* might spawn east of Guam (Fig. 4.3). A similar lack of sufficient sampling occurred in the French Polynesia region, where the 2 long latitudinal transects of stations in 2016 (Kuroki et al. 2020; Miller et al. 2022) may have been either too far west or too far east of the larger islands of the Tahiti region if those eels spawn locally as they might do at Samoa (Schabetsberger et al. 2021). The eastern transect passed close to the western side of the Marquesas Islands (Kuroki et al. 2020; Miller et al. 2022), where anguillid eels also live; however, this might have been the wrong season or spawning might have occurred to the east of the islands. Additional larval sampling with large nets is required during several seasons around the suspected spawning areas in the Indian Ocean, the southern margin of the WNP, and within the waters of French Polynesia to find the spawning areas of those species of anguillid eels.

4.6 Factors Determining Spawning Area Locations

Although not all anguillid spawning areas have been discovered, it appears that some general patterns have emerged after more has been learned about the locations of species other than the northern hemisphere temperate anguillids. To begin to understand why spawning areas occur in specific locations, several factors must be considered. First, anguillids are semelparous and each female spawns millions of eggs, which means that the spawning area must be able to support many larvae during the spawning season. Second, larvae need sufficient time to feed and grow until they reach large enough sizes for metamorphosis without excessive predation. The classic northern hemisphere long migrations to the westward-flowing equatorial currents provide all these conditions because the larvae are continuously transported out of the spawning areas in oligotrophic waters with theoretically lower numbers of predators, and the currents transport them towards the recruitment areas. Local spawning over tropical deep-water areas or basins with a high degree of eddies or other retention mechanisms could function similarly, such as in the Celebes Sea, Tomini Bay, and off West Sumatra. It is also possible that seasonally occurring currents, such as the South Equatorial Counter Current (SECC) in the WSP, might make a successful offshore spawning area that results in the larvae being distributed to many different areas in that particular region, even though it requires the adult eels to make a long spawning migration. Other tropical areas with weaker or eddy-dominated current flows, such as in French Polynesia, might be most suitable for local spawning because of the lack of regular current flows to transport larvae back to their growth habitats. Finding the locations of the remaining spawning areas and conducting more nuclear DNA population structure research (maternally inherited mtDNA is not suitable) will likely reveal these factors more clearly.

Presently, the Northern Hemisphere, Australia, and New Zealand species fit the model of long offshore migrations that are followed by larval drift in equatorial, western boundary, or other types of current flows; however, active swimming appears to be required to complete the larval period and reach coastal recruitment areas (Miller and Tsukamoto 2017). Tropical species seem likely to have local spawning and offshore migrations if the patterns of ocean currents can support larval transport to recruitment areas (i.e., the Northern Population of *A. marmorata* in the NEC and the spawning area northeast of Vanuatu). However, the factor of apparently genetically programmed swimming behavior (i.e., *A. rostrata*: west; *A. anguilla*: east) seems to be required to recruit successfully, and some species such as *A. bornensis* and *A. luzonensis* appear to preferentially recruit to some areas but not others that could be reached in the same currents used by other species with different geographic recruitment areas. These factors suggest that the evolution of spawning and recruitment areas might be more complex than presently imagined.

4.7 Summary and Future Research

After Schmidt's remarkably accurate estimation of the 2 Atlantic eel spawning-area locations, later surveys showed that both species spawn south of the temperature fronts that form in the Sargasso Sea. The Japanese eel spawning area was found to be in the NEC along the western side of a seamount ridge, where intensive surveys were led by Katsumi Tsukamoto, who along with his colleagues eventually collected all spawning-area life history stages (eggs, preleptocephali, and adults), and video-recorded the first anguillid eel in its spawning area. *Anguilla marmorata* (and probably *A. luzonensis*) was also found to spawn offshore in an overlapping area of the NEC. The same larval-survey techniques were then used to conduct internationally collaborative cruises to search for other Indo-Pacific anguillid eel spawning areas. Short-distance spawning migrations were found to occur in *A. bornensis* and *A. celebesensis*, and also likely occur in *A. interioris* and *A. bicolor bicolor* in the Indonesian Seas/eastern Indian Ocean region. Australia and New Zealand eels were found to migrate to the westward flows of the SEC, and other spawning locations were indicated by both larval sampling and PSAT tagging studies in the western region of the South Pacific. Despite a few larval survey efforts, the northern and western Indian Ocean and the central South Pacific spawning areas of several tropical anguillid species have not yet been found. The areas where spawning may occur in those ocean regions have been somewhat narrowed down and can be inferred based on current patterns and the occurrence patterns of known spawning areas. Future larval surveys that utilize that information can now be conducted in more limited areas and in specific seasons to discover these spawning areas. Nuclear DNA population studies will also be useful for understanding where these eels spawn and how they choose mates in areas where larvae might originate from different spawning locations. It is unclear however, if the future information that will be obtained from new studies will make the migrations and spawning areas of anguillid eels seem less mysterious, or more remarkable.

References

Aoyama J, Wouthuyzen S, Miller MJ, Inagaki T, Tsukamoto K (2003) Short–distance spawning migration of tropical freshwater eels. Biol Bull 204:104–108. https://doi.org/10.2307/1543500

Aoyama J, Wouthuyzen S, Miller MJ, Minegishi Y, Kuroki M, Suharti SR, Kawakami T, Sumardiharga KO, Tsukamoto K (2007) Distribution of leptocephali of the freshwater eels, genus *Anguilla*, in the waters off West Sumatra in the Indian Ocean. Environ Biol Fish 80:445–452. https://doi.org/10.1007/s10641-006-9143-z

Aoyama J, Watanabe S, Miller MJ, Mochioka N, Otake T, Yoshinaga T, Tsukamoto K (2014) Spawning sites of the Japanese eel in relation to oceanographic structure and the West Mariana Ridge. PLoS One 9:e88759. https://doi.org/10.1371/journal.pone.0088759

Aoyama J, Wouthuyzen S, Miller MJ, Sugeha HY, Kuroki M, Watanabe S, Syahailatua A, Tantu FY, Hagihara S, Triyanto OT, Tsukamoto K (2018) Reproductive ecology and biodiversity of freshwater eels around Sulawesi Island Indonesia. Zool Stud 57:e30. https://doi.org/10.6620/ZS.2018.57-30

Chow S, Kurogi H, Mochioka N, Kaji S, Okazaki M, Tsukamoto K (2009) Discovery of mature freshwater eels in the open ocean. Fish Sci 75:257–259. https://doi.org/10.1007/s12562-008-0017-5

Fukuba T, Miwa T, Watanabe S, Mochioka N, Yamada Y, Miller MJ, Okazaki M, Kodama T, Kurogi H, Chow S, Tsukamoto K (2015) A new drifting underwater camera system for observing spawning Japanese eels in the epipelagic zone along the West Mariana Ridge. Fish Sci 81:235–246. https://doi.org/10.1007/s12562-014-0837-4

Fukuda N, Kurogi H, Ambe D, Chow S, Yamamoto T, Yokouchi K, Shinoda A, Masuda Y, Sekino M, Saitoh K, Masujima M, Watanabe T, Mochioka N, Kuwada H (2018) Location, size and age at onset of metamorphosis in the Japanese eel *Anguilla japonica*. J Fish Biol 92:1342–1358. https://doi.org/10.1111/jfb.13590

Hanel R, Stepputtis D, Bonhommeau S, Castonguay M, Schaber M, Wysujack K, Vobach M, Miller MJ (2014) Low larval abundance in the Sargasso Sea: new evidence about reduced recruitment of the Atlantic eels. Naturwissenschaften 101:1041–1052. https://doi.org/10.1007/s00114-014-1243-6

Jellyman D, Tsukamoto K (2010) Vertical migrations may control maturation in migrating female *Anguilla dieffenbachii*. Mar Ecol Prog Ser 404:241–247. https://doi.org/10.3354/meps08468

Kimura S, Tsukamoto K (2006) The salinity front in the North Equatorial Current: a landmark for the spawning migration of the Japanese eel (*Anguilla japonica*) related to the stock recruitment. Deep-Sea Res Part II 53:315–325. https://doi.org/10.1016/j.dsr2.2006.01.009

Kleckner RC, McCleave JD (1988) The northern limit of spawning by Atlantic eels (*Anguilla* spp.) in the Sargasso Sea in relation to thermal fronts and surface water masses. J Mar Res 46:647–667

Kuroki M, Aoyama J, Miller MJ, Yoshinaga T, Shinoda A, Hagihara S, Tsukamoto K (2009) Sympatric spawning of *Anguilla marmorata* and *Anguilla japonica* in the western North Pacific Ocean. J Fish Biol 74:1853–1865. https://doi.org/10.1111/j.1095-8649.2009.02299.x

Kuroki M, Aoyama J, Miller MJ, Yoshinaga T, Watanabe S, Tsukamoto K (2012) Offshore spawning of the newly discovered anguillid species *Anguilla luzonensis* (Teleostei: Anguillidae) in the western North Pacific. Pacific Sci 66:497–507

Kuroki M, Miller MJ, Feunteun E, Sasal P, Pikering T, Han Y, Faliex E, Acou A, Dessier A, Schabetsberger R, Watanabe S, Kawakami T, Onda H, Higuchi T, Takeuchi A, Shimizu M, Hewavitharane CA, Hagihara S, Taka T, Kimura S, Mochioka N, Otake T, Tsukamoto K (2020) Distribution of anguillid leptocephali and possible spawning areas in the South Pacific Ocean. Prog Oceanogr 180:102234. https://doi.org/10.1016/j.pocean.2019.102234

Miller MJ, Tsukamoto K (2017) The ecology of oceanic dispersal and survival of anguillid leptocephali. Can J Fish Aquat Sci 74:958–971. https://doi.org/10.1139/cjfas-2016-0281

Miller MJ, Bonhommeau S, Munk P, Castonguay M, Hanel R, McCleave JD (2015) A century of research on the larval distributions of the Atlantic eels: a re-examination of the data. Biol Rev 90:1035–1064. https://doi.org/10.1111/brv.12144

Miller MJ, Westerberg H, Sparholt H, Wysujack K, Sørensen SR, Marohn L, Jacobsen MW, Freese M, Ayala DJ, Pohlmann J-D, Svendsen JC, Watanabe S, Andersen L, Møller PR, Tsukamoto K, Munk P, Hanel R (2019) Spawning by the European eel across 2000 km of the Sargasso Sea. Biol Lett 15:20180835. https://doi.org/10.1098/rsbl.2018.0835

Miller MJ, Shimizu M, Aoyama J, Watanabe S, Kuroki M, Feunteun E et al (2022) Distribution and abundance of leptocephali in the western South Pacific region during two large-scale sampling surveys. Prog Oceanogr 206:102853. https://doi.org/10.1016/j.pocean.2022.102853

Otake T, Miller MJ, Inagaki T, Minagawa G, Shinoda A, Kimura Y, Sasai S, Oya M, Tasumi S, Suzuki Y, Tsukamoto K (2006) Evidence for migration of metamorphosing larvae of *Anguilla japonica* in the Kuroshio. Coast Mar Sci 30:453–458. https://doi.org/10.15083/00040737

Schabetsberger R, Chang Y-L, Miller MJ (2021) Spawning migration and larval dispersal of tropical Pacific eels in the centre of their distribution ranges. Mar Ecol Prog Ser 670:167–184. https://doi.org/10.3354/meps13745

Schmidt J (1922) The breeding places of the eel. Phil Trans R Soc 211:179–208

Schmidt J (1935) Danish eel investigations during 25 years (1905–1930). The Carlsberg Foundation, Copenhagen, pp 1–16

Schoth M, Tesch F-W (1982) Spatial distribution of 0-group eel larvae (*Anguilla* sp.) in the Sargasso Sea. Helgol wissenshaft Meeresunter 35:309–320. https://doi.org/10.1007/BF02006139

Shinoda A, Aoyama J, Miller MJ, Otake T, Mochioka N, Watanabe S, Minegishi Y, Kuroki M, Yoshinaga T, Yokouchi K, Fukuda N, Sudo R, Hagihara S, Zenimoto K, Suzuki Y, Oya M, Inagaki T, Kimura S, Fukui A, Lee TW, Tsukamoto K (2011) Evaluation of the larval distribution and migration of the Japanese eel in the western North Pacific. Rev Fish Biol Fish 21:591–611. https://doi.org/10.1007/s11160-010-9195-1

Takeuchi A, Watanabe S, Yamamoto S, Miller MJ, Fukuba T, Miwa T, Okino T, Minamoto T, Tsukamoto K (2019) First use of oceanic environmental DNA to study the spawning ecology of the Japanese eel *Anguilla japonica*. Mar Ecol Prog Ser 609:187–196. https://doi.org/10.3354/meps12828

Takeuchi A, Higuchi T, Watanabe S, Yama R, Fukuba T, Okamura A, Miller MJ, Okino T, Miwa T, Tsukamoto K (2021) Several possible spawning sites of the Japanese eel determined from collections of their eggs and preleptocephali. Fish Sci 87:339–352. https://doi.org/10.1007/s12562-021-01519-4

Takeuchi A, Higuchi T, Kuroki M, Watanabe S, Miller MJ, Okino T, Miwa T, Tsukamoto K (2022) Environmental DNA detects a possible Japanese eel spawning event near a video-recorded anguillid eel in the open ocean. Mar Ecol Prog Ser 689:95–107. https://doi.org/10.3354/meps14038

Tsukamoto K (1992) Discovery of the spawning area for Japanese eel. Nature 356:789–791. https://doi.org/10.1038/356789a0

Tsukamoto K (2006) Spawning of eels near a seamount. Nature 439:929. https://doi.org/10.1038/439929a

Tsukamoto K, Otake T, Mochioka N, Lee T-W, Fricke H, Inagaki T, Aoyama J, Ishikawa S, Kimura S, Miller MJ, Hasumoto H, Oya M, Suzuki Y (2003) Seamounts, new moon and eel spawning: the search for the spawning site of the Japanese eel. Environ Biol Fish 66:221–229. https://doi.org/10.1023/A:1023926705906

Tsukamoto K, Chow S, Otake T, Kurogi H, Mochioka N, Miller MJ, Aoyama J, Kimura S, Watanabe S, Yoshinaga T, Shinoda A, Kuroki M, Oya M, Watanabe T, Hata K, Ijiri S, Kazato Y, Nomura K, Tanaka H (2011) Oceanic spawning ecology of freshwater eels in the western North Pacific. Nat Commun 2:1–9. https://doi.org/10.1038/ncomms1174

Tsukamoto K, Mochioka N, Miller MJ, Koyama S, Watanabe S, Aoyama J (2013) Video observation of an eel in the *Anguilla japonica* spawning area along the West Mariana ridge. Fish Sci 79:407–416. https://doi.org/10.1007/s12562-013-0611-z

Westerberg H, Miller MJ, Wysujack K, Marohn L, Freese M, Pohlmann J-D, Watanabe S, Tsukamoto K, Hanel R (2018) Larval abundance across the European eel spawning area: an analysis of recent and historic data. Fish Fish 19:890–902. https://doi.org/10.1111/faf.12298

Wouthuyzen S, Aoyama J, Sugeha YH, Miller MJ, Kuroki M, Minegishi Y, Suharti S, Tsukamoto K (2009) Seasonality of spawning by tropical anguillid eels around Sulawesi Island, Indonesia. Naturwissenschaften 96:153–158. https://doi.org/10.1007/s00114-008-0457-x

Chapter 5
Larval Transport

Shingo Kimura

Contents

In general, the larval stage of pelagic surface fish lasts a mere few weeks, and quickly undergo metamorphosis into juveniles, where swimming ability increases dramatically. In contrast, eels spend from 6 months to 2 years as leptocephalus larvae. During this period, they are transported long distances by ocean currents to join the distribution areas of their respective species. However, if they are not transported to the proper location, their subsequent survival, growth, and spawning returns can be negatively impacted, putting them at risk of death migration. Thus, successful or unsuccessful transport of larvae is a significant cause of annual fluctuations in abundance, and understanding this transport mechanism will lead to a better understanding of the early life history of eels and mechanisms of stock fluctuations.

5.1 Larval Transport Process of the Japanese Eel

The spawning area of the Japanese eel is located in the North Equatorial Current (NEC) to the west of the Mariana Islands, about 3000 km away from their juvenile growth areas in East Asia (Japan, China, Taiwan, and Korea) (Fig. 5.1). The discovery of the spawning area of the Japanese eel in 1991 (Tsukamoto 1992) marked the beginning of several decades of effort to understand the ecologies of

S. Kimura (✉)
Graduate School of Frontier Sciences/Atmosphere and Ocean Research Institute, The University of Tokyo, Chiba, Japan
e-mail: s-kimura@aori.u-tokyo.ac.jp

K. Tsukamoto et al. (eds.), *Eel Science*, Fisheries Science Series,
https://doi.org/10.1007/978-981-99-5692-0_5

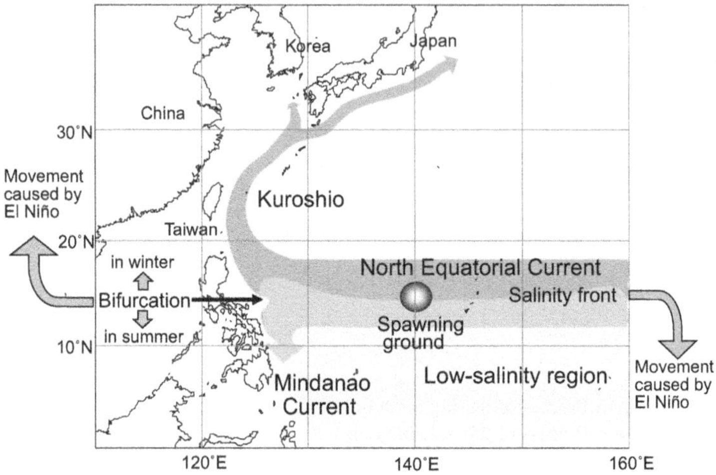

Fig. 5.1 Distributions of ocean circulation and low-salinity region in the western North Pacific related to larval transport and spawning migration of the Japanese eel. An arrow on the right indicates the directions of the salinity front movement associated with El Niño. Averaged bifurcation latitude is indicated by an arrow on the left. The bifurcation latitude changes seasonally, around 13–14°N in summer and 15–16°N in winter. El Niño cause the bifurcation latitudes to shift northward

the spawning area and larval phase of the Japanese eel (Kimura et al. 1994; Tsukamoto et al. 2011; Shinoda et al. 2011), which included research on the feeding ecology of the larvae (Kimura and Tsukamoto 2006; Miyazaki et al. 2011). A salinity front located around 15°N in NEC is considered a landmark for the endpoint of spawning migration of the Japanese eel (Kimura et al. 1994, 2001; Kimura and Tsukamoto 2006). In the far western tropical Pacific, the salinity front represented by salinity 34.5 is generated by 2 water masses: the southern low-salinity water diluted by precipitation and the northern high-salinity water caused by excessive evaporation. The main flow of the westward current of NEC occurs between approximately 10°N and 18°N, but the latitude of the salinity front can vary within the latitude and among years. Spawning usually occurs south of the front (Kimura et al. 2001; Kimura and Tsukamoto 2006). In some months during the spawning season, there is no distinct front, even if lower salinity is present; however, spawning occurs within low-salinity water (Aoyama et al. 2014). In other years, a distinct front formed in the south, and spawning appeared to be restricted to the south of the front. Variations in upper-layer water masses might indicate different biological communities. Furthermore, isotopic signatures of particulate organic matter (POM) related to the salinity front would cause differences in spawning latitude, because POM is estimated to be the diet of leptocephalus larvae (Kimura and Tsukamoto 2006).

The NEC forms part of a subtropical circulation system that circles the North Pacific Ocean from the Kuroshio Current - North Pacific Current - California Current - North Equatorial Current. Because it is different from the western boundary current like the Kuroshio Current, the NEC has a velocity of approximately 30 cm/s at the surface, and it takes several months for leptocephalus larvae to reach the eastern Philippine Sea area. During this period, owing to the strong vertical stratification, the larvae are transported through waters with extremely low biological production, where nutrient supply from the lower layers is scarce, and the water depth at which primary production at its maximum is approximately 120 m, which is considerably deeper than that in the main current region of the Kuroshio Current.

The carbon-nitrogen stable isotope ratios of suspended POM, which is the food source for leptocephalus larvae in NEC, are low in nitrogen and high in carbon at shallow depths, and their characteristics differ around the chlorophyll-a maximum depth. Vertically different food isotope ratios lead to differences in larval assimilation, resulting in differences in feeding depth among species at different developmental stages. Japanese eel larvae may feed more on POM from 50 to 100 m depths (Miyazaki et al. 2011), and leptocephali of *Anguilla* spp., including the Japanese eel, were more abundant near the top of the thermocline at 70–100 m at night (Onda et al. 2017). This suggests that leptocephalus larvae feed near the top of the thermocline. These results are related to the diurnal vertical migration of larvae, which stay on the surface at night and dive deeper during the day, indicating that leptocephalus larvae actively select depth and are transported through horizontal flow environments that vary with depth.

The larvae that reach the eastern Philippine Sea must transfer to the Kuroshio Current. If they are mistakenly swept toward the Mindanao Current, which is the opposite of the Kuroshio Current, they will undergo death migration. A conceptual model for the appropriate migration involves a transport mechanism that incorporates vertical migration and Ekman transport by trade winds (i.e., while at the surface, they are more likely to be transported to the right relative to the wind direction in the northern hemisphere and to a more northerly path due to the influence of the earth's rotation by trade winds) (Kimura et al. 1994); a three-dimensional numerical simulation model based on this mechanism has been introduced by Kimura et al. (1999). The larval period of Japanese eels is estimated to be approximately 160 days (Kuroki et al. 2014) which corresponds to the results estimated by several numerical simulations.

5.2 Larval Transport Process of the Atlantic Eels

There are 2 species of eels in the North Atlantic: the European eel *A. anguilla*, which inhabits rivers along the European and North African coasts; and the American eel *A. rostrata*, which lives along the North American Atlantic coast and West Indies. Their larvae have been intensively collected in a large water temperature gradient near the subtropical convergence zone in the Sargasso Sea (25–30°N), particularly in the high-temperature and high-salinity water mass to the south of the convergence zone (Kleckner and McCleave 1988). This suggests that parent eels stop their spawning migration and are encouraged to spawn when they experience rapid changes in water temperature due to the water temperature front, along with differences in water quality and odor in the water masses distributed north and south of this water temperature front. In other words, the water temperature front likely played an important role in spawning migration. This is consistent with the fact that in Japanese eels distributed in East Asia, the site water mass variations between the north and south of the ocean front is a landmark for spawning.

European and American eels, which have the same spawning area in the Atlantic, are transported by the Gulf Stream from the Sargasso Sea via the Florida Current. European eels are also transported by the North Atlantic Current to Europe and North Africa. In contrast, estimating from the known location of past leptocephalus larval collections, the role of the Gulf Stream is limited to the European eel. Thus, there may be an alternate trans-Atlantic route via the eastward counter current north of the subtropical convergence zone and the Azores Current (Miller et al. 2015).

The larval period of European eels in the Atlantic Ocean is estimated to last 180–350 days based on otolith dairy ring analyses (Lecomte-Finiger 1992; Wang and Tzeng 2000; Kuroki et al. 2008) compared to that of the Japanese eel. Since the spawning area in the Sargasso Sea is far from juvenile growth areas and the distance is twice as long as that of the Japanese eel, the significant variation in the larval period between the 2 species is reasonable. Numerical simulations using hydrodynamic models have also estimated that European eels have a 2-year long larval-transport period (Kettle and Haines 2006; Bonhommeau et al. 2010; Zenimoto et al. 2011). Nevertheless, the 3 eel species, including the American eel, can suffer from high mortality over long distances and time to reach the juvenile growth area.

5.3 Effects of Global Environmental Change

The success of the Japanese eel larval transfer from NEC to the Kuroshio Current was significantly reduced when El Niño occurred. This is because the bifurcation where the NEC splits into the Kuroshio and Mindanao Currents in the eastern Philippine Sea shifts significantly to the north during El Niño events. Weakening of eastward trade winds due to the interaction between the atmosphere and ocean causes these events. In addition, the bifurcation latitude shifts north and south around 13–14°N in summer and 15–16°N in winter owing to seasonal wind fluctuations. The northward shift in bifurcation latitude during El Niño events did not result in a northward shift in spawning areas, which would result in more larvae being transported to the Mindanao Current. Numerical simulations based on models of Pacific Ocean circulation with a 1/10-degree grid of latitude/longitude incorporating the diurnal vertical migration of leptocephalus larvae showed a double probability of larval transport toward the Mindanao Current during El Niño years (Kim et al. 2007; Zenimoto et al. 2009), and Hsiung et al. (2018) reported similar results. In addition, the location of the salinity front of NEC is thought to be a landmark for spawning migration that shifts significantly south during El Niño events and spurs larval transport toward the Mindanao Current (Kimura et al. 1994). This idea of the larval migration mechanism in response to El Niño occurrences is supported by the fact that glass eel catches on the Japanese coast decrease during El Niño events (Kimura et al. 2001; Kimura and Tsukamoto 2006).

To evaluate the effects of the fluctuating NEC bifurcation position and salinity front during El Niño events on the success rate of Japanese eel larvae entering the Kuroshio Current, otoliths of glass eels collected from the estuary in Taiwan during 1972–2013 were used for growth analysis based on the number of otolith daily increments and body length. Furthermore, numerical simulations based on the results of a high-resolution climate model, the Model for Interdisciplinary Research on Climate (MIROC), have been conducted (Hsiung et al. 2018). In this study, the impact of ENSO on the onset of La Niña was also evaluated, and comparisons were made for 3 time periods, including years when neither phenomena occurred. The results showed that during El Niño, larvae experienced slower currents due to the southward shift of the salinity front, and the bifurcation of NEC tended to move northward, which may have increased the distance between the spawning areas and its bifurcation position. Moreover, the proportion of larvae entering the Kuroshio Current was reduced owing to the increased flow of NEC into the Mindanao Current (Fig. 5.2, bottom panel). Analyses based on numerical simulations and otolith daily ring increment analysis revealed that the transport period to reach the Kuroshio Current was longer during El Niño (Fig. 5.2, top panel), and the body length of migrating glass eels tended to increase. This indicates that a longer larval transport period also increases the time required for metamorphosis.

The transport and dispersal processes of Japanese eel larvae in El Niño, non-El Niño, and La Niña years estimated by numerical simulations (Kim et al. 2007; Zenimoto et al. 2009) also showed results similar to those of These results clearly indicate that many larvae were transported to the Mindanao Current region and taken into the Indonesian region during El Niño years, thereby indicating that death migration does not contribute to Japanese eel reproduction. However, this condition may allow good larval transport in other tropical eel species, such as the Indo-Pacific eel *A. marmorata*.

However, when La Niña occurs (the opposite event of El Niño), the number of larvae migrating to the Japanese coast is slightly lower than that in normal years (Zenimoto et al. 2009; Hsiung et al. 2018). However, because the decrease in La Niña years is not as significant as that in El Niño years, the impact on the recruitment of Japanese eel stocks is considered to be limited in La Niña years. This result is consistent with those of the otolith daily ring increment analysis by Hsiung et al. (2018).

Recently, it has been proposed that the Philippines–Taiwan Oscillation (Hsu et al. 2017) and subtropical/tropical wind stress curl (Chang et al. 2018) may also influence larval transport. The combination of these factors may contribute to the further understanding of the Japanese eel larval transport mechanism.

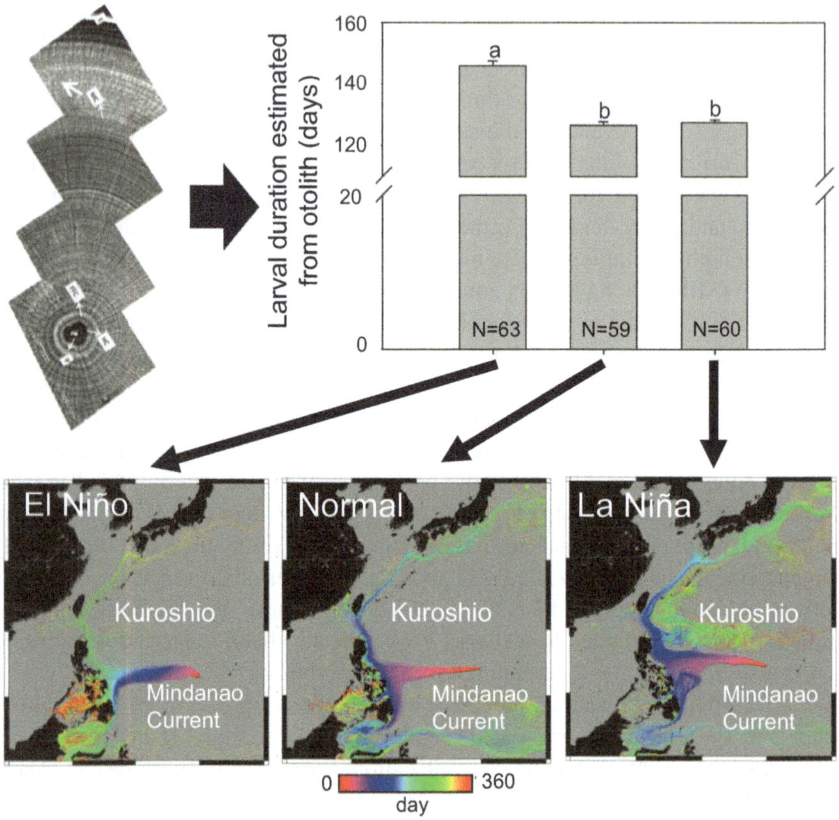

Fig. 5.2 Transport and dispersal processes of the Japanese eel larvae in El Niño, normal (non-El Niño and non-La Niña) and La Niña years estimated from numerical simulations (bottom panel) with larval durations for each climate period estimated from otolith daily ring increments (top panels) (modified from Hsiung et al. (2018))

5.4 Large Time-Scale Climate Change

Since the mid-1970s, Japanese, American, and European eels stocks have declined. The causes of this are overfishing (glass eels and parents), deterioration of the river environment (revetments, estuary weirs, water quality, and dams), and global environmental changes (regime shift, El Niño, and global warming). Various eel species living in both oceans began to decrease almost simultaneously; thus, the effects of global warming should be considered to affect dispersal processes. The marine environments of the Pacific and Atlantic oceans are closely linked to the climatic conditions of their respective oceans, and the pressure pattern can be representative of the interannual variability of these conditions. In the Pacific Ocean, the pressure difference between Darwin and Tahiti is known as the Southern Oscillation Index, and in the Atlantic Ocean, the pressure difference between the Azores and Iceland is known as the North Atlantic Oscillation Index, which has been studied in relation to various marine organisms other than eels. The Southern Oscillation Index, which is an indicator of El Niño, has shown large fluctuations since 1976 and has not been observed in the past. In contrast, the North Atlantic Oscillation Index has a long-term cyclical trend, with its positive and negative values reversing after 1976. Although the data for the past 130 years showed no correlated fluctuations between the 2 oceans, 1976 can be regarded as a year in which both the Atlantic and Pacific oceans underwent major changes, known as a climatic regime shift (Kimura and Tsukamoto 2006). In the Atlantic, the water temperature front structure in the Sargasso Sea changed, suggesting a correlation between the North Atlantic Oscillation Index and the number of glass eels migrating into Europe (Knights 2003; Friedland et al. 2007). These may be considered possible reasons for the synchronized fluctuations in the long-term declining trends of eel species in the temperate Pacific and Atlantic oceans. According to the updated current data set of the Southern Oscillation Index and the North Atlantic Oscillation Index, both datasets show a large temporal change around 2010, but they recovered to the same level as the previous trend. It is difficult to assess the effect on larval transport and spawning migration because the timescale of this temporal change is very short. Therefore, the effect of long decadal-scale changes on larval migration should be monitored continuously in the future.

In addition, variations in oceanic environmental conditions under global warming could significantly influence larval and juvenile transport of the Japanese eel. Numerical simulations are useful tools to detect this influence on future larval migrations, even if uncertain assumptions are set in the models. Under these conditions, a numerical simulation of larval transport of the Japanese eel at different swimming speeds, based on the Intergovernmental Panel on Climate Change A1B climate-warming scenario, was conducted (Hsiung and Kimura 2019). According to the comparison of results before and after global warming, the model predicted a 14–26% decline in particles recruited into the Kuroshio Current and a 7–12% increase in particles entrained in the Mindanao Current. The drifting time from the spawning area to the Kuroshio Current was extended by approximately 35 days, which prolonged the larval time in the pelagic environment of the open ocean and potentially affected the recruitment success of this species.

These estimations have significant uncertainties, and there is a lack of data and analysis to assess the factors quantitatively; in particular, the separation of the effects of anthropogenic disturbances, such as overfishing and habitat degradation, should be considered. Under strong demands for the conservation of eel species, many trials to predict their prospects are required to perform any conservation procedures.

References

Aoyama J, Watanabe S, Miller MJ, Mochioka N, Otake T, Yoshinaga T, Tsukamoto K (2014) Spawning sites of the Japanese eel in relation to oceanographic structure and the West Mariana Ridge. PLoS One 9:e88759. https://doi.org/10.1371/journal.pone.0088759

Bonhommeau S, Castonguay M, Rivot E, Sabatié R, Le Pape O (2010) The duration of migration of Atlantic *Anguilla* larvae. Fish Fish 11:289–306. https://doi.org/10.1111/j.1467-2979.2010.00362.x

Chang YL, Miyazawa Y, Miller MJ, Tsukamoto K (2018) Potential impact of ocean circulation on the declining Japanese eel catches. Sci Rep 8:5496. https://doi.org/10.1038/s41598-018-23820-6

Friedland KD, Miller MJ, Knights B (2007) Oceanic changes in the Sargasso Sea and declines in recruitment of the European eel. ICES J Mar Sci 64:519–530. https://doi.org/10.1093/icesjms/fsm022

Hsiung K-M, Kimura S (2019) Impacts of global warming on larval and juvenile transport of Japanese eels (*Anguilla japonica*). Deep-Sea Res II 169:104685. https://doi.org/10.1016/j.dsr2.2019.104685

Hsiung K-M, Kimura S, Han Y-S, Takeshige A, Iizuka Y (2018) Effect of ENSO events on larval and juvenile duration and transport of Japanese eel (*Anguilla japonica*). PLoS One 13: e0195544. https://doi.org/10.1371/journal.pone.0195544

Hsu A-C, Xue H, Chai F, Xiu P, Han Y-S (2017) Variability of the Pacific North Equatorial Current and its implications on Japanese eel (*Anguilla japonica*) larval migration. Fish Oceanogr 26: 251–267. https://doi.org/10.1111/fog.12189

Kettle AJ, Haines K (2006) How does the European eel (*Anguilla anguilla*) retain its population structure during its larval migration across the North Atlantic Ocean? Can J Fish Aquat Sci 63: 90–106. https://doi.org/10.1139/f05-198

Kim H, Kimura S, Shinoda A, Kitagawa T, Sasai Y, Sasaki H (2007) Effect of *El Niño* on migration and larval transport of the Japanese eel (*Anguilla japonica*). ICES J Mar Sci 64:1387–1395. https://doi.org/10.1093/icesjms/fsm091

Kimura S, Tsukamoto K (2006) The salinity front in the north equatorial current: a landmark for the spawning migration of the Japanese eel (*Anguilla japonica*) related to the stock recruitment. Deep-Sea Res II 53:315–325. https://doi.org/10.1016/j.dsr2.2006.01.009

Kimura S, Tsukamoto K, Sugimoto T (1994) A model for the larval migration of the Japanese eel: roles of the trade winds and salinity front. Mar Biol 119:185–190. https://doi.org/10.1007/BF00349555

Kimura S, Döös K, Coward AC (1999) Numerical simulation to resolve the downstream migration of the Japanese eel. Mar Ecol Prog Ser 186:303–306. https://doi.org/10.3354/meps186303

Kimura S, Inoue T, Sugimoto T (2001) Fluctuation in distribution of low-salinity water in the north equatorial current and its effect on the larval transport of the Japanese eel. Fish Oceanogr 10:51–60. https://doi.org/10.1046/j.1365-2419.2001.00159.x

Kleckner RC, McCleave JD (1988) The northern limit of spawning by Atlantic eels (*Anguilla* spp.) in the Sargasso Sea in relation to thermal fronts and surface-water masses. J Mar Res 46:647–667

Knights B (2003) A review of the possible impacts of long-term oceanic and climate changes and fishing mortality on recruitment of anguillid eels of the northern hemisphere. Sci Total Environ 310:237–244. https://doi.org/10.1016/S0048-9697(02)00644-7

Kuroki M, Kawai M, Jonsson B, Aoyama J, Miller MJ, Noakes DLG, Tsukamoto T (2008) Inshore migration and otolith microstructure/microchemistry of anguillid glass eels recruited to Iceland. Environ Biol Fish 83:309–325. https://doi.org/10.1007/s10641-008-9341-y

Kuroki M, Miller MJ, Tsukamoto K (2014) Diversity of early life-history traits in freshwater eels and the evolution of their oceanic migrations. Can J Zool 92:749–770. https://doi.org/10.1139/cjz-2013-0303

Lecomte-Finiger R (1992) Growth history and age at recruitment of European glass eel (*Anguilla anguilla*) as revealed by otolith microstructure. Mar Biol 114:205–210. https://doi.org/10.1007/BF00349520

Miller MJ, Bonhommeau S, Munk P, Castonguay M, Hanel R, McCleave JD (2015) A century of research on the larval distributions of the Atlantic eels: a re-examination of the data. Biol Rev 90:1035–1064. https://doi.org/10.1111/brv.12144

Miyazaki S, Kim H, Zenimoto K, Kitagawa T, Miller MJ, Kimura S (2011) Stable isotope analysis of two species of anguilliform leptocephali (*Anguilla japonica* and *Ariosoma major*) relative to their feeding depth in the north equatorial current region. Mar Biol 158:2555–2564. https://doi.org/10.1007/s00227-011-1756-x

Onda H, Miller MJ, Takeshige A, Miyake Y, Kuroki M, Aoyama J, Kimura S (2017) Vertical distribution and assemblage structure of leptocephali in the north equatorial current region of the western Pacific. Mar Ecol Prog Ser 575:119–136. https://doi.org/10.3354/meps12198

Shinoda A, Aoyama J, Miller MJ, Otake T, Mochioka N, Watanabe S, Minegishi Y, Kuroki M, Yoshinaga T, Yokouchi K, Fukuda N, Sudo R, Hagihara S, Zenimoto K, Suzuki Y, Oya M, Inagaki T, Kimura S, Fukui A, Lee TW, Tsukamoto K (2011) Evaluation of the larval distribution and migration of the Japanese eel in the western North Pacific. Rev Fish Biol Fish 21:591–611. https://doi.org/10.1007/s11160-010-9195-1

Tsukamoto K (1992) Discovery of the spawning area for Japanese eel. Nature 356:789–791. https://doi.org/10.1038/356789a0

Tsukamoto K, Chow S, Otake T, Kurogi H, Mochioka N, Miller MJ, Aoyama J, Kimura S, Watanabe S, Yoshinaga T, Shinoda A, Kuroki M, Oya M, Watanabe T, Hata K, Ijiri S, Kazeto Y, Nomura K, Tanaka H (2011) Oceanic spawning ecology of freshwater eels in the western North Pacific. Nat Commun 2:179. https://doi.org/10.1038/ncomms1174

Wang CH, Tzeng WN (2000) The timing of metamorphosis and growth rates of American and European eel leptocephali: a mechanism of larval segregative migration. Fish Res 46:191–205. https://doi.org/10.1016/S0165-7836(00)00146-6

Zenimoto K, Kitagawa T, Miyazaki S, Sasai Y, Sasaki H, Kimura S (2009) The effects of seasonal and interannual variability of oceanic structure in the western Pacific north equatorial current on larval transport of the Japanese eel (*Anguilla japonica*). J Fish Biol 74:1878–1890. https://doi.org/10.1111/j.1095-8649.2009.02295.x

Zenimoto K, Sasai Y, Sasaki H, Kimura S (2011) Estimation of larval duration in *Anguilla* spp., based on cohort analysis, otolith microstructure, and Lagrangian simulations. Mar Ecol Prog Ser 438:219–228. https://doi.org/10.3354/meps09255

Chapter 6
Glass Eel Recruitment

Nobuto Fukuda and Akira Shinoda

Contents

Japanese eel larvae that hatch in the west Mariana Ridges are transported via the North Equatorial Current (NEC) and the western boundary current, known as the Kuroshio Current (KC), to an offshore region of East Asia. After metamorphosing from leptocephali to glass eels, they leave the KC and migrate into their growth habitats with brackish and fresh water. The inshore migration behavior of glass eels in the open ocean remains unknown. However, it is known that when they approach coastal areas, various environmental factors in estuarine areas strongly influence their behavior. The abundance and timing of glass eels migrating upstream in estuaries also vary greatly from year to year. This section outlines the mysterious ecology of glass eels from the open ocean to estuaries.

Nobuto Fukuda and Akira Shinoda contributed equally to this work.

N. Fukuda
Nikko Field Station, Fisheries Technology Institute, Japan Fisheries Research and Education Agency, Tochigi, Japan

A. Shinoda (✉)
Department of Biology, Tokyo Medical University, Tokyo, Japan
e-mail: shinoda@tokyo-med.ac.jp

K. Tsukamoto et al. (eds.), *Eel Science*, Fisheries Science Series,
https://doi.org/10.1007/978-981-99-5692-0_6

6.1 Metamorphosis from Leptocephalus to Glass Eel

The leptocephali of the Japanese eel are passively transported from their spawning grounds to the KC via the NEC. In addition to this main route, some larvae pass through mesoscale eddies to the north of the NEC and enter the KC off Taiwan (Fukuda et al. 2018). Leptocephali at the metamorphic stage were collected in the upstream region of the KC and mesoscale eddies adjacent to the KC (Otake et al. 2006; Shinoda et al. 2011; Fukuda et al. 2018). This suggests that metamorphosis begins around the area where they enter the KC.

The size of wild leptocephalus at the onset of metamorphosis was assumed to be approximately 68 mm in length (Fukuda et al. 2018). Contraction (12.5% on average) in length during metamorphosis has been observed in artificially produced leptocephali. The length of leptocephali at the onset of metamorphosis was calculated from the length of wild glass eels in estuaries (approximately 60 mm in the mean).

To learn the trigger for leptocephalus to begin metamorphosis, an experiment was conducted using artificially produced leptocephali (Okamura et al. 2012). The threshold size for metamorphosis was a total length of 50–55 mm. Smaller larvae did not begin metamorphosis in the absence of food, whereas larvae reaching the threshold size were induced to undergo metamorphosis by starvation. This study suggests that food deprivation is a cue for metamorphosis onset.

Leptocephali with higher growth rates during transport reach the fully grown size at the leptocephalus stage earlier, and thus, they appear to begin metamorphosis earlier (Tsukamoto 1990; Cheng and Tzeng 1996; Shinoda 2004). The otolith microstructure analyses of glass eels suggested that individuals with higher growth rates during the leptocephalus stage begin metamorphosis 1–2 months earlier than those with lower growth rates, even when they hatched in the same month (Shinoda 2004). Otolith microstructure and microchemistry analyses also indicate that the metamorphosis of leptocephali into glass eels may take approximately 20 days (Shinoda 2004).

During metamorphosis, leptocephali transform from a laterally compressed leaf-like body to a cylindrical body. Before metamorphosis, the body of the leptocephali is buoyant because it is filled with large amounts of stored transparent gelatinous glycosaminoglycans. However, there is a sudden reduction in buoyancy during metamorphosis (Tsukamoto et al. 2009) owing to the loss of the glycosaminoglycan matrix (which causes body water content reduction) and the formation of the muscle, cartilage, and bone. As a result, the specific gravity of glass eels becomes heavier than that of the surrounding seawater (Tsukamoto et al. 2009).

6.2 Offshore Transportation Routes of Glass Eels to Taiwan, China, and Japan

Recruiting glass eels of Japanese eels have been found in a wide range of latitudes (Fig. 6.1), from Luzon Island in the Philippines at the southern limit (Yoshinaga et al. 2014) to Hokkaido in Japan at the northern limit (Morita and Kuroki 2021; Kasai et al. 2021).

Historically, the leptocephali of Japanese eels have rarely been collected in the sea shelf, which is similar to European eels (Tesch 1980), and few glass eels have been collected in the offshore area of the KC and the sea shelf (Otake et al. 2006; Shinoda et al. 2011). The pathways of glass eels on the East Asian continental shelf

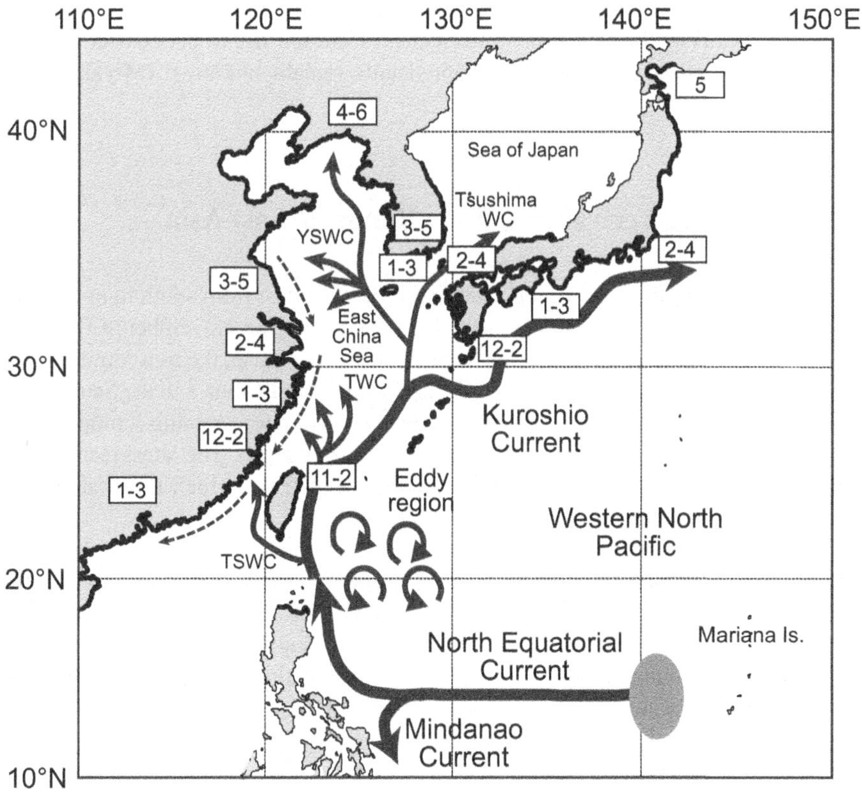

Fig. 6.1 Map of the western North Pacific region showing oceanic currents and recruitment locations of glass eels. The bold lines along the coast indicate locations with glass eel recruitment. The numbers in boxes indicate the main month of glass eel recruitment at each location (Morita and Kuroki (2021) on Hokkaido and Han (2011) on other locations). The gray circle represents the spawning area of Japanese eels. (This figure was adapted from Shinoda et al. (2011), Han (2011), and Han et al. (2019)). *TSWC* Taiwan Strait Warm Current, *TWC* Taiwan Warm Current, *YSWC* Yellow Sea Warm Current, *Tsushima WC* Tsushima Warm Current

were assumed to have 5 main recruitment blocks: (1) the main KC, (2) the Taiwan Strait Warm Current, (3) the Taiwan Warm Current, (4) the Yellow Sea Warm Current, and (5) the branch of the Yellow Sea Warm Current (Fig. 6.1), owing to evaluations based on glass eel otolith increments, drifting oceanic simulation, and fishing seasons in each location (Han et al. 2019).

A study in Lake Hamana, located in central Japan, showed that the catch of glass eels was positively influenced by the nearshore positions of the KC in the 1960s–1980s, indicating that recruitment is influenced by KC position; however, this tendency has been lost in recent years (Miyake et al. 2020). Another study in southern Japan suggested that a cyclonic mesoscale eddy induced between the KC and the estuary when the KC was displaced offshore may enhance the recruitment of glass eels into the estuary (Aoki et al. 2018). The opposing effect of KC positions on the glass eel recruitment in these 2 studies may be explained by the regional difference forming the KC intrusion, which may transport glass eels into the study area. The details of the recruitment mechanism from the KC to the coastal area are unclear, as glass eel behavior and other details remain unknown (Miyake et al. 2020).

6.3 Glass Eel Arrival in a Wide Range of East Asia

The recruitment of glass eels occurs along the Pacific coast from south to north with time lags (toward the downstream of the KC; Fig. 6.1), from November to February in Taiwan and December to April in Chiba Prefecture in Japan, the most downstream area of the KC. In areas far from the KC, such as mainland China and the Seto Inland Sea in Japan, the recruitment period is delayed by a few months compared to locations at the same latitude near the KC (Tsukamoto 1990). The latest recruitment occurs from April to June at the mouth of the Yalu River, which flows along the border between China and North Korea (Han 2011).

A possible factor that determined recruitment location is the duration of the larval phase before metamorphosis. Studies using otolith microstructure analyses showed that larval durations tended to be longer in glass eels caught at higher latitudes (Cheng and Tzeng 1996; Fukuda et al. 2018). This suggests that individuals with shorter larval durations tend to recruit into southern latitude areas such as Taiwan and southern China, while those with longer larval durations were likely transported northward with the KC and recruited into northern areas such as the Yalu River in China and Chiba Prefecture in Japan.

Time-series fluctuations in the recruitment of glass eels have been evaluated in Taiwan, the country nearest to the spawning grounds of Japanese eels. In the offshore area of Yilan, batch-like arrival waves of glass eels with a 1-month periodicity were observed, all of which were in the early pigmentation stage and similar in age (approximately 150–160 days), suggesting that the cohorts reflect those spawned every new moon period (Han et al. 2016). Glass eels caught in Taiwan had longer larval durations and longer body lengths during El Niño years, likely because it may prolong the time needed for the larvae to enter the KC from their spawning ground (Hsiung et al. 2018). Furthermore, environmental factors within the spawning area also influence the body lengths of glass eels recruited to Taiwan (Hsiung et al. 2022).

The leptocephalus duration and oceanic currents determine the dispersal locations up to the glass eel phase, whereas temperatures determine the timing of upstream migration at each location (Han 2011). The recruitment season generally starts in southern East Asia (i.e., Taiwan) in November and in northern areas (i.e., China/Korea) in April of the following year, with a lag of approximately 5 months, whereas daily growth rings in otoliths differed by only 1–2 months in the mean leptocephalus stage between southern and northern East Asian samples (Han 2011). Glass eel stages in northern areas should last approximately 3 months; however, the durations estimated from otolith growth rings ranged between 7 and 40 days, which is considerably less than the expected difference (Han 2011). In rearing experiments, glass eels in water temperatures below 10 °C ceased otolith growth and remained in the early pigmentation stage (Fukuda et al. 2009). The time lag in recruitment can be accounted for by a longer leptocephalus stage combined with a low-temperature-driven delay to upstream migration in winter (Han 2011).

6.4 Yearly Variation in Recruitment Peak of Glass Eels

The recruitment peak of the glass eels at each location varied from year to year. A quantitative study on the recruitment during 1991–2003 on Tanegashima Island, located in southern Japan, found that glass eels were usually present from October to May, with the peak occurring between January and March. Rarely, their earliest peak appeared in November and their latest in April (Shinoda 2004). Fukuda et al. (2016) also reported that glass eel recruitment in Lake Hamana occurred from November to May during 2003–2005, peaking in January. However, in the estuary of the Sagami River, located in central Japan, the highest peak recruitment was observed in early summer (June) in the 2009–2010 and 2010–2011 seasons, which occurred after the end of commercial fishing season (Aoyama et al. 2012). Year-round surveys of glass eels were also conducted at 6 locations in Japan from 2013 to 2018, showing that peak recruitment varied depending on the region and year, and all peaks except the 2017–2018 season were within the commercial fishing seasons (Japan Fisheries Agency 2019). The delayed peak recruitment in the 2009–2010, 2010–2011, and 2017–2018 seasons corresponded to years with low glass eel commercial catches in Japan (Fig. 6.2). This implies that a potential cause of low-commercial catches is peak recruitment delayed outside the fishing season. At present, the causes of annual variations in recruitment peaks have not been clarified. Thus, breakthrough studies are required to reveal yearly fluctuations in the peak spawning time, larval growth, and survival conditions during transportation from spawning grounds to estuaries.

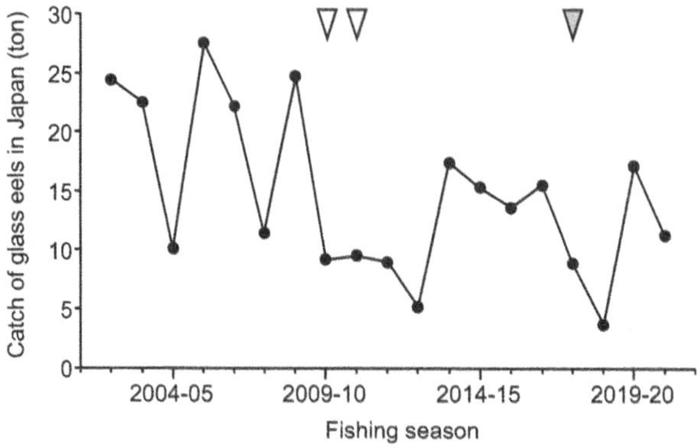

Fig. 6.2 Catch of glass eels in Japan. The data are based on statistics from the Fisheries Agency of Japan, which is calculated as "inputs into aquaculture ponds" minus "imports of glass eels". The fishing season "20XX-XX + 1" refers to the input season which starts from November 1, 20XX to October 31, 20XX + 1. Open inverted triangle: delayed peak recruitment observed in the Sagami River (Aoyama et al. 2012), closed inverted triangle: delayed peak recruitment observed in the national glass eel monitoring (Japan Fisheries Agency 2019)

6.5 Glass Eel Behavior and Their Mysterious Ecology in the Offshore Area

Extensive studies on the behavior of European eels at the glass eel stage have been conducted in relation to olfactory cues, salinity, and temperature (Cresci 2020). However, few studies have been conducted on the behavior of Japanese glass eels.

Some behavioral experiments have shown that Japanese eels at the glass eel stage have strong preferences for odors in inland waters (Fukuda et al. 2019a; Kumai and Kuroki 2021). The attractive odorant substances were not removed by glass fiber filtration but were diminished by aeration and absorbed by activated carbon, suggesting that the attractants are likely present in biodegradable and/or volatile dissolved organic matter (Fukuda et al. 2019b). Under the same odorous conditions, glass eels prefer colder waters (Fukuda et al. unpublished data); attraction to low-salinity (freshwater) has also been observed (Fukuda et al. 2019a, b; Kumai and Kuroki 2021).

In the KC region, Japanese glass eels first need to detrain from the KC and move toward the coasts. Model simulations for American eels have shown that directional swimming significantly improves the success rates of larvae reaching the continental shelf (Rypina et al. 2014). Olfactory cues are unlikely to be effective in the offshore areas because they are far from the estuaries. *In situ* experiments revealed that European eels at the glass eel stage can use the Earth's magnetic field and possibly lunar cues for orientation in offshore areas (Cresci 2020). Furthermore, other experiments showed that Japanese eels at the glass eel stage can also use geomagnetic cues for orientation (Nishi et al. 2018). In the KC area, Japanese glass eels may swim toward the coasts using geomagnetic cues or celestial cues and detrain from the main KC, as assumed in European eels.

Another key of the potential biotic factor influencing shoreward transport is the vertical distribution of glass eels. An in-depth study of the depth distribution of glass eels in the KC would enhance our understanding of the biophysical mechanisms driving inshore migration from the KC; however, this is a difficult issue to assess because catch rates of glass eels are typically low in the KC (Miyake et al. 2020). The rapid decrease in buoyancy during metamorphosis from leptocephali to glass eels may be related to the depth distributions of glass eels that must swim toward coastal waters (Tsukamoto et al. 2009). In European eels, metamorphosing leptocephali are known to occur at 300–600 m depths during daytime and 35–125 m at night off the continental shelf; however, glass eels were not collected off or on the shelf (Tesch 1980). Similarly, only a few glass eels have been collected in historical surveys of Japanese eels, and their vertical distribution on or off the shelf remains unknown.

6.6 Upstream Migration in Coastal Areas and Estuaries

Olfactory cues are likely to play a central role in the eels navigating coastal waters to find estuaries. As mentioned above, glass eels have a strong preference for odors in inland waters. After crossing the continental shelf, they arrive at shallow coastal waters. Rivers discharge cold odorous freshwater into coastal waters. Glass eels would follow a gradient of odor, salinity, and water temperature, and are attracted to estuarine areas.

After arriving in estuaries, the catch of glass eels fluctuates with various time-scale periodicities, such as monthly (lunar cycle), semi-monthly (spring-neap tidal cycle), daily (day-night cycle), and tidal cycles (flood-ebb tidal cycle) (Tzeng 1985; Fukuda et al. 2016). Basically, because glass eels are photophobic, they are collected only at night. In estuarine areas where tidal influences are large, glass eels use tidal flow to effectively move upstream (Fukuda et al. 2016); this mechanism is called Selective Tidal Stream Transport (STST) and has been reported in both European and American eels (Harrison et al. 2014). During low tides, glass eels remain close to the sea floor to prevent seaward transport, and when the tide rises, they ascend to the water columns and move upstream with tidal flow. The rhythmic movements of European glass eels are likely caused by changes in odor. A temperature increase at the beginning of the flood tide may also be a cue for the ascending behavior of Japanese glass eels (Fukuda et al. 2016). Rhythmic patterns synchronized to the tidal period were observed as an endogenous circa-tidal rhythm in laboratory experiments of American eels (Wippelhauser and McCleave 1988). This ability to memorize the tide also plays a role in orientation with respect to the Earth's magnetic field, which can be utilized for orientation in estuaries (Cresci 2020).

Because glass eels migrate with flood tidal flow, the catch of glass eels in estuaries is higher in spring tides with large tidal amplitudes than in neap tides.

Regardless of the full or new moon phase, a greater catch of glass eels has been observed during spring tides in some rivers in Taiwan and Korea (Tzeng 1985; Hwang et al. 2014) and in the Yangtze River estuary in China (Guo et al. 2017). Increased catches in response to large tidal amplitudes have also been reported for other *Anguilla* species (Harrison et al. 2014).

However, there are cases in which glass eel abundance increases only during the new moon phase and not during the full moon phase. The increase in abundance strictly during the new moon phase has been observed not only in Japanese eels but also in other temperate eels, such as European and New Zealand eels (Harrison et al. 2014). Furthermore, a survey of tropical eels on Sulawesi Island also showed that glass eels rarely appear on the shore during the full moon phase (Sugeha et al. 2001). A possible reason for the lack of an increase in abundance during the full moon phase is behavior restricted by the moon's illumination, wherein high moonlight intensity and low turbidity can induce glass eels to reside at the bottom (Harrison et al. 2014).

High sensitivity to light at the early glass eel stage may inhibit their migration behavior. Tzeng (1985) showed that glass eel abundance increased only during the new moon phase in coastal waters, whereas that in a river increased during the full moon phase. This may be because the pigmentation stage in glass eels is advanced in the river compared to that in coastal waters (Tzeng 1985). In laboratory experiments, developed European glass eels were less influenced by light than unpigmented ones (Bardonnet et al. 2005); thus, glass eels with advanced pigmentation stages might occasionally be active, even during the full moon phase.

Glass eel recruitment is associated with complex environmental factors such as oceanic conditions of the KC; water conditions (odor, salinity, temperature, and turbidity) in coastal waters and estuaries; effects of celestial bodies including tidal effects, illumination, and geomagnetism; and ontogenetic factors from metamorphosis and glass eel development. The combination and interaction of these factors form a unique recruitment ecology at each location, resulting in large fluctuations in the timing and abundance of glass eel recruitment from year to year, even at the same location. Long-term monitoring at a wide range of locations is necessary to gain a better understanding of the interactions between glass eel recruitment and environmental factors.

References

Aoki K, Yamamoto T, Fukuda N, Yokouchi K, Kurogi H, Setou T, Kuroda H, Kameda T, Takafuji K, Tokeshi T (2018) Enhanced local recruitment of glass eel *Anguilla japonica* in Oyodo River, Miyazaki and offshore environmental conditions in 2002. Fish Sci 84:777–785. https://doi.org/10.1007/s12562-018-1207-4

Aoyama J, Shinoda A, Yoshinaga T, Tsukamoto K (2012) Late arrival of *Anguilla japonica* glass eels at the Sagami River estuary in two recent consecutive year classes: ecology and socio-economic impacts. Fish Sci 78:1195–1204. https://doi.org/10.1007/s12562-012-0544-y

Bardonnet A, Bolliet V, Belon V (2005) Recruitment abundance estimation: role of glass eel (*Anguilla anguilla* L.) response to light. J Exp Mar Biol Ecol 321:181–190. https://doi.org/10.1016/j.jembe.2005.02.004

Cheng PW, Tzeng WN (1996) Timing of metamorphosis and estuarine arrival across the dispersal range of the Japanese eel *Anguilla japonica*. Mar Ecol Prog Ser 131:87–96. https://doi.org/10.3354/meps131087

Cresci A (2020) A comprehensive hypothesis on the migration of European glass eels (*Anguilla anguilla*). Biol Rev 95:1273–1286. https://doi.org/10.1111/brv.12609

Fukuda N, Kuroki M, Shinoda A, Yamada Y, Okamura A, Aoyama J, Tsukamoto K (2009) Influence of water temperature and feeding regime on otolith growth in *Anguilla japonica* glass eels and elvers: does otolith growth cease at low temperatures? J Fish Biol 74:1915–1933. https://doi.org/10.1111/j.1095-8649.2009.02287.x

Fukuda N, Aoyama J, Yokouchi K, Tsukamoto K (2016) Periodicities of inshore migration and selective tidal stream transport of glass eels, *Anguilla japonica*, in Hamana Lake, Japan. Environ Biol Fish 99:309–323. https://doi.org/10.1007/s10641-016-0475-z

Fukuda N, Kurogi H, Ambe D, Chow S, Yamamoto T, Yokouchi K, Shinoda A, Masuda Y, Sekino M, Saitoh K, Masujima M, Watanabe T, Mochioka N, Kuwada H (2018) Location, size and age at onset of metamorphosis in the Japanese eel *Anguilla japonica*. J Fish Biol 92:1342–1358. https://doi.org/10.1111/jfb.13590

Fukuda N, Yokouchi K, Yamamoto T, Kurogi H, Yada T (2019a) Salinity and odor preferences of Japanese eel during the first year of post-recruitment growth in saline water. J Ethol 37:93–99. https://doi.org/10.1007/s10164-018-0572-x

Fukuda N, Yokouchi K, Yamamoto T, Kurogi H, Yada T (2019b) Insight into Japanese eel divergence into seaestuaries and rivers. Eels: biology, monitoring, management, culture and exploitation. 5m Publishing, Sheffield, pp 145–157

Guo H, Zhang X, Zhang Y, Tang W, Wu J (2017) Effects of environmental variables on recruitment of *Anguilla japonica* glass eels in the Yangtze Estuary, China. Fish Sci 83:333–341. https://doi.org/10.1007/s12562-017-1071-7

Han Y-S (2011) Temperature-dependent recruitment delay of the Japanese glass eel *Anguilla japonica* in East Asia. Mar Biol 158:2349–2358. https://doi.org/10.1007/s00227-011-1739-y

Han Y-S, Wu C-R, Iizuka Y (2016) Batch-like arrival waves of glass eels of *Anguilla japonica* in offshore waters of Taiwan. Zool Stud 55:36. https://doi.org/10.6620/ZS.2016.55-36

Han Y-S, Hsiung K-M, Zhang H, Chow L-Y, Tzeng W-N, Shinoda A, Yoshinaga T, Hur S-P, Hwang S-D, Iizuka Y, Kimura S (2019) Dispersal characteristics and pathways of Japanese glass eel in the East Asian continental shelf. Sustainability 11:2572. https://doi.org/10.3390/su11092572

Harrison AJ, Walker AM, Pinder AC, Briand C, Aprahamian MW (2014) A review of glass eel migratory behaviour, sampling techniques and abundance estimates in estuaries: implications for assessing recruitment, local production and exploitation. Rev Fish Biol Fish 24:967–983. https://doi.org/10.1007/s11160-014-9356-8

Hsiung K-M, Kimura S, Han Y-S, Takeshige A, Iizuka Y (2018) Effect of ENSO events on larval and juvenile duration and transport of Japanese eel (*Anguilla japonica*). PLoS One 13: e0195544. https://doi.org/10.1371/journal.pone.0195544

Hsiung K-M, Ma C, Ko C-Y, Tseng Y-H, Kuo Y-C, Han Y-S (2022) Effects of environmental factors within the spawning area and migration routes on the length of *Anguilla japonica* glass eels recruited to Taiwan. Mar Ecol Prog Ser 683:109–121. https://doi.org/10.3354/meps13931

Hwang SD, Lee TW, Choi IS, Hwang SW (2014) Environmental factors affecting the daily catch levels of *Anguilla japonica* glass eels in the Geum River estuary, South Korea. J Coast Res 30: 954–960. https://doi.org/10.2112/JCOASTRES-D-13-00144.1

Japan Fisheries Agency (2019) Annual report 2018–19 on research programs for the recruitment and inhabiting status of Japanese eels. Chuo Suisan Kenkyujo, Yokohama; in Japanese

Kasai A, Yamazaki A, Ahn H, Yamanaka H, Kameyama S, Masuda R, Azuma N, Kimura S, Karaki T, Kurokawa Y, Yamashita Y (2021) Distribution of Japanese eel *Anguilla japonica* revealed by environmental DNA. Front Ecol Evol 9:621461. https://doi.org/10.3389/fevo.2021.621461

Kumai Y, Kuroki M (2021) Salinity, freshwater and agricultural water preferences of glass eels of the Japanese eel *Anguilla japonica* collected in southern Japan. J Fish Biol 99:288–292. https://doi.org/10.1111/jfb.14715

Miyake Y, Tellier MA, Takeshige A, Itakura H, Yoshida A, Kimura S (2020) Past and lost influence of the Kuroshio on estuarine recruitment of *Anguilla japonica* glass eels. J Oceanogr 76:259–270. https://doi.org/10.1007/s10872-020-00543-9

Morita K, Kuroki M (2021) Japanese eel at the northern edge: glass eel migration into a river on Hokkaido, Japan. Ichthyol Res 68:217–221. https://doi.org/10.1007/s10228-020-00771-5

Nishi T, Archdale MV, Kawamura G (2018) Behavioural evidence for the use of geomagnetic cue in Japanese glass eel *Anguilla japonica* orientation. Ichthyol Res 65:161–164. https://doi.org/10.1007/s10228-017-0587-2

Okamura A, Yamada Y, Mikawa N, Horie N, Tsukamoto K (2012) Effect of starvation, body size, and temperature on the onset of metamorphosis in Japanese eel (*Anguilla japonica*). Can J Zool 90:1378–1385. https://doi.org/10.1139/cjz-2012-0146

Otake T, Miller MJ, Inagaki T, Minagawa G, Shinoda A, Kimura Y, Sasai S, Oya M, Tasumi S, Suzuki Y, Uchida M, Tsukamoto K (2006) Evidence for migration of metamorphosing larvae of *Anguilla japonica* in the Kuroshio. Cost Mar Sci 30:453–458. https://doi.org/10.15083/00040737

Rypina II, Llopiz JK, Pratt LJ, Lozier MS (2014) Dispersal pathways of American eel larvae from the Sargasso Sea. Limnol Oceanogr 59:1704–1714. https://doi.org/10.4319/lo.2014.59.5.1704

Shinoda A (2004) The ecology of inshore migration of the Japanese eel, *Anguilla japonica*. PhD dissertation. The University of Tokyo, Tokyo (in Japanese)

Shinoda A, Aoyama J, Miller MJ, Otake T, Mochioka N, Watanabe S, Minegishi Y, Kuroki M, Yoshinaga T, Yokouchi K, Fukuda N, Sudo R, Hagihara S, Zenimoto K, Suzuki Y, Oya M, Inagaki T, Kimura S, Fukui A, Lee TW, Tsukamoto K (2011) Evaluation of the larval distribution and migration of the Japanese eel in the western North Pacific. Rev Fish Biol Fish 21:591–611. https://doi.org/10.1007/s11160-010-9195-1

Sugeha HY, Arai T, Miller MJ, Limbong D, Tsukamoto K (2001) Inshore migration of the tropical eels *Anguilla* spp. recruiting to the Poigar River estuary on North Sulawesi Island. Mar Ecol Prog Ser 221:233–243. https://doi.org/10.3354/meps221233

Tesch F-W (1980) Occurrence of eel *Anguilla anguilla* larvae west of the European continental shelf, 1971–1977. Environ Biol Fish 5:185–190. https://doi.org/10.1007/BF00005354

Tsukamoto K (1990) Recruitment mechanism of the eel, *Anguilla japonica*, to the Japanese coast. J Fish Biol 36:659–671. https://doi.org/10.1111/j.1095-8649.1990.tb04320.x

Tsukamoto K, Yamada Y, Okamura A, Kaneko T, Tanaka H, Miller MJ, Horie N, Mikawa N, Utoh T, Tanaka S (2009) Positive buoyancy in eel leptocephali: an adaptation for life in the ocean surface layer. Mar Biol 156:835–846. https://doi.org/10.1007/s00227-008-1123-8

Tzeng WN (1985) Immigration timing and activity rhythms of the eel, *Anguilla japonica*, elvers in the estuary of northern Taiwan, with emphasis on environmental influences. Bull Jpn Soc Fish Oceanogr 47:11–28

Wippelhauser GS, McCleave JD (1988) Rhythmic activity of migrating juvenile American eels *Anguilla rostrata*. J Mar Biol Assoc UK 68:81–91. https://doi.org/10.1017/S0025315400050116

Yoshinaga T, Aoyama J, Shinoda A, Watanabe S, Azanza RV, Tsukamoto K (2014) Occurrence and biological characteristics of glass eels of the Japanese eel *Anguilla japonica* at the Cagayan River of Luzon Island, Philippines in 2009. Zool Stud 53:1–6. https://doi.org/10.1186/1810-522X-53-13

Chapter 7
Spawning Migration

Takatoshi Higuchi

Contents

Freshwater eels are catadromous fish that migrate between oceanic spawning areas and growth habitats in fresh, brackish, or coastal waters. After establishing themselves in growth habitats, they physiologically transform into silver eels (the early stage of sexual maturity) and migrate to the oceanic spawning area. Based on the geographic location of their growth habitat, anguillid eels can be classified into temperate and tropical species, and are known to migrate long and short distances, respectively. Their spawning migration is an important life-history event for reproductive success. Although many aspects of spawning migration behaviors of these species remain shrouded in mystery, recent telemetry and numerical simulation research have greatly advanced their research. This chapter concerns the technical aspects, current status, and perspective of eel spawning migration research.

7.1 Technical Status of Telemetry-Based Research for Silver Eels

Once silver eels reach maturity in their growth habitats, they return to their spawning areas in a phenomenon called spawning migration. Life-history studies on spawning migration of anguillid eels has yielded limited results. Telemetry-based research has been conducted to investigate the behavioral ecology of anguillid eels. Telemetry

T. Higuchi (✉)
Atmosphere and Ocean Research Institute, The University of Tokyo, Chiba, Japan
e-mail: t-higuchi@aori.u-tokyo.ac.jp

K. Tsukamoto et al. (eds.), *Eel Science*, Fisheries Science Series,
https://doi.org/10.1007/978-981-99-5692-0_7

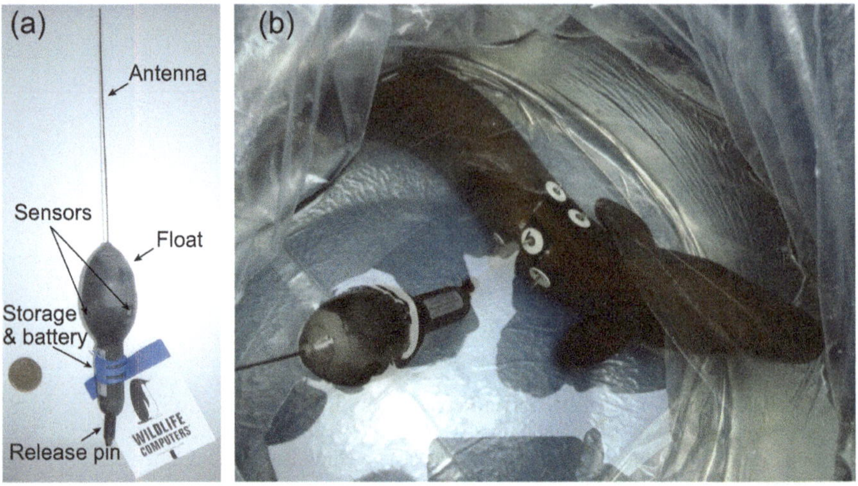

Fig. 7.1 A pop-up satellite archival tag (PSAT) and the tagged *A. japonica*. (**a**) In most PSAT tracking of *A. japonica*, the MiniPAT (Wildlife Computers Inc., 12.4 cm length, 3.8 cm diameter, weight in air: 60 g) have been used. (**b**) The PSAT was surgically attached to the dorsal side of the lateral musculature by the method of Manabe et al. (2011), and released from the southern part of Japan (Higuchi et al. 2021)

has been used primarily in rivers and estuaries for yellow eels (see Chap. 8) but has since been used to study the downstream and spawning migration behaviors of silver eels. In rivers and coastal waters, "passive tracking," in which acoustic receivers are installed on the river or seabed, enables long-term tracking, but the spatial coverage is limited by the number and array of receivers. Therefore, passive tracking is not suitable for tracking the spawning migration of silver eels, which move toward the open sea; thus, "active tracking" is used to monitor the migration route of individuals in real time using vessels. The oceanic spawning migration of silver eels was first studied by tracking a small number of individuals for short time periods (<1 week) over several kilometers using boats and acoustic tags (Tesch 1989). Since then, the number of telemetry studies on anguillid eels has increased annually. The introduction of data storage tags (DSTs), which are tags with sensors that store recorded data (environmental or physiological information), remedied the drawback of acoustic tracking (Wahlberg et al. 2014). DSTs with floats were developed, allowing the tags to drift with ocean currents at the surface and then be found and returned to researchers (Thorstad et al. 2013). However, this method requires the physical retrieval of tags to obtain data, and the recovery rate is extremely low for tracking silver eels migrating into the open ocean (Westerberg et al. 2014).

To overcome the drawbacks of acoustic tags and DSTs, pop-up satellite archival tags (PSATs; Fig. 7.1) were developed to transmit data stored during tracking to the Advanced Research and Global Observation Satellite (ARGOS) system after tracking. PSAT is a relatively large tag (length: >10 cm, diameter: approximately 3 cm) that enables large-scale migration studies of large fish, such as tuna and sword fish.

The PSAT collects experienced environmental data of the fish, such as light, depth, and water temperature. After a preprogrammed period, the slightly positive buoyant tag detaches from the fish and pops up to the sea surface. The amount of data received depends on the observation range of the satellite (time and duration of pop-up location passage and weather conditions), type of PSAT product, and battery capacity.

The first PSAT tracking of anguillid eels was conducted in New Zealand on the largest anguillid species, silver-phase New Zealand longfin eels *A. dieffenbachii* (Jellyman and Tsukamoto 2002). Following the development of smaller PSAT tags, 21 satellite tracking studies (including the first *A. dieffenbachii* tracking) were conducted on 9 anguillid species (Table 7.1). To conserve battery power and data storage space, tags were programmed to record up to 32 sets of observations of water temperature (0.05 °C resolution) and depth (±0.5 m) per day, and data were stored in 4 time bins each day to track *A. dieffenbachii* (Jellyman and Tsukamoto 2002, 2005, 2010; Watanabe et al. 2020) and the first Japanese eel *A. japonica* (Manabe et al. 2011). The other PSAT tracking experiments improved the battery and data storage capacity and provided time series data (i.e., depth and temperature) for 2–15 min intervals. PSATs cannot directly observe the horizontal migration pathway; therefore, the trajectories must be reconstructed using auxiliary data. Traditionally, light intensity data have been used for longitudinal estimation. However, migrating silver eels avoid the photic zone, making the tag's light sensor useless and making this method inapplicable (Manabe et al. 2011). Recorded vertical migration profiles provide estimates of sunrise and sunset (Westerberg et al. 2014) and the maximum swimming depths combined with bathymetry depth data may be used to identify possible swimming locations. Vertical profiles of the experienced water temperature of eels can be compared with environmental observations or modeling to improve the accuracy of tracking pathway estimation (Chang et al. 2020a; Higuchi et al. 2021). Current telemetry technologies allow us to study parameters that are otherwise unknown, such as vertical migration, migration pathways, and prey events, during the spawning migration of anguillid eels. Particle tracking simulations based on telemetry-based surveys have also been used to estimate the spawning migration pathways of silver eels (Chang et al. 2016).

7.2 Spawning Migration of Anguillid Eels

The first tracking with acoustic tags in silver eels reported the results of several hours of tracking the European eel *Anguilla anguilla* in the Elbe River and the North Sea during their early spawning migrations (Tesch 1989). In these studies, acoustic tags with pressure sensors were used primarily to investigate the vertical movement of eels, and the swimming depths of eels during the early stages of spawning migration were reported for the first time. The development of small PSATs has allowed eels to be successfully tracked in the open ocean for weeks to months, revealing the

Table 7.1 PSAT-based research on oceanic spawning migration of freshwater eels

Species	Region of major growth habitat; Tracking area	References
Anguilla dieffenbachii	Temperate; South Pacific	Jellyman and Tsukamoto (2002)
A. dieffenbachii	Temperate; South Pacific	Jellyman and Tsukamoto (2005)
A. anguilla	Temperate; North Atlantic	Aarestrup et al. (2009)
A. dieffenbachii	Temperate; South Pacific	Jellyman and Tsukamoto (2010)
A. japonica	Temperate; North Pacific	Manabe et al. (2011)
A. marmorata	Tropical; South Pacific	Schabetsberger et al. (2013)
A. megastoma	Tropical; South Pacific	
A. obscura	Tropical; South Pacific	
A. anguilla	Temperate; North Atlantic	Westerberg et al. (2014)
A. rostrata	Temperate; North Atlantic	Béguer-Pon et al. (2015)
A. marmorata	Tropical; South Pacific	Schabetsberger et al. (2015)
A. megastoma	Tropical; South Pacific	
A. anguilla	Temperate; North Atlantic	Wysujack et al. (2015)
A. anguilla	Temperate; North Atlantic	Amilhat et al. (2016)
A. anguilla	Temperate; North Atlantic	Righton et al. (2016)
A. rostrata	Temperate; North Atlantic	Béguer-Pon et al. (2017)
A. japonica	Temperate; North Pacific	Chen et al. (2018)
A. marmorata	Tropical; South Pacific	
A. bicolor bicolor	Tropical; South Pacific	
A. japonica	Temperate; North Pacific	Higuchi et al. (2018)
A. marmorata	Tropical; South Pacific	Schabetsberger et al. (2019)
A. megastoma	Tropical; South Pacific	
A. dieffenbachii	Temperate; South Pacific	Watanabe et al. (2020)
A. marmorata	Tropical; South Pacific	Chang et al. (2020a)
A. megastoma	Tropical; South Pacific	
A. japonica	Temperate; North Pacific	Higuchi et al. (2021)
A. marmorata	Tropical; South Pacific	Schabetsberger et al. (2021)
A. megastoma	Tropical; South Pacific	
A. australis australis	Temperate; South Pacific	Koster et al. (2021)
A. anguilla	Temperate; North Atlantic	Wright et al. (2022)

mysteries of their ocean spawning migrations. PSAT telemetry studies have revealed parts of the migration routes and vertical migration behavior of several eel species.

Through telemetry studies, several detailed vertical profiles were obtained using telemetry tracking (Table 7.1). These species commonly exhibit diel vertical migration (DVM) that swim in the shallow (100–400 m) and deep (500–800 m) layers during the night and day, respectively (Table 7.2). However, swimming depth varies according to species, individuals, and the tracking area. Intensive research has been

Table 7.2 Examples of the mean swimming depths and experienced temperatures (mean ± SD) during day and night of migrating silver eels (*Anguilla* spp.) tagged with PSATs

	Daytime		Nighttime		
Species	Depth (m)	Temp. (°C)	Depth (m)	Temp. (°C)	Reference
A. anguilla	564 ± 125	10.12 ± 0.89	282 ± 138	11.68 ± 0.48	Aarestrup et al. (2009)
A. marmorata	631	6	175	23	Schabetsberger et al. (2013)
A. megastoma	743	5.6	186	22.9	Schabetsberger et al. (2013)
A. obscura	312	15.1	226	19.7	Schabetsberger et al. (2013)
A. rostrata	618 ± 16	12	141 ± 14	24	Béguer-Pon et al. (2015)
A. bicolor pacifica	602.0 ± 77.3	7.5 ± 1.1	165.0 ± 75.6	20.0 ± 3.8	Chen et al. (2018)
A. dieffenbachii	928	5.6	336	19.8	Watanabe et al. (2020)
A. japonica	787.6 ± 54.87	5.2 ± 0.3	267.3 ± 52.6	18.2 ± 3.0	Higuchi et al. (2021)
A. australis australis	~ 700–900	6–8	~100–300	15–20	Koster et al. (2021)

conducted on the characteristics of *A. japonica* DVMs in the western North Pacific. In this section, an overview of DVMs for anguillid eels is provided, and the detailed behavioral characteristics of *A. japonica* are described in the next section. The influence of the lunar cycle on night-swimming depths was first reported in the South Pacific, the Indo-Pacific eel *A. marmorata*, and Polynesian longfin eel *A. megastoma*. These eels were released off Vanuatu and have been found to swim deeper during the full moon than during the new moon (Schabetsberger et al. 2013). Subsequently, this characteristic has been reported in several temperate species, including *A. japonica*, the Australian shortfin eel *A. australis australis* in eastern Australia, and *A. dieffenbachii* in New Zealand (Table 7.1). Large-scale ascending and descending behaviors have also been linked to the timing of sunset and sunrise in many species (Westerberg et al. 2014). To change their swimming depth in response to sunlight and moonlight, eels are known to increase their eye diameter and the number of rod cells responsible for light reception in the retina when they begin their spawning migration (see Chaps. 9 and 13). In addition, the expression levels of the "deep-sea rod opsin" gene (perceiving ocean blue light, approximately 480 nm wavelength) and the "freshwater rod opsin" gene (perceiving freshwater green light, approximately 500 nm wavelength) increase and decrease, respectively, in their retina at the start of spawning migration (Zhang et al. 2000). This allows migrating eels to have better vision under low light conditions at night as well as at deep water depths. The ecological significance of these DVMs is not fully understood, but a likely hypothesis is that there is a trade-off between predator avoidance (reducing predation risk from marine predators by migrating at minimal light levels) and thermoregulation, which controls the metabolic rate and gonad maturation (Aarestrup et al. 2009; Jellyman and Tsukamoto 2010; Higuchi et al. 2021). During oceanic migration, eels experience large temperature differences (>20 °C) depending on the ocean area. Recent laboratory experiments have suggested that the thermal shock associated with the DVM results in high energy consumption requirements (Trancart et al. 2015). These studies suggest that there is a significant benefit from DVM as a tradeoff between the associated energy loss and gain from a reduced risk of predation.

Based on the depth and temperature data recorded by the PSATs or the estimated surfacing position, spawning migration pathways that could not be directly observed by the PSAT were estimated. Jellyman and Tsukamoto (2002) reconstructed a 2250 km migration path from *A. dieffenbachii* tracking data; however, the surfacing location of the tag was over 2000 km from the spawning area estimated by Jellyman and Bowen (2009). Subsequent tracking experiments reported the migration process of tagged eels toward the estimated spawning areas north of New Zealand (see Chap. 4; Jellyman and Tsukamoto 2010). Reanalysis of tracking data confirmed the northward process of spawning migration based on changes in empirical water temperatures (Watanabe et al. 2020). Similar to *A. dieffenbachii*, the process of northward migration toward the spawning area was reconstructed in *A. australis australis*, which is distributed in the South Pacific (Koster et al. 2021). In 2014, a 2400 km migration route of a single American eel *A. rostrata* released from the coast of Nova Scotia, Canada, was reconstructed, providing the first direct evidence of silver eels migrating to the Sargasso Sea (Béguer-Pon et al. 2015). In *A. anguilla*, spawning migration from the North Sea was reconstructed as a southward pathway off the Faroe Islands or the Celtic Sea (Righton et al. 2016; Verhelst et al. 2022). Research on individuals from the Mediterranean led to the reconstruction of the westward migration through the Strait of Gibraltar, thus providing evidence that eels from the Mediterranean contribute to spawning migration (Amilhat et al. 2016). The latest PSAT tracking indicated that eels migrate eastward toward known spawning areas (Wright et al. 2022). All 3 reconstructed routes in *A. anguilla* suggest returning to the spawning area in the Sargasso Sea via the Azores. However, the results of particle-tracking simulations have proposed a new hypothesis in which Atlantic eels, *A. anguilla* and *A. rostrata*, spawn along the mid-Atlantic Ridge rather than in the Sargasso Sea, as previously thought (Chang et al. 2020b). Therefore, spawning migration routes must be considered with both possible spawning locations. Durif et al. (2022) showed that geomagnetism can explain the spawning migration routes of eels, as the reconstructed routes of Atlantic species were consistent with the horizontal gradient of the geomagnetic field calculated by the IGRF model (Thébault et al. 2015). They proposed a scenario in which the orientation of the silver eel spawning migration was determined by the imprinting of geomagnetic information experienced during the leptocephalus period. In contrast, Westerberg et al. (2014) proposed the hypothesis that genetically inherited compasses and geomagnetic maps determine orientation during spawning migration. Although it has been confirmed that eels have magnetic senses (Nishi et al. 2004) and there seems to be geomagnetism in their spawning migrations, the mechanism behind their utilization of geomagnetic fields remains a matter of debate.

Telemetry surveys have revealed that migrating silver eels are preyed upon by marine mammals and fish, and predation during spawning migration may be a major factor in their mortality. Predation by marine mammals and endothermic fish was determined by the rapid increase in temperature recorded by the tags (36 °C for marine mammals and 26 °C for endothermic animals). Telemetry studies of *A. anguilla* in the North Atlantic have confirmed its predation by cetaceans and sharks (Westerberg et al. 2014; Amilhat et al. 2016). Predation events were also detected in the open ocean, indicating that anguillid eels are at risk of predator attack during spawning migration (Righton et al. 2016). Predations by the porbeagle shark *Lamna nasus* and Atlantic bluefin tuna *Thunnus thynnus* have been detected in the tracking of *A. rostrata* (Béguer-Pon et al. 2015). Eels are presumed to be preyed upon by the great white shark (*Carcharodon carcharias*) or Pacific bluefin tuna *Thunnus orientalis* around the continental shelf of the East China Sea and by swordfish *Xiphias gladius* in the open ocean; however, predation events have not been clearly identified (Manabe et al. 2011; Higuchi et al. 2021). PSAT tracking experiments of these 3 anguillid species show high prey rates (*A. anguilla*: <60%, Amilhat et al. 2016; *A. rostrata*: <80%, Béguer-Pon et al. 2012; *A. japonica*: <40%, Higuchi et al. 2021). In biologging, the "5% (or 3%) rule" is commonly used, which states that the size of a data logger that does not affect animal behavior is 3–5% of the weight of the target animal. The size of the PSAT for eels far exceeds this rule, and thus, the possibility that the PSAT increases the eel's prey rates cannot be ruled out.

7.3 Spawning Migration of *A. japonica* in the Western North Pacific

The spawning area of *A. japonica* has been identified in the western Mariana Islands based on the collection of pre-feeding larvae (preleptocephali), fertilized eggs, and spawning-conditioned adults (see Chap. 4). Leptocephali are transported by the North Equatorial Current and the Kuroshio to their growth habitats in East Asia (see Chap. 5). After growing as yellow eels for 5–10 years or more, they enter the initial phase of sexual maturity in the fall or winter and migrate to spawning areas in the open ocean. However, since migrating *A. japonica* have never been collected in the open ocean, their migration route from growth to spawning habitats and behavior has not been clarified. In general, most telemetry studies on anguillid eels target the pathway of spawning migration, while studies on *A. japonica* have been actively conducted to understand the behavioral mechanism of DVMs. Aoyama et al. (2002) tracked the rapid migration from estuarine to coastal areas. In this study, artificially mature eels were subjected to active tracking using acoustic tags. The results suggest that silver eels may choose to travel to deeper water and move offshore immediately after their downstream migration. This feature is similar to the spawning migration

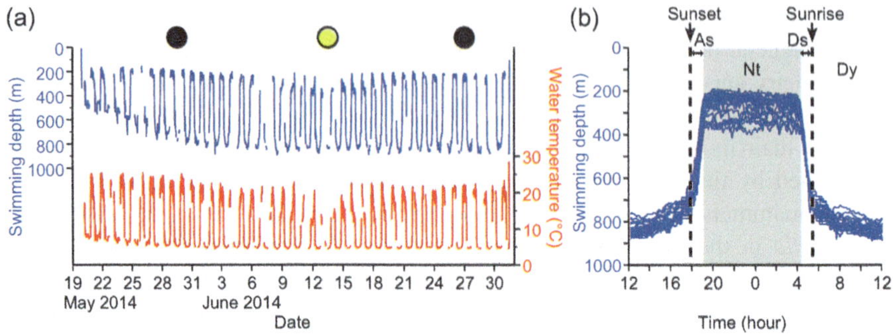

Fig. 7.2 PSAT-based tracking records of an *A. japonica* in the western North Pacific. (**a**) The swimming depths (upper) and experienced water temperatures (below). The yellow and black circles show the day of full moon and new moon, respectively. (**b**) Swimming depth data are separated into 4 phases of daytime (Dy), ascent (As), nighttime (Nt), and descent (Ds). Modified from Higuchi et al. (2018) with permission of the authors

behavior of *A. anguilla* in the European coastal area, which moves vertically between the surface and bottom toward deeper offshore depths (Tesch 1989) and may represent a common orientation method for anguillid species. This bathymetry-based orientation with surface-bottom oscillation was also observed in PSAT tracking of silver eels in coastal areas (Higuchi et al. 2021).

In recent telemetry studies, the behavioral characteristics of DVMs of *A. japonica* have been analyzed in detail by dividing them into 4 time segments: nighttime (the period during which they swim in the upper layer from the end of their ascent to the start of their descent), descent (the period during which they continuously dive from dawn to sunrise), daytime (the period during which they swim in the deeper layer from the end of their descent to the start of their ascent), and ascent (the period during which they continuously rise from sunset to dusk) (Higuchi et al. 2018, 2021; Fig. 7.2). When the moon altitude is higher, the nighttime swimming depth of the DVM is deeper (see the last section). The swimming depths were shallower layers with higher chlorophyll concentrations, suggesting that their nighttime behavior was regulated in response to the moonlight intensity transmitted through the water. Because nighttime swimming depth is determined by the light environment, the experienced water temperature may vary depending on the area and season in which the eels migrate. In contrast, eels tend to swim deeper at higher sun altitudes, suggesting that swimming depth changes during the day in response to sunlight reaching a daily maximum depth at sunset (Higuchi et al. 2018). The daily maximum depths were also found to peak at a fixed water temperature (approximately 5 °C) during the entire spawning migration period. Higuchi et al. (2021) reported that eels generally started to descend from shallow to deeper layers during the period from astronomical (sun altitude ≈ -18) to nautical dawn (sun altitude ≈ -12), and they finished descending during the period from civil dawn (sun altitude ≈ -6) and sunrise (sun altitude ≈ 0). After the daytime behavioral period, the eels started to ascend from deeper to shallower layers during the period from sunset (sun altitude $\approx$

0) and civil dusk (sun altitude ≈ -6), and they reached the shallower layer during the period from nautical (sun altitude ≈ -12) to astronomical dusk (sun altitude ≈ -18). Astronomical dawn is generally defined as the time when sunlight begins to influence a completely dark night; thus, eels might use the illumination change as a cue for the start of their descent. The observed pattern of irradiance changes, which the solar irradiance increases/decreases rapidly from astronomical twilights to sunrise/sunset, and increases/decreases slowly during the higher solar elevation period (Spitschan et al. 2016), was similar to the pattern of swimming depth changes in the DVMs of *A. japonica*. Based on the above behavioral characteristics, Higuchi et al. (2021) proposed a hypothesis for the DVM mechanism of silver-phase *A. japonica*, stating that they have increased visual sensitivity to blue light; thus, they select swimming depths to experience a constant light environment during their spawning migration, and the maximum swimming depths are determined physiologically by the water temperature.

In general, pelagic fishes and zooplankton are known to function with DVM to avoid predators during the day and maximize foraging at night. In a PSAT-based tracking study, individuals that did not exhibit standard DVM were preyed upon by endothermic fishes, suggesting that eel diving behavior seen during DVM may have a predator avoidance effect (Higuchi et al. 2021). However, a study comparing stable isotopes of silver eels during their spawning migration with those of spawning-condition adults in their spawning area suggested that eels may not feed during their ocean spawning migration (Chow et al. 2010). Therefore, the reason for ascending to a shallow layer at night for foraging, which has been observed in many pelagic fish, is not applicable to anguillid eels. It is known that oogenesis progresses to the oil droplet or previtellogenic stages when *A. japonica* begins spawning migration in winter (Utoh et al. 2004). This suggests that *A. japonica* underwent progressive oogenesis during spawning migration, which agrees with reports that the gonadosomatic index increases slightly in *A. japonica* when they experience water temperature changes in the laboratory (Mikawa et al. 2019). Thus, the DVM in *A. japonica* may play a role in visual predator avoidance and oogenesis promotion.

To evaluate the possible migration routes of *A. japonica*, Tsukamoto (2009) reviewed 3 hypotheses: (1) eels move eastward with the northern Kuroshio in the initial phase of their spawning migration from Japan, and then migrates southward to their spawning area; (2) eels cross the Kuroshio and migrate directly toward their spawning area from any part of their species range; and (3) eels reach their spawning area by swimming in the upstream direction in relation to the Kuroshio and the North Equatorial Current.

However, no evidence of a true spawning migration pathway has been provided. Most PSAT- or acoustic-based telemetry studies in anguillid eels are released from coastal waters, but even PSATs that have relatively long tracking periods cannot cover the entire process of their spawning migration, which extends several thousand kilometers (except for *A. rostrata* in the North Atlantic). Therefore, tracking experiments are being conducted in various areas, including coastal, pelagic, and spawning areas, to reveal all spawning migration routes and behaviors of *A. japonica* (Higuchi et al. 2021; Fukuda et al. 2022). Piecing together the fragmentary evidence of spawning migration pathways provided results that support the "Kuroshio Following Hypothesis" of Tsukamoto (2009). The particle tracking simulation involving a silver-phase *A. japonica* also appears to be consistent with this hypothesis (Chang et al. 2016). However, the orientation mechanisms that determine spawning migration paths remain unclear. It has been suggested that geomagnetism may be related to the orientation of spawning migration in *A. japonica*, as in other anguillid species (Nishi et al. 2004; Durif et al. 2022). However, acoustic tracking in the open ocean suggests that *A. japonica* may use a solar compass for meridional orientation in combination with multiple environmental cues (Fukuda et al. 2022).

7.4 Perspective

Recent telemetry studies have accumulated behavioral information on the spawning migration of freshwater eels (Table 7.2). Although the association between DVM and lunar phase has been reported for several species, the factors determining DVM characteristics are limited to their correspondence with the moon/sun altitudes in *A. japonica* (Higuchi et al. 2021) and the estimation of underwater irradiance in *A. marmorata* (Schabetsberger et al. 2013). The analysis methods used for *A. japonica* and *A. marmorata* could be applied to DVM analysis of other eel species, and interspecific comparisons could reveal the ecological significance and evolutionary background of DVM in anguillids.

The spawning migration of *A. japonica* begins in colder temperate regions in winter and ends in warmer tropical regions in summer; therefore, it is expected that the experienced water temperature at night will increase as spawning migration progresses (Fig. 7.3). An increase in experienced water temperature during the nighttime has also been observed in *A. dieffenbachii* in New Zealand and may be common to temperate eels. In contrast, tropical eels, which migrate relatively short distances, have both growth and spawning habitats in the tropics, and therefore, the long-term changes in experienced water temperature during their spawning migration might be small or absent (e.g., *A. megastoma* in Chang et al. 2020a). Oogenesis of a tropical eel, *A. celebesensis* captured in Indonesia, has been found to reach the midvitellogenic stage at the beginning of their spawning migration, giving it an advantage over *A. japonica* at the onset of spawning migration (see Chap. 13). Based on the above background, a 2-step scenario for the oogenesis of temperate eels can be speculated: (Step 1) Eels in temperate zones experience warm waters in the early stages of their spawning migration, which leads to a gradual progression of oogenesis. (Step 2) Subsequently, oogenesis may progress rapidly as they experience higher water temperatures in tropical waters. In contrast, because tropical eels inhabit the tropics during their entire spawning migration, gonadal maturation may be completed even over short migrations.

A silver-phase *A. japonica* collected in the spawning area is considered unsuitable for swimming because the maturity index reaches 47.8% and the

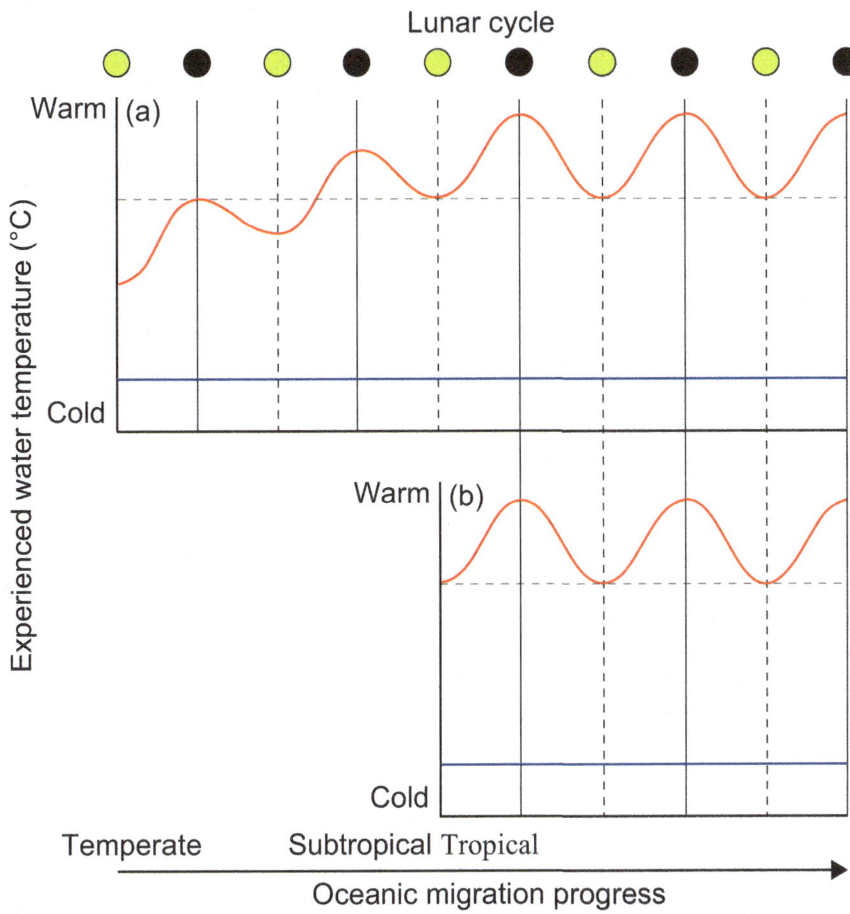

Fig. 7.3 Model for experienced temperature variations in (**a**) temperate and (**b**) tropical eels. Red wavy lines and blue straight lines show simplified night and day experience water temperatures, respectively. The yellow and black circles show the day of full moon (vertical dashed lines) and new moon (vertical solid lines), respectively. The horizontal gray dashed lines of the same water temperature are to facilitate intercomparisons of the experienced water temperatures between temperate and tropical eels

abdomen expands (Tsukamoto et al. 2011). If *A. japonica*, which migrates long distances, undergoes body-shape changes midway through spawning migration, their swimming ability may be reduced, and the risk of predation may increase. Therefore, it is favorable to accelerate oogenesis in the tropics or subtropics closer to the spawning area, rather than in temperate zones where they are in the process of spawning migration. This scenario may be reasonable based on the origin of the genus *Anguilla* and the locations of their current spawning areas are in the tropics. Artificial maturation and fertilization by exogenous hormone injection have been successful in several anguillid species, but artificial seedling production by spontaneous maturation without exogenous hormones has not been successful (Oliveira and Hable 2010). By reproducing the circadian and long-term water temperature variations that silver eels experience in their natural spawning migration in an artificial environment, a spontaneous maturation-inducing technique without exogenous hormones could be established.

Geomagnetic and solar compasses have been proposed as possible orientation methods for spawning migration of freshwater eels. Durif et al. (2022) argued that silver eel orientations were determined by the imprinting of geomagnetic features when they hatched from eggs at each spawning site. However, because the spawning location varies with hydrographic structures (i.e., salinity fronts), geomagnetic and/or solar compass orientation would not be sufficient to complete spawning migrations. Mouritsen (2018) considered that large-scale migration can be divided into long-, mid-, and near-range phases based on their progression and distance to the goal, each using different environmental cues for orientation. Similarly, eels may use different cues to orient themselves in stages as spawning migration progresses. Geomagnetic and solar compasses would correspond to the long-range phase of this theory. A comprehensive migration model is needed to understand the navigation mechanisms of anguillid eels, including hypotheses about orientation during the mid- (or near-) range phase, which have not yet been tested. The migration of marine organisms is the result of active swimming and passive transport by ocean currents. It is expected that the current hypotheses on possible orientation methods will be tested using bio-physical modeling or other methods to clarify the mysterious navigation mechanism of anguillid eels.

References

Aarestrup K, Økland F, Hansen MM, Righton D, Gargan P, Castonguay M, Bernatchez L, Howey P, Sparholt H, Pedersen MI, McKinley RS (2009) Oceanic spawning migration of the European eel (*Anguilla anguilla*). Science 325:1660. https://doi.org/10.1126/science.1178120

Amilhat E, Aarestrup K, Faliex E, Simon G, Westerberg H, Righton D (2016) First evidence of European eels exiting the Mediterranean Sea during their spawning migration. Sci Rep 6:1–9. https://doi.org/10.1038/srep21817

Aoyama J, Sasai S, Miller MJ, Shinoda A, Nakamura A, Kawazu K, Tsukamoto K (2002) A preliminary study of the movements of yellow and silver eels, *Anguilla japonica*, in the estuary of the Fukui River, Japan, as revealed by acoustic tracking. Hydrobiologia 470:31–36. https://doi.org/10.1023/A:1015604906154

Béguer-Pon M, Benchetrit J, Castonguay M, Aarestrup K, Campana SE, Stokesbury MJW, Dodson J (2012) Shark predation on migrating adult American eels (*Anguilla rostrata*) in the Gulf of St. Lawrence. PLoS One 7:e46830. https://doi.org/10.1371/journal.pone.0046830

Béguer-Pon M, Castonguay M, Shan S, Benchetrit J, Dodson JJ (2015) Direct observations of American eels migrating across the continental shelf to the Sargasso Sea. Nat Commun 6:1–9. https://doi.org/10.1038/ncomms9705

Béguer-Pon M, Shan S, Castonguay M, Dodson JJ (2017) Behavioural variability in the vertical and horizontal oceanic migrations of silver American eels. Mar Ecol Prog Ser 585:123–142. https://doi.org/10.3354/meps12380

Chang Y-L, Miyazawa Y, Béguer-Pon M (2016) Simulating the oceanic migration of silver Japanese eels. PLoS One 11:e0150187. https://doi.org/10.1371/journal.pone.0150187

Chang Y-LK, Olmo GD, Schabetsberger R (2020a) Tracking the marine migration routes of South Pacific silver eels. Mar Ecol Prog Ser 646:1–12. https://doi.org/10.3354/meps13398

Chang Y-LK, Feunteun E, Miyazawa Y, Tsukamoto K (2020b) New clues on the Atlantic eels spawning behavior and area: the Mid-Atlantic Ridge hypothesis. Sci Rep 10:1–12. https://doi.org/10.1038/s41598-020-72916-5

Chen SC, Chang CR, Han YS (2018) Seaward migration routes of indigenous eels, *Anguilla japonica*, *A. marmorata*, and *A. bicolor pacifica*, via satellite tags. Zool Stud 57:e21. https://doi.org/10.6620/zs.2018.57-21

Chow S, Kurogi H, Katayama S, Ambe D, Okazaki M, Watanabe T, Ichikawa T, Kodama M, Aoyama J, Shinoda A, Watanabe S, Tsukamoto K, Miyazaki S, Kimura S, Yamada Y, Nomura K, Tanaka H, Kazeto Y, Hata K, Handa T, Tawa A, Mochioka N (2010) Japanese eel *Anguilla japonica* do not assimilate nutrition during the oceanic spawning migration: evidence from stable isotope analysis. Mar Ecol Prog Ser 402:233–238. https://doi.org/10.3354/meps08448

Durif CM, Stockhausen HH, Skiftesvik AB, Cresci A, Nyqvist D, Browman HI (2022) A unifying hypothesis for the spawning migrations of temperate anguillid eels. Fish Fish 23:358–375. https://doi.org/10.1111/faf.12621

Fukuda N, Yamamoto T, Yokouchi K, Kurogi H, Okazaki M, Miyake Y, Watanabe T, Chow S (2022) Active swimming and transport by currents observed in Japanese eels (*Anguilla japonica*) acoustically tracked in the western North Pacific. Sci Rep 12:1–14. https://doi.org/10.1038/s41598-022-05880-x

Higuchi T, Watanabe S, Manabe R, Kaku T, Okamura A, Yamada Y, Miller MJ, Tsukamoto K (2018) Tracking *Anguilla japonica* silver eels along the West Mariana Ridge using pop-up archival transmitting tags. Zool Stud 57:e24. https://doi.org/10.6620/ZS.2018.57-24

Higuchi T, Watanabe S, Manabe R, Tanimoto A, Miller MJ, Kojima T, Tsukamoto K (2021) Horizontal and vertical migration behavior of silver-phase Japanese eels in coastal, pelagic and spawning areas observed by pop-up satellite archival tags. J Exp Mar Biol Ecol 542-543: 151587. https://doi.org/10.1016/j.jembe.2021.151587

Jellyman D, Bowen M (2009) Modelling larval migration routes and spawning areas of anguillid eels of New Zealand and Australia. Am Fish Soc Symp 69:255–274. https://doi.org/10.47886/9781934874080.ch17

Jellyman D, Tsukamoto K (2002) First use of archival transmitters to track migrating freshwater eels *Anguilla dieffenbachii* at sea. Mar Ecol Prog Ser 233:207–215. https://doi.org/10.3354/meps233207

Jellyman D, Tsukamoto K (2005) Swimming depths of offshore migrating longfin eels *Anguilla dieffenbachii*. Mar Ecol Prog Ser 286:261–267. https://doi.org/10.3354/meps286261

Jellyman D, Tsukamoto K (2010) Vertical migrations may control maturation in migrating female *Anguilla dieffenbachii*. Mar Ecol Prog Ser 404:241–247. https://doi.org/10.3354/meps08468

Koster WM, Aarestrup K, Birnie-Gauvin K, Church B, Dawson D, Lyon J, O'Connor J, Righton D, Rose D, Westerberg H, Stuart I (2021) First tracking of the oceanic spawning migrations of Australasian short-finned eels (*Anguilla australis*). Sci Rep 11:1–13. https://doi.org/10.1038/s41598-021-02325-9

Manabe R, Aoyama J, Watanabe K, Kawai M, Miller MJ, Tsukamoto K (2011) First observations of the oceanic migration of Japanese eel, from pop-up archival transmitting tags. Mar Ecol Prog Ser 437:229–240. https://doi.org/10.3354/meps09266

Mikawa N, Yamada Y, Horie N, Okamura A, Utoh T, Tanaka S, Tsukamoto K (2019) A preliminary experiment regarding the natural induction of gonadal development in female Japanese eels without hormone treatment. Aquac Res 50:3749–3754. https://doi.org/10.1111/are.14337

Mouritsen H (2018) Long-distance navigation and magnetoreception in migratory animals. Nature 558:50–59. https://doi.org/10.1038/s41586-018-0176-1

Nishi T, Kawamura G, Matsumoto K (2004) Magnetic sense in the Japanese eel, *Anguilla japonica*, as determined by conditioning and electrocardiography. J Exp Biol 207:2965–2970. https://doi.org/10.1242/jeb.01131

Oliveira K, Hable WE (2010) Artificial maturation, fertilization, and early development of the American eel (*Anguilla rostrata*). Can J Zool 88:1121–1128. https://doi.org/10.1139/Z10-081

Righton D, Westerberg H, Feunteun E, Økland F, Gargan P, Amilhat E, Metcalfe J, Lobon-Cervia J, Sjöberg N, Simon J, Acou A, Vedor M, Walker A, Trancart T, Brämick U, Aarestrup K (2016) Empirical observations of the spawning migration of European eels: the long and dangerous road to the Sargasso Sea. Sci Adv 2:e1501694. https://doi.org/10.1126/sciadv.1501694

Schabetsberger R, Økland F, Aarestrup K, Kalfatak D, Sichrowsky U, Tambets M, Dall'Olmo G, Kaiser R, Miller PI (2013) Oceanic migration behaviour of tropical Pacific eels from Vanuatu. Mar Ecol Prog Ser 475:177–190. https://doi.org/10.3354/meps10254

Schabetsberger R, Økland F, Kalfatak D, Sichrowsky U, Tambets M, Aarestrup K, Gubili C, Sarginson J, Boufana B, Jehle R, Dall'Olmo G, Miller MJ, Scheck A, Kaiser R, Quartly G (2015) Genetic and migratory evidence for sympatric spawning of tropical Pacific eels from Vanuatu. Mar Ecol Prog Ser 521:171–187. https://doi.org/10.3354/meps11138

Schabetsberger R, Scheck A, Kaiser R, Leaana R, Gubili C, Økland F (2019) Oceanic migration behaviour of Pacific eels from Samoa. Fish Manag Ecol 26:53–56. https://doi.org/10.1111/fme.12298

Schabetsberger R, Chang YLK, Miller MJ (2021) Spawning migration and larval dispersal of tropical Pacific eels (*Anguilla* spp.) in the centre of their distribution ranges. Mar Ecol Prog Ser 670:167–184. https://doi.org/10.3354/meps13745

Spitschan M, Aguirre GK, Brainard DH, Sweeney AM (2016) Variation of outdoor illumination as a function of solar elevation and light pollution. Sci Rep 6:1–14. https://doi.org/10.1038/srep26756

Tesch F-W (1989) Changes in swimming depth and direction of silver eels (*Anguilla anguilla L.*) from the continental shelf to the deep sea. Aquat Living Resour 2:9–20. https://doi.org/10.1051/alr:1989002

Thébault E, Finlay CC, Alken P, Beggan CD, Canet E, Chulliat A, Langlais B, Lesur V, Lowes FJ, Manoj C, Rother M, Schachtschneider R (2015) Evaluation of candidate geomagnetic field models for IGRF-12. Earth Planets Space 67:1–23. https://doi.org/10.1186/s40623-015-0273-4

Thorstad EB, Rikardsen AH, Alp A, Økland F (2013) The use of electronic tags in fish research – an overview of fish telemetry methods. Turkish J Fish Aquat Sci 13:881–896. https://doi.org/10.4194/1303-2712-v13_5_13

Trancart T, Tudorache C, Van den Thillart GEEJM, Acou A, Carpentier A, Boinet C, Gouchet G, Feunteun E (2015) The effect of thermal shock during diel vertical migration on the energy required for oceanic migration of the European silver eel. J Exp Mar Biol Ecol 463:168–172. https://doi.org/10.1016/j.jembe.2014.12.004

Tsukamoto K (2009) Oceanic migration and spawning of anguillid eels. J Fish Biol 74:1833–1852. https://doi.org/10.1111/j.1095-8649.2009.02242.x

Tsukamoto K, Chow S, Otake T, Kurogi H, Mochioka N, Miller MJ, Aoyama J, Kimura S, Watanabe S, Yoshinaga T, Shinoda A, Kuroki M, Oya M, Watanabe T, Hata K, Ijiri S, Kazeto Y, Monura K, Tanaka H (2011) Oceanic spawning ecology of freshwater eels in the western North Pacific. Nat Commun 2:1–9. https://doi.org/10.1038/ncomms1174

Utoh T, Mikawa N, Okamura A, Yamada Y, Tanaka S, Horie N, Akazawa A, Oka HP (2004) Ovarian morphology of the Japanese eel in Mikawa Bay. J Fish Biol 64:502–513. https://doi.org/10.1111/j.0022-1112.2004.00317.x

Verhelst P, Reubens J, Coeck J, Moens T, Simon J, Van Wichelen J, Westerberg H, Wysujack K, Righton D (2022) Mapping silver eel migration routes in the North Sea. Sci Rep 12:1–10. https://doi.org/10.1038/s41598-021-04052-7

Wahlberg M, Westerberg H, Aarestrup K, Feunteun E, Gargan P, Righton D (2014) Evidence of marine mammal predation of the European eel (*Anguilla anguilla L.*) on its marine migration. Deep Sea Res I: Oceanogr Res Pap 86:32–38. https://doi.org/10.1016/j.dsr.2014.01.003

Watanabe S, Higuchi T, Noshiro M, Manabe R, Miller MJ, Jellyman DJ, Tsukamoto K (2020) Reexamination of the spawning migration of *Anguilla dieffenbachii* in relation to water temperature and the lunar cycle. NZ J Mar Freshw Res 54:131–147. https://doi.org/10.1080/00288330.2019.1614075

Westerberg H, Sjöberg N, Lagenfelt I, Aarestrup K, Righton D (2014) Behaviour of stocked and naturally recruited European eels during migration. Mar Ecol Prog Ser 496:145–157. https://doi.org/10.3354/meps10646

Wright RM, Piper AT, Aarestrup K, Azevedo JMN, Cowan G, Don A, Gollock M, Ramallo SR, Velterop R, Walker A, Westerberg H, Righton D (2022) First direct evidence of adult European eels migrating to their breeding place in the Sargasso Sea. Sci Rep 12:1–6. https://doi.org/10.1038/s41598-022-19248-8

Wysujack K, Westerberg H, Aarestrup K, Trautner J, Kurwie T, Nagel F, Hanel R (2015) The migration behaviour of European silver eels (*Anguilla anguilla*) released in open ocean conditions. Mar Freshw Res 66:145–157. https://doi.org/10.1071/MF14023

Zhang H, Futami K, Horie N, Okamura A, Utoh T, Mikawa N, Tamada Y, Tanaka S, Okamoto N (2000) Molecular cloning of fresh water and deep-sea rod opsin genes from Japanese eel *Anguilla japonica* and expressional analyses during sexual maturation. FEBS Lett 469:39–43. https://doi.org/10.1016/s0014-5793(00)01233-3

Chapter 8
Behavior

Hikaru Itakura

Contents

In contrast to the large-scale oceanic migration between spawning grounds in the open ocean and continental habitats, eels exhibit sedentary behavior in rivers, lakes, and estuaries, where they spend a large part of their life. The emerging biotelemetry of free-ranging fish represents a promising approach for better observation of the cryptic sedentary behavior of eels over an extensive period (Béguer-Pon et al. 2018). This chapter describes the behavior of yellow eels, which is mainly revealed by acoustic telemetry, including their activity, feeding behavior, movement, homing, and habitat utilization.

8.1 Diel Activity

It is widely known that yellow eels of various anguillid eel species exhibit nocturnal and crepuscular behaviors, and thus are most active during night and twilight, respectively (*Anguilla rostrata*, Hedger et al. 2010; Béguer-Pon et al. 2015, *A. anguilla*, Ovidio et al. 2013; Walker et al. 2014; Verhelst et al. 2017, *A. japonica*, Itakura et al. 2018; Noda et al. 2020, *A. mossambica*, *A. bengalensis*, *A. marmorata*, Hanzen et al. 2021, *A. australis*, *A. dieffenbachii*, Jellyman and Sykes 2003). During the day, they shelter in mud, sand, rocks, vegetation, and other potential refuges (Aoyama et al. 2005). Furthermore, results of an acoustic

H. Itakura (✉)
Atmosphere and Ocean Research Institute, The University of Tokyo, Chiba, Japan
e-mail: hikaruitakura@aori.u-tokyo.ac.jp

K. Tsukamoto et al. (eds.), *Eel Science*, Fisheries Science Series,
https://doi.org/10.1007/978-981-99-5692-0_8

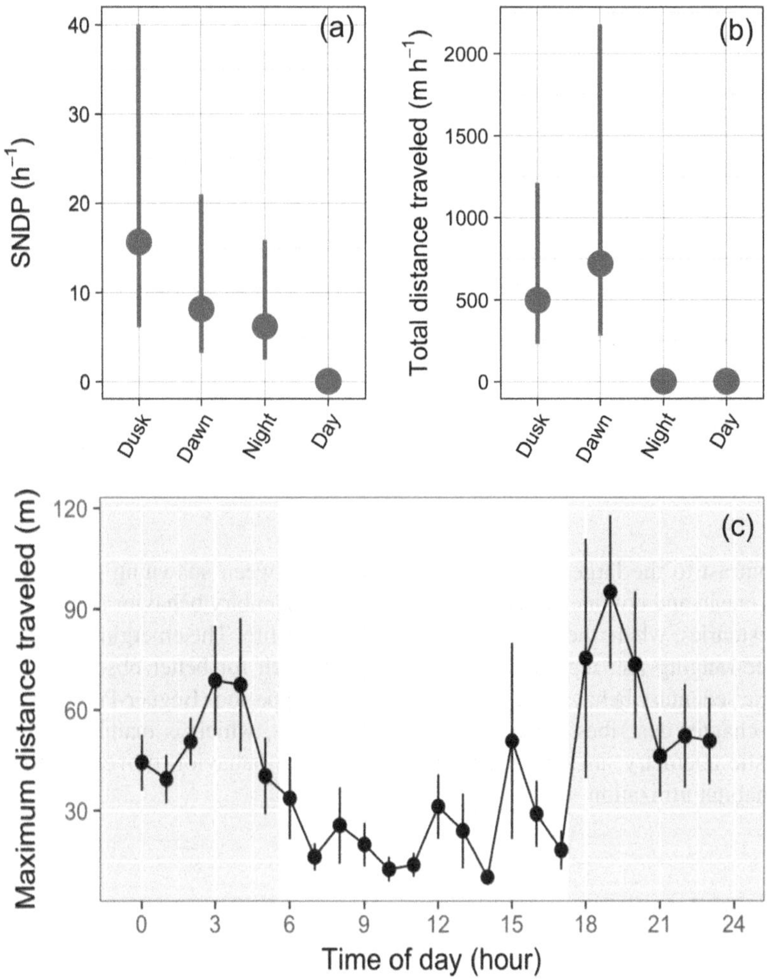

Fig. 8.1 Diel activity and movement patterns of tagged yellow-phase Japanese eels as inferred by acoustic telemetry. (**a**) the standardized number of detected positions (SNDP) and (**b**) the total distance traveled at time of day. (**c**) the hourly maximum distance traveled. Shadings represents dark periods (dawn, dusk, and night) (modified from Itakura et al. 2018, 2022)

biotelemetry study of Japanese eels in a brackish lake indicated that 98% of eel positions, which are often used as an activity index, are detected between sunset and sunrise (Itakura et al. 2022). It is possible that eels feed primarily during the dark times of the day. Fishers say that the best time to angle eels is just after sunset. Indeed, biotelemetry studies have shown that eel activity is greatest during dusk and dawn compared to that during nighttime and daytime, indicating strong crepuscular behavior for feeding (Fig. 8.1) (Jellyman and Sykes 2003; Itakura et al. 2022).

Because eels rest (non-feeding) during the day, their appetite is likely to be strongest at dusk when they start to forage, which may explain the high activity at that time.

Such diel patterns of eel behavior likely vary depending on environmental conditions and habitat type, which can alter light intensity. The masking effects of light alter normal diel activity patterns by markedly decreasing or increasing the activity levels of fish (Reebs 2002). Although eels are active during dark times of the day, anecdotes by anglers and fishers in Japan have long indicated that eels can be caught during the day when it is overcast or when rain reduces the light intensity. Notably, the diurnal activity of eels has been observed during darker conditions when eels extend their night activity into the morning, suggesting that overcast skies reduce light intensity and prolong nocturnal activity during the daytime (Itakura et al. 2022). Furthermore, it has been reported that while eels in shallow waters display exclusive nocturnal and crepuscular behaviors (Itakura et al. 2022), eels in deeper and turbid habitats exhibit partial nocturnal behavior, showing some activity during the day (Walker et al. 2014; Itakura et al. 2018; Piper et al. 2022). In deep environments, bottom habitats may be dark enough to encourage eels to be mobile during the day, which might explain their increased activity levels (Itakura et al. 2022). In contrast, light levels would be high at the bottom in shallow habitats compared with deeper habitats, which likely results in a clear contrast in light conditions between day and night. Thus, eel activity in shallow habitats may be more sensitive to light conditions than in deeper habitats.

The nocturnal and crepuscular behaviors of eels may be related to their foraging behavior, which is based on olfaction rather than visual cues, as olfaction is more important under dark conditions (Barbin 1998). These behaviors may also be related to reducing the risk of predation by potential predators such as piscivorous birds and fishes, which could prey on smaller eels. In shallow environments, eels are likely exposed to high risks of predation compared to those in deeper habitats; thus, crepuscular and nocturnal behaviors are necessary to reduce predation risk. In contrast, deep environments, where predation risks are likely to be relatively low, could allow eels to be somewhat mobile during the day (Itakura et al. 2022).

Eels seem to change their feeding habitats at night and resting habitats during the day, suggesting diel habitat shifts (Jellyman and Sykes 2003; Itakura et al. 2022). In lacustrine habitats, Japanese eels rest nearshore during the day, with space utilization expanding offshore at dusk and night, and returning to nearshore areas at dawn (Itakura et al. 2022). Because offshore areas are generally deeper than nearshore areas, the depth distribution of eels may also change during different times of the day in relation to changes in the distance to the shore. Such diel changes in depth distribution have also been reported for other eel species in New Zealand (Jellyman and Sykes 2003). These results showed that eels occupy deeper areas at night than during the day, and this movement pattern could be attributed to food availability.

8.2 Seasonal Activity

Temperate eels exhibit seasonal activity patterns, which are likely driven by a combination of multiple factors but may be partly associated with seasonal changes in water temperature. In general, eels actively feed and grow from spring to autumn months when the water temperature is relatively high, whereas they tend to be less active or torpor (inactive, cease feeding, and have a saltatory depression of basal metabolism) during winter months when temperatures fall below 3–12 °C (potentially depending on species, habitat, and latitude) (Walsh et al. 1983; Ovidio et al. 2013; Westerberg and Sjöberg 2015; Itakura et al. 2018; Rohtla et al. 2022). During the torpor period at low temperatures, eels shelter in substrate refuges, such as mud, either in the same habitat where they spend other periods or they migrate to other habitats for overwintering (Tomie et al. 2017). It has been reported that Japanese eels stop feeding when the water temperature falls below their optimum feeding temperature range (13.2–25 °C) (Matsui 1972). Indeed, Japanese eels inhabiting freshwater habitats are active from spring to autumn when the water temperature is above 13 °C and become less active or dormant during winter when the water temperature falls below 13 °C (Itakura et al. 2018). Reductions in activity at low temperatures have been commonly observed among species. *Anguilla anguilla* are most active in spring (at 12 °C), and less active during winter, reaching a minimum at a temperature of 3 °C (Ovidio et al. 2013). The activities of New Zealand eels decrease when the water temperature falls below 6 °C (Jellyman 1991), with no significant differences in activities between varying temperatures above this level (Jellyman and Sykes 2003). In addition to the interspecific differences, the water temperature range at which eels cease activity during winter may differ within eel species, because the temperature that eels experience in their habitats varies depending on the distributional latitude. Intraspecific variation in such thermal niches has not been well evaluated, which is an urgent issue due to global warming.

The seasonal activity patterns of eels may differ among habitat locations where they inhabit. It appears that eels inhabiting saline habitats may also be mobile during the winter. Specifically, Japanese eels in saline habitats have been reported to feed throughout the year (Katahira et al. 2016) and had longer feeding durations than eels in freshwater habitats, which was attributed to warmer water temperatures in saline habitats during winter (Kaifu et al. 2013a). This longer feeding duration can partially contribute to the higher growth rate of eels in saline habitats than in freshwater habitats (Kaifu et al. 2013a). Moreover, although eels tend to remain in a particular habitat throughout the season, some individuals change their habitat by season (see below for more details) (Thibault et al. 2007; Hedger et al. 2010; Béguer-Pon et al. 2015). The seasonal migration of eels among different habitats has been poorly investigated, and it is unclear how this migration contributes to their growth. In addition, seasonal activity of eels may be influenced not only by seasonal changes in water temperature, but also by other factors such as photoperiod (Rohtla et al. 2022), food availability, and body condition; therefore, further studies considering multiple factors are required.

8.3 Feeding Behavior

Yellow eels are well-known as facultative polytrophic predators that feed on a range of prey taxa (Dörner et al. 2009; Kaifu et al. 2013b; Itakura et al. 2015). For example, Japanese eels feed on aquatic organisms, such as small fish, crustaceans (shrimps, crabs, and crayfish), polychaetes, bivalves, aquatic insects (dragonfly and mayfly larvae), leeches, and terrestrial organisms, such as oligochaetes (earthworms), insects, and their larvae (Fig. 8.2) (Kaifu et al. 2013b; Itakura et al. 2015; Itakura et al. 2021; Wakiya and Mochioka 2021). Yellow eels are generally carnivorous and are considered one of the highest-order predators in riverine freshwater ecosystems (Itakura et al. 2020).

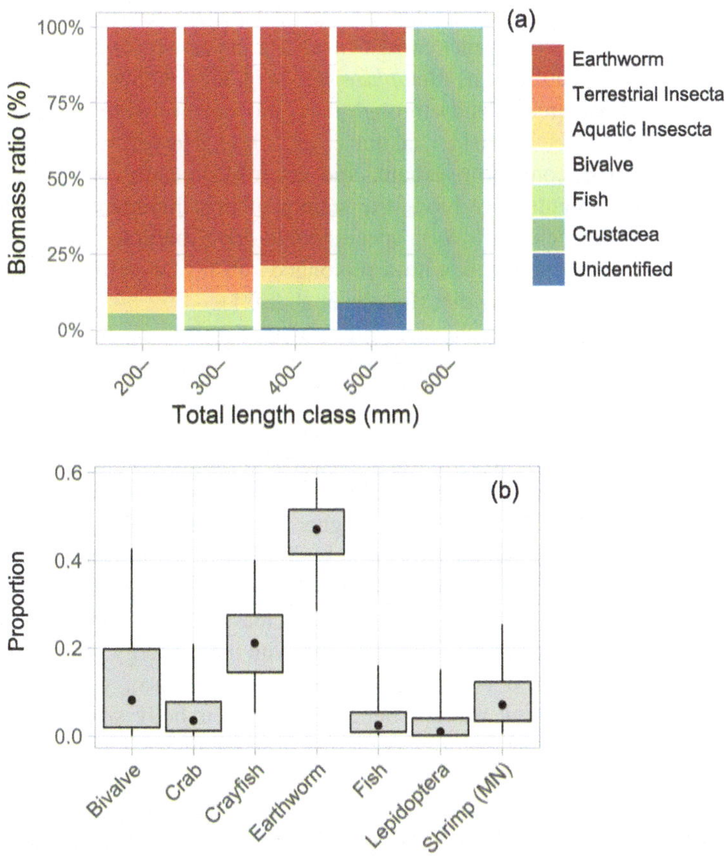

Fig. 8.2 Diet of yellow-phase Japanese eels, collected in natural shore sites of freshwater areas in the Tone River watershed, Japan. (**a**) stomach contents of eels, shown as biomass ratio (%) per total length class. (**b**) prey contributions to the diet of eels estimated by the mixing model based on carbon and nitrogen stable isotopic ratios (modified from Itakura et al. 2021)

The diet of eels varies depending on their environment. A study in the Kojima Bay-Asahi River system, Japan, reported that Japanese eels inhabiting brackish bay areas and freshwater river waters mainly ate mud shrimp *Upogebia major* and crayfish *Procambarus clarkii*, respectively (Kaifu et al. 2013a, b). Studies also showed that the annual food consumption of yellow eels was higher in brackish waters than that in freshwater, which may lead to a higher annual growth rate of eels in brackish waters than in freshwater.

Eels also feed on terrestrial organisms. Japanese eels inhabiting the upper reaches of a river mostly feed on terrestrial organisms such as cockroaches *Epilampridae* spp., centipedes *Scolopendra* spp., and earthworms *Megascolecidae* spp., as well as amphibious Japanese freshwater crabs *Geothelphusa dehaani*. However, the contribution of terrestrial resources to the eel diet is likely much lower in the lower reaches of a river, where eels primarily eat shore crabs that feed on sea algae (Wakiya and Mochioka 2021). Another study conducted in freshwater areas of the lower reaches of the Tone River, Japan, showed that the diet of Japanese eels inhabiting natural shoreline areas, where riverbanks consist of vegetation or exposed soil, depended largely on terrestrial earthworms *Metaphire* spp. (Fig. 8.2), which are supplied to a river as a pulsed resource subsidy through mass movement driven by rainfall into river waters (Itakura et al. 2015; Itakura et al. 2021). However, eels inhabiting revetment areas with concrete riverbanks did not feed on earthworms, suggesting that shoreline revetments may block the supply of this important allochthonous resource subsidy. Diverse earthworm species could drive multiple pulsed subsidies across seasons and provide eels with a prolonged subsidy, thereby enhancing the long-term contribution to their diet (Itakura et al. 2021).

The diet of the yellow eels changes with growth. Smaller eels feed primarily on annelids, including earthworms and insects with relatively small and soft bodies, while larger individuals mainly consume fish and crustaceans with relatively large and hard bodies, indicating an ontogenetic diet shift (Michel and Oberdorff 1995; Itakura et al. 2021). Therefore, the tropic positions of larger eels tend to be higher than those of smaller eels, suggesting that eels transition from feeding on prey animals with lower trophic positions to higher ones as they grow (Itakura et al. 2021). In general, gape size, relative to body size, has a significant effect on the available size of prey animals (Brönmark and Hansson 1998), and handling time for feeding decreases as the predator's total length increases (Werner 1974). Assuming that larger prey animals provide a predatory fish with more energy, there may be an optimal prey size for each body size of the predator. Hence, yellow eels may selectively eat suitable prey animals depending on their body size, which might allow them to maximize the energy they derive from feeding.

As noted above, eels utilize a wide variety of prey taxa at a water system or the population scales. However, at the individual level or at each feeding event, eels tend to feed exclusively on a single prey species (Dörner et al. 2009; Kaifu et al. 2013b). Because some eels move among different environments with completely different prey compositions (i.e., brackish water and freshwater), it is assumed that they may flexibly change their food habits depending on the environment. In fact, it has been reported for *A. anguilla* that although eels tend to feed exclusively on a single prey species at each feeding event, they change their target prey species in response to the density of the prey species (Dörner et al. 2009). In addition to such flexible change in feeding habits according to various habitats and prey abundance, by feeding exclusively on a single prey species that is abundant in quantity at each feeding event, eels could use prey resources in habitats effectively, with minimal effort.

8.4 Movement, Home Range, and Homing Behavior

It is challenging to generalize the features of movement and home range of eels, because their estimates may differ depending on study designs (i.e., tracking methods (active or passive), physical and statistical tools, and spatiotemporal scale and coverage) (Béguer-Pon et al. 2018), habitat scale and type (i.e., lake, estuary, river, and stream), and large individual variations. However, numerous telemetry studies have shown that tagged eels are continuously detected in areas close to the capture points, suggesting that their travel distances are generally short, and that they tend to remain at a specific location for several months to years. In general, the home range size of eels appears to increase with body size (Thibault et al. 2007; Barry et al. 2016; Hanzen et al. 2021), although some studies have found no significant relationship between them (Walker et al. 2014; Itakura et al. 2018). For Japanese eels, the annual home range and linear distance traveled of yellow eels living in a freshwater area of a river have been estimated to be <0.33 km^2 and <1.1 km, respectively (Itakura et al. 2018) (Fig. 8.3), whereas in a brackish lake tagged Japanese eels tended to be continuously distributed within ~1 km of the release point (Noda et al. 2020). The home range of yellow eels has been estimated to be <1.0 km^2 (Barry et al. 2016) or several meters to kilometers (Ovidio et al. 2013; Walker et al. 2014; Verhelst et al. 2017) for *A. anguilla*, <19 km for *A. rostrata* (Thibault et al. 2007; Hedger et al. 2010; Béguer-Pon et al. 2015), and <~l km for *A. australis* and *A. dieffenbachii* (Jellyman and Sykes 2003; Crook et al. 2014; Jellyman and Crow 2016). For *A. marmorata*, a 2-year mark-recapture experiment conducted in a small river of the Amami-Oshima Island, Japan, has shown that 50% of recaptured eels were recaptured from the original section of capture (i.e., <10 m distances moved), with distance moved by another half of individuals being 46.5 ± 72.5 m (median: 20 m, range: 10–380 m) (Itakura and Wakiya 2020) (Fig. 8.4), whereas eels stayed for 25% of a 1-year tracked period within 1-km study areas near the headwaters of a tributary of the Cagayan River, Philippines (Piper et al. 2022). In the Thukela River, South Africa, it was reported that the home range and linear distance traveled for *A. marmorata* and *A. bengalensis* were <0.04 km^2 and <2.3 km, respectively (Hanzen et al. 2021). Furthermore, the home range of eels tended to be bordered by shorelines, and they primarily moved along shorelines with movements to opposite shorelines being rarely observed, demonstrating their limited habitat use with an emphasis on nearshore areas (Jellyman and Sykes 2003; Thibault et al. 2007; Itakura et al. 2018, 2022). The small home range size and limited habitat-use patterns of yellow eels indicate strong

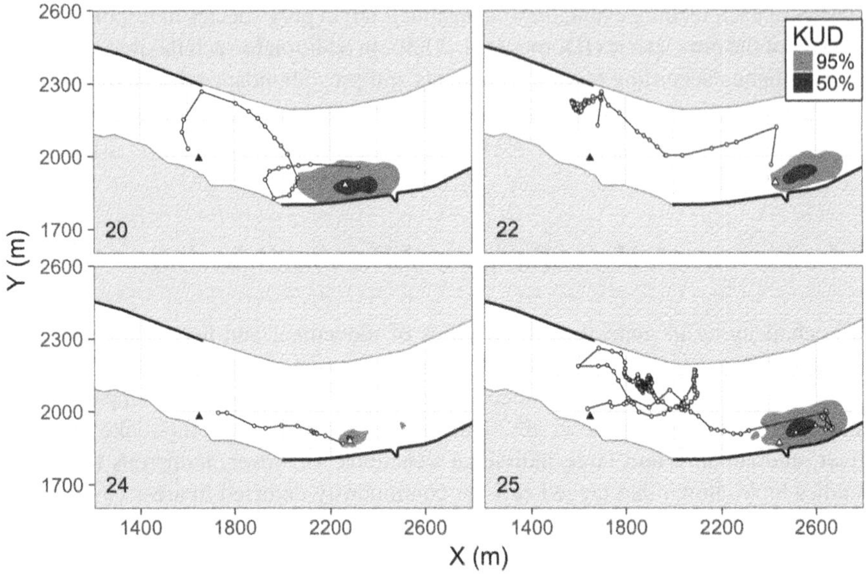

Fig. 8.3 Trajectories of the homing behaviour and home range (95% kernel utilization distribution [KUD]) and core area (50% KUD) for tagged Japanese eels. White and black triangles indicate the capture and release points, respectively. The numbers in the bottom left corner of each panel represent the individual tag number (modified from Itakura et al. 2018)

fidelity to specific sites, which contrasts with the long distances they travel during their upstream and downstream migration phases in rivers and spawning migrations in the ocean. Feeding within such a 'familiar' small home range could allow easier access to food for eels.

The strong site fidelity of the yellow eels was also evidenced by their homing abilities. Homing abilities in specific areas of rivers, lakes, and estuaries have been documented at various spatial scales, ranging from hundreds of meters within a watershed to hundreds of kilometers outside the watershed (*A. rostrata*, Parker 1995; Lamothe et al. 2000; Thibault et al. 2007; Béguer-Pon et al. 2015, *A. japonica*, Itakura et al. 2018, *A. anguilla*, Tesch 1967, and *A. australis*, Jellyman et al. 1996). For example, a transport-release experiment conducted within a freshwater river area demonstrated that Japanese eels that were displaced outside their home range (~800 m away from their capture sites) returned to their respective capture sites within 13 days (average: 6.2 days, range: 162 min to 13.3 days) after release (Itakura et al. 2018) (Fig. 8.3). Similarly, *A. rostrata* can return to their original areas, as reported by transport-release experiments among adjacent freshwater ponds (distance transported: 1.6 km) (Lamothe et al. 2000) and between tidal freshwater and brackish water areas (10–17 km) (Parker 1995). Larger-scale experiments showed that *A. anguilla* that were captured in rivers and then transplanted into offshore areas returned to their original capture areas, including those transported 180 km away (Tesch 1967). In addition to homing behavior within continental habitats, eels

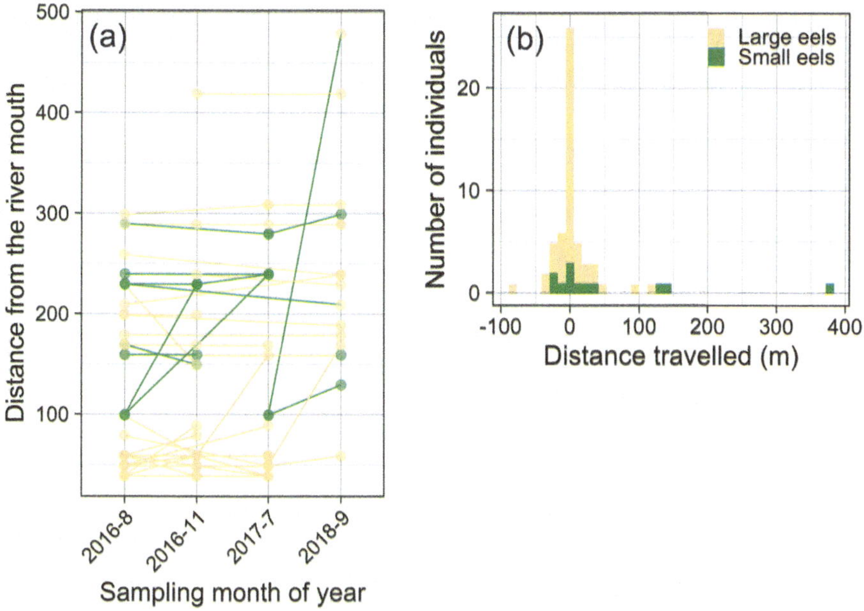

Fig. 8.4 Movement of recaptured A. marmorata in the Oganeku River, Amami-Oshima Island, Japan. (**a**) capture and recapture lacations during each sampling survey connected by lines for each eel. (**b**) histogram of distance traveled of eels (modified from Itakura and Wakiya 2020)

perform large-scale natal homing, which extends thousands of kilometers from continental nurseries to spawning grounds in the open ocean. Eels may therefore possess the ability to accurately determine orientation and return to specific areas regardless of distance, habitat, or life history. The ecological benefits of eel homing abilities during their continental phase remain poorly understood; therefore, further research is needed. However, homing abilities could have merit for eels when they encounter significant ecological disturbances, such as flood events, which significantly disturb riverine ecosystems. When eels encounter flood events, eels may evacuate to other habitats (i.e., tributaries) or be transported to the lower reaches of rivers. By possessing homing abilities, eels could return to their original homes from other habitats after floods have passed.

Other interesting migration behaviors of yellow eels have been identified by telemetry studies, one of which involves extensive seasonal migration between river (freshwater) and estuarine (brackish water) habitats tens to hundreds of kilometers away. It has been demonstrated for *A. rostrata* that some tagged eels living in freshwater rivers migrated to brackish estuaries during summer, probably for feeding (Thibault et al. 2007; Hedger et al. 2010; Béguer-Pon et al. 2015), and some of the eels that migrated to estuaries returned to rivers for overwintering (Thibault et al. 2007). Moreover, some yellow eels migrate among the rivers of different watersheds (Walker et al. 2014; Kume et al. 2021). Furthermore, daily recurring migration between river and estuarine habitats has been previously reported, wherein eels moved to a river at night, followed by a return to an estuary at dawn (Hedger et al. 2010; Noda et al. 2020). Importantly, these migration behaviors between different habitats are likely performed by some of the eels inhabiting each watershed. Therefore, future studies are required to investigate the environmental and biological factors that can drive these unique migration behaviors.

References

Aoyama J, Shinoda A, Sasai S, Miller MJ, Tsukamoto K (2005) First observations of the burrows of *Anguilla japonica*. J Fish Biol 67:1534–1543. https://doi.org/10.1111/j.1095-8649.2005.00860.x

Barbin GP (1998) The role of olfaction in homing and estuarine migratory behavior of yellow-phase American eels. Can J Fish Aquat Sci 55:564–575. https://doi.org/10.1139/f97-274

Barry J, Newton M, Dodd JA, Hooker OE, Boylan P, Lucas MC, Adams CE (2016) Foraging specialisms influence space use and movement patterns of the European eel *Anguilla anguilla*. Hydrobiologia 766:333–348. https://doi.org/10.1007/s10750-015-2466-z

Béguer-Pon M, Castonguay M, Benchetrit J, Hatin D, Legault M, Verreault G, Mailhot Y, Tremblay V, Dodson JJ (2015) Large-scale, seasonal habitat use and movements of yellow American eels in the St. Lawrence River revealed by acoustic telemetry. Ecol Freshw Fish 24: 99–111. https://doi.org/10.1111/eff.12129

Béguer-Pon M, Dodson JJ, Castonguay M, Jellyman D, Aarestrup K, Tsukamoto K (2018) Tracking anguillid eels: five decades of telemetry-based research. Mar Freshw Res 69:199–219. https://doi.org/10.1071/MF17137

Brönmark C, Hansson L-A (1998) Food web interactions in freshwater ecosystems. In: The biology of lakes and ponds. Oxford University Press, Oxford, pp 187–235

Crook DA, Macdonald JI, Morrongiello JR, Belcher CA, Lovett D, Walker A, Nicol SJ (2014) Environmental cues and extended estuarine residence in seaward migrating eels (*Anguilla australis*). Freshw Biol 59:1710–1720. https://doi.org/10.1111/fwb.12376

Dörner H, Skov C, Berg S, Schulze T, Beare DJ, Van Der Velde G (2009) Piscivory and trophic position of *Anguilla anguilla* in two lakes: importance of macrozoobenthos density. J Fish Biol 74:2115–2131. https://doi.org/10.1111/j.1095-8649.2009.02289.x

Hanzen C, Lucas MC, O'Brien G, Calverley P, Colleen DT (2021) Spatial ecology of freshwater eels in South Africa: implications for conservation. Hydrobiologia 848:2579–2593. https://doi.org/10.1007/s10750-021-04581-2

Hedger RD, Dodson JJ, Hatin D, Caron F, Fournier D (2010) River and estuary movements of yellow-stage American eels *Anguilla rostrata*, using a hydrophone array. J Fish Biol 76:1294–1311. https://doi.org/10.1111/j.1095-8649.2010.02561.x

Itakura H, Wakiya R (2020) Habitat preference, movements and growth of giant mottled eels, *Anguilla marmorata*, in a small subtropical Amami-Oshima Island river. PeerJ 8:1–28. https://doi.org/10.7717/peerj.10187

Itakura H, Kaino T, Miyake Y, Kitagawa T, Kimura S (2015) Feeding, condition, and abundance of Japanese eels from natural and revetment habitats in the Tone River, Japan. Environ Biol Fish 98:1871–1888. https://doi.org/10.1007/s10641-015-0404-6

Itakura H, Miyake Y, Kitagawa T, Kimura S (2018) Site fidelity, diel and seasonal activities of yellow-phase Japanese eels (*Anguilla japonica*) in a freshwater habitat as inferred from acoustic telemetry. Ecol Freshw Fish 27:737–751. https://doi.org/10.1111/eff.12389

Itakura H, Wakiya R, Gollock M, Kaifu K (2020) Anguillid eels as a surrogate species for conservation of freshwater biodiversity in Japan. Sci Rep 10:8790. https://doi.org/10.1038/s41598-020-65883-4

Itakura H, Miyake Y, Kitagawa T, Sato T, Kimura S (2021) Large contribution of pulsed subsidies to a predatory fish inhabiting large stream channels. Can J Fish Aquat Sci 78:144–153. https://doi.org/10.1139/cjfas-2020-0004

Itakura H, Miyake Y, Wakiya R, Kimura S (2022) Environmental influences on late-summer individual Japanese eel diel activity and space utilization patterns in a shallow western Japan brackish lake. Fish Sci 88:29–43. https://doi.org/10.1007/s12562-021-01560-3

Jellyman DJ (1991) Factors affecting the activity of two species of eel (*Anguilla* spp.) in a small New Zealand lake. J Fish Biol 39:7–14. https://doi.org/10.1111/j.1095-8649.1991.tb04336.x

Jellyman DJ, Crow SK (2016) Population size, growth and movements of *Anguilla australis* in a small lake. J Fish Biol 88:2157–2174. https://doi.org/10.1111/jfb.12962

Jellyman DJ, Sykes JRE (2003) Diel and seasonal movements of radio-tagged freshwater eels, *Anguilla* spp., in two New Zealand streams. Environ Biol Fish 66:143–154. https://doi.org/10.1023/A:1023691604088

Jellyman DJ, Glova GJ, Todd PR (1996) Movements of shortfinned eels, *Anguilla australis*, in Lake Ellesmere, New Zealand: results from mark-recapture studies and sonic tracking. N Z J Mar Freshw Res 30:371–381. https://doi.org/10.1080/00288330.1996.9516724

Kaifu K, Miller MJ, Yada T, Aoyama J, Washitani I, Tsukamoto K (2013a) Growth differences of Japanese eels *Anguilla japonica* between fresh and brackish water habitats in relation to annual food consumption in the Kojima Bay-Asahi River system, Japan. Ecol Freshw Fish 22:127–136. https://doi.org/10.1111/eff.12010

Kaifu K, Miyazaki S, Aoyama J, Kimura S, Tsukamoto K (2013b) Diet of Japanese eels *Anguilla japonica* in the Kojima bay-Asahi river system, Japan. Environ Biol Fish 96:439–446. https://doi.org/10.1007/s10641-012-0027-0

Katahira H, Mizuno K, Nagasawa K (2016) Year-round infections and complicated demography of a food-transmitted parasite *Heliconema anguillae* implying the feeding activity of Japanese eels in saline habitats. Fish Sci 82:863–871. https://doi.org/10.1007/s12562-016-1017-5

Kume M, Nakayama N, Iwasaki Y, Hori T, Watanabe S, Terashima Y, Medo A, Arai N, Yamashita Y, Mitamura H (2021) River to river: first evidence of eel movement between distant rivers via the sea. Environ Biol Fish 104:529–533. https://doi.org/10.1007/s10641-021-01090-y

Lamothe PJ, Gallagher M, Chivers DP, Moring JR (2000) Homing and movement of yellow-phase American eels in freshwater ponds. Environ Biol Fish 58:393–399. https://doi.org/10.1023/A:1007639615834

Matsui I (1972) Eel biology cultivation techniques. Koseisha-Koseikaku, Tokyo

Michel P, Oberdorff T (1995) Feeding habits of fourteen european freshwater fish species. Cybium 19:5–46. https://sfi-cybium.fr/fr/feeding-habits-fourteen-european-freshwater-fish-species. Accessed 4 June 2023

Noda T, Wada T, Mitamura H, Kume M, Komaki T, Fujita T, Sato T, Narita K, Yamada M, Matsumoto A, Hori T, Takagi J, Kutzer A, Arai N, Yamashita Y (2020) Migration, residency and habitat utilisation by wild and cultured Japanese eels (*Anguilla japonica*) in a shallow brackish lagoon and inflowing rivers using acoustic telemetry. J Fish Biol 98:507–525. https://doi.org/10.1111/jfb.14595

Ovidio M, Seredynski AL, Philippart JC, Nzau Matondo B (2013) A bit of quiet between the migrations: the resting life of the European eel during their freshwater growth phase in a small stream. Aquat Ecol 47:291–301. https://doi.org/10.1007/s10452-013-9444-1

Parker SJ (1995) Homing ability and home range of yellow-phase American eels in a tidally dominated estuary. J Mar Biol Assoc United Kingdom 75:127–140. https://doi.org/10.1017/S0025315400015241

Piper A, Belen A, Villanueva B, Gollock M (2022) Residence, activity patterns and behaviour of the giant mottled eel Anguilla marmorata in two freshwater protected areas in The Philippines. Aquat Conserv Mar Freshw Ecosyst 32:1606–1617. https://doi.org/10.1002/aqc.3870

Reebs SG (2002) Plasticity of diel and circadian activity rhythms in fishes. Rev Fish Biol Fish 12:349–371. https://doi.org/10.1023/A:1025371804611

Rohtla M, Moland E, Skiftesvik AB, Thorstad EB, Bosgraaf S, Olsen EM, Browman HI, Durif CMF (2022) Overwintering behaviour of yellow-stage European eel (*Anguilla anguilla*) in a natural marine fjord system. Estuar Coast Shelf Sci 276:108016. https://doi.org/10.1016/j.ecss.2022.108016

Tesch F-W (1967) Homing of eels (*Anguilla anguilla*) in the southern North Sea. Mar Biol 1:2–9. https://doi.org/10.1007/BF00346688

Thibault I, Dodson JJ, Caron F (2007) Yellow-stage American eel movements determined by microtagging and acoustic telemetry in the St Jean river watershed, Gaspé, Quebec, Canada. J Fish Biol 71:1095–1112. https://doi.org/10.1111/j.1095-8649.2007.01584.x

Tomie JPNA, Cairns DKB, Hobbs RSC, Desjardins MC (2017) American eel (*Anguilla rostrata*) substrate selection for daytime refuge and winter thermal sanctuary. Mar Freshw Res 68:95–105. https://doi.org/10.1071/MF15102

Verhelst P, Reubens J, Pauwels I, Buysse D, Aelterman B, Van Hoey S, Goethals P, Moens T, Coeck J, Mouton A, Van Hoey S, Goethals P, Moens T, Coeck J, Mouton A (2017) Movement behaviour of large female yellow European eel (*Anguilla anguilla* L.) in a freshwater polder area. Ecol Freshw Fish 27:471–480. https://doi.org/10.1111/eff.12362

Wakiya R, Mochioka N (2021) Contrasting diets of the Japanese eel, *Anguilla japonica*, in the upper and lower areas of Tsuchikawa-Gawa River, Kagoshima, Japan. Ichthyol Res 68:145–151. https://doi.org/10.1007/s10228-020-00755-5

Walker AM, Godard MJ, Davison P (2014) The home range and behaviour of yellow-stage European eel *Anguilla anguilla* in an estuarine environment. Aquat Conserv Mar Freshw Ecosyst 24:155–165. https://doi.org/10.1002/aqc.2380

Walsh PJ, Foster GD, Moon TW (1983) The effects of temperature on metabolism of the American eel *Anguilla rostrata* (LeSueur): compensation in the summer and torpor in the winter. Physiol Zool 56:532–540. https://doi.org/10.1086/physzool.56.4.30155876

Werner EE (1974) The fish size, prey size, handing time relation in several sunfishes and some implications. J Fish Res Board Canada 31:1531–1536. https://doi.org/10.1139/f74-186

Westerberg H, Sjöberg N (2015) Overwintering dormancy behaviour of the European eel (*Anguilla anguilla* L.) in a large lake. Ecol Freshw Fish 24:532–543. https://doi.org/10.1111/eff.12165

Part III
Physiology

Chapter 9
Senses and Nervous System

Naoyuki Yamamoto and Hanako Hagio

Contents

Animals must adapt to internal and external environments and properly respond to the situations to survive, grow, and reproduce. Sensory organs play pivotal roles in obtaining information from the environment, and the nervous system is indispensable for integrating different types of sensory information and determining appropriate behavioral responses, such as escape, aggression, and reproductive behavior. The central nervous system (brain and spinal cord) receives environmental information from sensory organs and sends motor commands to the muscles via the peripheral nervous system. Eels, like other fish, possess several sorts of sensory organs and the nervous system to adapt in response to environmental conditions. In this chapter, we provide an outline of the sensory organs and nervous system of teleost fish, focusing on the features seen in Japanese eels.

9.1 Senses

Teleost fish generally possess the same sensory systems as us humans for various types of environmental information, such as olfaction, vision, auditory sense, sense of balance, taste, and visceral sense. However, we do not possess a lateral line

N. Yamamoto (✉)
Graduate School of Bioagricultural Sciences, Nagoya University, Aichi, Japan
e-mail: nyama@nuagr1.agr.nagoya-u.ac.jp

H. Hagio
Graduate School of Bioagricultural Sciences, Nagoya University, Aichi, Japan

Institute for Advanced Research, Nagoya University, Aichi, Japan

K. Tsukamoto et al. (eds.), *Eel Science*, Fisheries Science Series,
https://doi.org/10.1007/978-981-99-5692-0_9

sensory system, and a specific pH sense of the external environment may only be present in teleosts (among vertebrates).

9.1.1 Olfaction

The olfactory organs of teleosts respond strongly to amino acids, bile acids, sex steroid hormones and their metabolites, and prostaglandins. Odorants are detected by the olfactory epithelium on the wall of the olfactory lamellae. The olfactory epithelium is equipped with olfactory receptor cells, which are sensory neurons with axons. The nasal cavity of eels accompanies a tube-shaped anterior naris rostrally and a round posterior naris at the caudal end of the cavity (Fig. 9.1a). The morphology of olfactory lamellae present in the olfactory chamber is diverse in teleosts; depending on the species, it can be cup-shaped, cup-shaped accompanying a plate called olfactory lamella, or disk-shaped with radially arranged olfactory lamellae. The olfactory lamellae of the Japanese eel are large and obround disk-shaped. A rostro-caudally oriented septum is present along the midline, from which numerous olfactory lamellae originate (Fig. 9.1b). Therefore, the surface area of the olfactory epithelium is enormous. Numerous olfactory receptor cells in the epithelium should result in high olfactory sensitivity of eels. An electrophysiological study of the

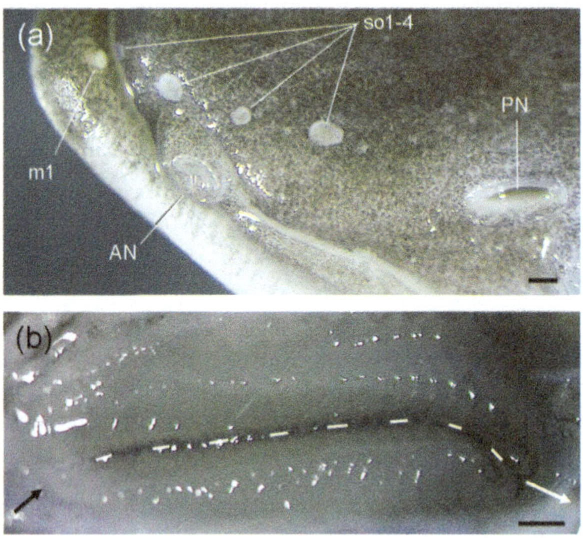

Fig. 9.1 Olfactory organ of Japanese eel (48 cm in total length). (**a**) Dorsal view of anterior naris (AN) and posterior naris (PN). Rostral is to the left. (**b**) Olfactory lamellae exposed by excising the skin covering the olfactory lamellae. The same specimen as shown in (**a**). Black arrow indicates the direction of water flow into the olfactory chamber; white arrow indicates the direction of outflow; white broken line indicates the septum. m1, lateral line pore 1 of mandibular canal; so1–4, lateral line pores 1–4 of superior orbital canal. (Modified from Yamamoto and Hagio (2019b))

olfactory epithelium of eels reported response thresholds to amino acids on the order of 10^{-7}–10^{-8} M, similar to those in other teleosts (Silver 1982). However, it is possible that much higher sensitivities are achieved by the summation of sensory information at the level of the olfactory bulb, the primary olfactory center.

9.1.2 Vision

Specialized visual organs such as the eyes of archerfish and telescopefish are frequently encountered in teleosts. The eyes of the Japanese eel are covered by thick, transparent skin, which protects the eyes when the fish digs into the sand or whisks into a crevice between rocks. Other than this, no particular features are observed in the gross anatomy of eyes. The eyes of yellow eels in the river are small relative to their body size, while silver eels in the sea possess much larger eyes. This anatomical change is an adaptation to the dim environment of the mesopelagic zone, through which the silver eels travel to the spawning site. An androgen present in teleosts, 11-keto-testosterone, is known to induce eye enlargement in Japanese eels; the blood concentration of this steroid hormone increases through the maturation of eels (Sudo et al. 2012). Visual cells also undergo changes during maturation. Both cone cells (serving for color vision in bright environments) and rod cells (serving for scotopic vision) are present in the retina of yellow eels, whereas only rod cells are found in silver eels. The molecular species of opsins, which serve for the reception of photons, also change during maturation (Wood et al. 1992). Thus, dramatic changes occur in the visual system of Japanese eels, reflecting the shift in life history from the yellow eel growing in rivers and coastal areas to the silver eel migrating to the spawning site.

9.1.3 Lateral Line Sense

The lateral line sense is the sensation of the flow and vibration of water at low frequencies, and most aquatic vertebrates possess this sense. The sensory endorgans for lateral line sense are small protuberances called neuromasts. There are two types of neuromasts: the canal neuromast and the free neuromast, which are located in the lateral line under the skin and on the body surfaces, respectively. Water enters and exits pores (lateral line pores) along the lateral line canal in response to the flow and vibration of water. Neuromasts are accompanied by a projection called the cupula. Water flow tilts the cupula and underlying cilia of hair cells, which results in excitation or inhibition of hair cells depending on the direction of the tilt. The lateral line canal system of the Japanese eel is composed of the supraorbital, infraorbital, and mandibular canals that are present in the head and the trunk canal running along the lateral side of the body. There are 4 lateral line pores in the supraorbital and infraorbital canals, and 9 in the mandibular canal (Fig. 9.2). Lateral line canals are

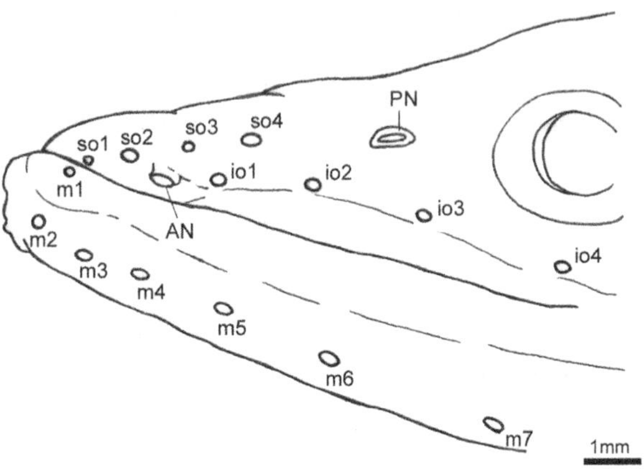

Fig. 9.2 Drawing of lateral line pores of Japanese eel (total length: 48 cm). Rostral is to the left. AN, anterior naris; io1–4, lateral line pores 1–4 of inferior orbital canal; m1–7, lateral line pores 1–7 of mandibular canal; PN, posterior naris; so1–4, lateral line pores 1–4 of superior orbital canal. Modified from Yamamoto and Hagio (2019b)

lacking in the Japanese eel at places where temporal and preopercular canals are present in other species of teleosts. Instead, there are free neuromasts at these places. Free neuromasts are also present on the surface of the trunk but at lower densities. In general, free neuromasts respond to vibrations at relatively lower frequencies, while canal neuromasts detect vibrations at relatively higher frequencies (Uematsu et al. 2013).

9.1.4 Auditory and Vestibular Senses

Both auditory and vestibular senses are detected by the inner ear. The sensory organs in the inner ear of teleosts comprise otolith organs that detect sound, gravity, and linear acceleration, and semicircular canals that detect rotations of the head. A "stone" (otolith) mainly containing calcium carbonate is housed within the thin membranous wall of otolith organs, and hair cells beneath the otolith are stimulated by the relative movement of otolith by the inertia. Within the inner ear of the Japanese eel, the utricle (otolith: lapillus) is the rostral-most otolith organ, the saccule (otolith: sagitta) is located ventrocaudal to the utricle, and the lagena (otolith: asteriscus) is located caudal to the saccule (Fig. 9.3). In goldfish, it has been suggested that the saccule receives auditory stimuli and the other otolith organs detect linear accelerations. However, functional segregation of otolith organs may not be so clear-cut, and further studies are necessary regarding this point for eels. The sagitta is the largest otolith in the Japanese eel and is used by researchers to determine age in days and years (Kuroki et al. 2008). The periods when an eel stayed in freshwater or seawater during its life span can also be determined by measuring the concentration of trace elements in the otoliths.

Semicircular canals are tubular half circles, and comprise anterior, posterior, and horizontal semicircular canals. The canals run along 3 planes that are perpendicular to each other (Fig. 9.3). Therefore, rotations in any 3-dimensional direction could be

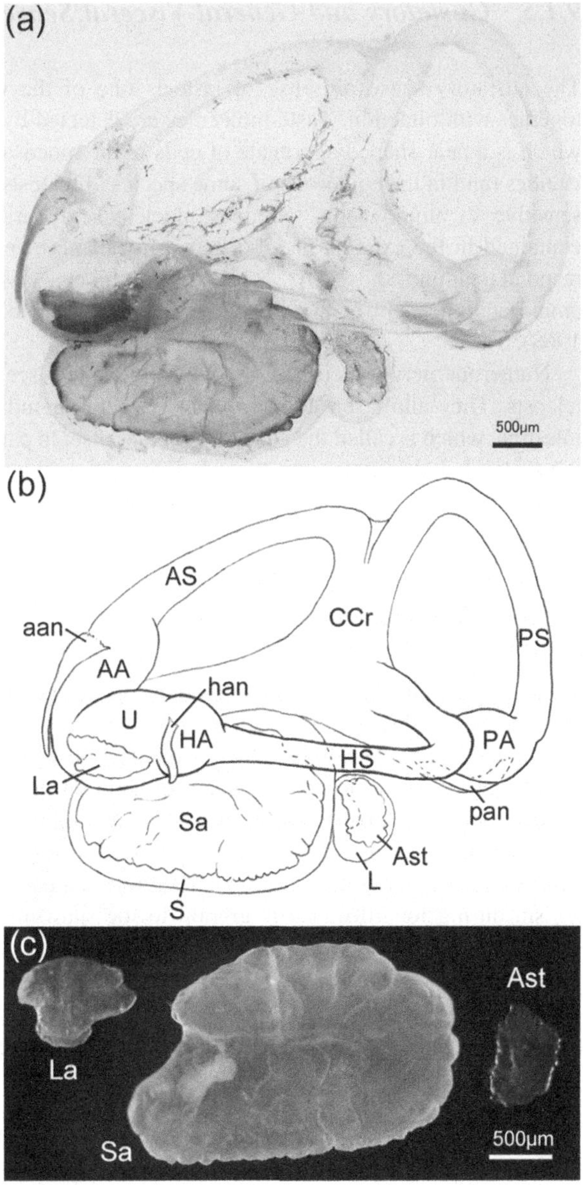

Fig. 9.3 Inner ear of Japanese eel (total length: 48 cm). (**a**) Lateral view of left inner ear excised from the head. Rostral is to the left. (**b**) Line drawing of (**a**). (**c**) Lateral views of otoliths removed from the membranous inner ear. *AA* anterior ampulla; *aan* anterior ampullary nerve, *AS* anterior semicircular canal, *Ast* asteriscus, *CCr* common crus, *HA* horizontal ampulla, *han* horizontal ampullary nerve, *HS* horizontal semicircular canal, *L* lagena, *La* lapillus, *PA* posterior ampulla, *pan* posterior ampullary nerve, *PS* posterior semicircular canal, *S* saccule, *Sa* sagitta, *U* utricle. (Modified from Yamamoto and Hagio (2019b))

detected by the 3 canals. Rotations are detected by a protuberance and hair cells beneath it that are housed in the bulge of each canal, called the crista ampullaris. The protuberance is tilted by the relative movement of liquid (inner lymph) within the canal by inertia.

9.1.5 Gustatory and General Visceral Senses

The gustatory sense or sense of taste is one of the important chemical senses, together with olfaction. Taste molecules are detected by taste cells in the taste bud, which is a pear-shaped aggregate of cells in the mucosa of the oral and pharyngeal cavities (and in the epidermis of some species of teleosts). Taste buds of teleosts are sensitive to amino acids, peptides, nucleic acid derivatives, and organic acids contained in the extracts of food items. In Japanese eels, taste buds on the palate respond to amino acids at 10^{-8}–10^{-9} M (Yoshii et al. 1979). Eels are attracted to the amino acids present in the extract of Japanese littleneck clams (Hashimoto et al. 1968).

Numerous nerve fibers are also present in the viscera of vertebrates, including teleosts. They allow sensation from the viscera, including the heart, stomach, and intestine, which is called the general visceral sense. In particular, sensory inputs from the stomach and intestine can provide information about the amount and quality of food items being digested and should be important for control of feeding behavior.

9.1.6 Magnetosense

There is a hypothesis that salmonids make use of geomagnetism when they return from the sea to their home area (olfaction is believed to mediate the final phase of finding the home river). In fact, there are reports that cells in the olfactory epithelium detect magnetism, which is conveyed to the brain through the trigeminal nerve. A previous study on Japanese eels reported behavioral responses to magnetism and the presence of a sensory system detecting magnetism in the olfactory chamber or nearby structures (Nishi et al. 2018). Therefore, Japanese eels might utilize geomagnetism during long-distance migration to the spawning site in the West Mariana Ridge.

9.1.7 Other Senses

There are several types of receptors in the skin of mammals, such as Merkel's disk, Pacinian corpuscle, and Clause's corpuscle, which are comprised of fibers and associated structures and a free nerve ending (not associated with specialized structures). These receptors are involved in the reception of various types of mechanical/physical stimuli, such as touch and vibration. In the skin of teleosts, only free nerve endings have been found, which respond to tactile stimuli (i.e., escape behavior in response to the touch of the skin).

Recently, a novel sense has been revealed in sea catfish. The barbels of the Japanese sea catfish, *Plotosus japonicus* can detect changes in pH with amazingly precise sensitivity (Caprio et al. 2014). Their taste cells might be extremely sensitive to acids involved in pH sensitivity; however, the specific receptor cell type involved in this sense remains unknown, and it is possible that a similar sensory system is also present in eels. Whether such a precise sense of pH is present in other vertebrates remains to be investigated, although free nerve endings and taste cells of mammals are known for more vague responses to low pH (acid).

A third type of chemical sense, distinct from olfaction and taste, is present in a number of vertebrate taxa. Individual cells (not groups) that are similar to taste cells in the taste buds are present on the skin and fins, and are called solitary chemosensory cells. To the best of our knowledge, investigations on the presence of solitary chemosensory cells have not been performed in Japanese eels.

9.2 Nervous System

9.2.1 Central Nervous System

Similar to other teleosts, the brain of the Japanese eel comprises the telencephalon, diencephalon, mesencephalon (midbrain), hindbrain, and medulla oblongata, in this order from rostral to caudal (Fig. 9.4). Neurons, which transmit and process information, possess a long process called axon (nerve fiber). Within the central

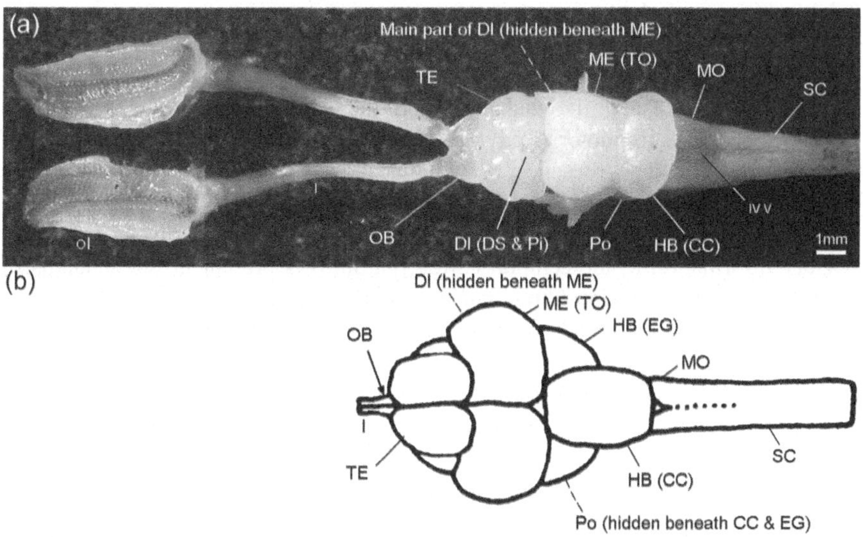

Fig. 9.4 Dorsal views of the brain of Japanese eel (**a**) and thread-sail filefish (**b**). Rostral is to the left. *CC* corpus cerebelli, *DI* diencephalon, *DS* dorsal sac, *EG* eminentia granularis, *HB* hindbrain, *ME* mesencephalon, *MO* medulla oblongata, *OB* olfactory bulb, *ol* olfactory lamellae, *Pi* pineal organ, *Po* pons, *SC* spinal cord, *TE* telencephalon, *TO* optic tectum, *I* olfactory nerve, *IV V* fourth ventricle (modified from Yamamoto and Hagio (2019a))

nervous system, there are nuclei containing tightly-gathered cell bodies (the main body of neurons), and nerve tracts, through which axons run in groups. There are a series of connected fluid-filled (cerebrospinal fluid) chambers in the central part of the brain, which are called ventricles.

9.2.1.1 Telencephalon

The telencephalon is composed of the telencephalon proper (cerebrum, cerebral hemisphere) and the olfactory bulb. The olfactory bulb receives olfactory information from the olfactory epithelium and is, hence, the primary olfactory center. The olfactory bulb is quite large in the Japanese eels, likely due to their nocturnal life and highly olfaction-dependent foraging, in sharp contrast to the small olfactory bulb of a diurnal filefish (Fig. 9.4). The telencephalon proper comprises dorsal and ventral areas, which correspond to the pallium and subpallium of mammals, respectively. The telencephalon proper of teleosts was previously called "the olfactory lobe" based on the assumption that it only receives olfactory information from the olfactory bulb. However, it is now known that all major sensory information reaches the dorsal telencephalon of goldfish (Yamamoto et al. 2007). Our preliminary data suggest that visual information reaches the dorsal telencephalon in Japanese eels, which is likely to be true for other senses.

9.2.1.2 Diencephalon

The diencephalon of teleosts comprises the epithalamus (pineal organ, parapineal organ, and dorsal sac), the thalamus, and the hypothalamus. The pineal and parapineal organs are photoreceptive and the former is involved in the control of circadian rhythm. The presence of parapineal organ has been reported in the European eel (Borg et al. 1983), but may be absent in the Japanese eel (Mukuda and Ando 2003; Yamamoto, unpubl. observation). A group of nuclei that relay sensory information to the cerebral cortex (pallium) is present in the thalamus of mammals. In contrast, teleost brain nuclei, which are situated in a diencephalic zone traditionally called the thalamus, play insignificant roles in relaying sensory information. Rather, a group of diencephalic nuclei called the preglomerular complex serves as the sensory relay station to the dorsal telencephalon (pallial homologue). The preglomerular complex is also present in Japanese eels, comprised of moderately developed nuclei. In general, the hypothalamus of teleosts is involved in the control of behavior and hormone secretion from the pituitary. Similar to other teleosts, neurons secreting various neurohormones and neuromodulators, such as gonadotropin-releasing hormone (GnRH) and dopamine, are present in the hypothalamus of eels. The hypothalamus of eels quite likely regulates various behaviors such as escape and aggression, as in other teleosts. In teleosts and cartilaginous fish, the lateral part of the hypothalamus forms a large bulge called the inferior lobe. However, the inferior lobe of eels is relatively small.

9.2.1.3 Mesencephalon

The optic tectum is a primary visual center occupying the dorsal part of the mesencephalon. Although the optic tectum of the Japanese eel is a 6-layered structure, as in other teleosts, it is rather small (Fig. 9.4; compare with filefish). This likely reflects the nocturnal life of Japanese eels, in which vision is not so much useful. The optic tectum of teleosts controls eye movement and vision-dependent behavior, and sends visual information to the dorsal telencephalon via the preglomerular complex (Yamamoto and Ito 2002). Ventral to the optic tectum is a moderate protuberance called the torus semicircularis with the mesencephalic ventricle in between. The torus is clearly laminated in the Japanese eel. In teleosts, this structure mainly receives lateral line and auditory information from the medulla oblongata and sends it to the dorsal telencephalon mediated by the preglomerular complex (Yamamoto et al. 2007). Oculomotor and trochlear nuclei, both of which are cranial nerve nuclei controlling eye movements, are present in the ventromedial zone of the mesencephalon, called the mesencephalic tegmentum. In the border zone between the diencephalon and the mesencephalon, the nucleus of the medial longitudinal fasciculus is present in the tegmentum. This nucleus sends axons to the spinal cord in eels, as in other teleosts, and is likely involved in initiating and controlling swimming (Bosch and Roberts 2001).

9.2.1.4 Hindbrain

The cerebellum and pons occupy the dorsal and ventral regions of the hindbrain, respectively. There is a common misconception that teleosts lack the pons, stemming from the fact that they do not possess pontine nuclei (relay nuclei of telencephalic inputs to the cerebellum), as seen in mammals. However, the pons is present in teleosts, and a number of nuclei are present in this brain region, such as the secondary gustatory nucleus, principal trigeminal nucleus, and trigeminal motor nucleus. The cerebellum is composed of the corpus cerebelli, which can be seen in the dorsal view (Fig. 9.4a), the valvula cerebelli that protrudes rostrally into the mesencephalic ventricle, eminentia granularis forming a bulge lateral to the corpus, and the caudal lobe or the caudal-most component. The cerebellum of the Japanese eel is relatively large, which may reflect the control of elaborate movement of the very long trunk of the eel.

9.2.1.5 Medulla Oblongata

The fourth ventricle, or the ventricle of the hindbrain and medulla oblongata, is large and forms a deep recess (Fig. 9.4a). The rostral zone of the medulla oblongata receives lateral line, auditory, and equilibrium senses. In Japanese eels, the equilibrium-receptive region is well-developed. Primary gustatory centers, or facial, glossopharyngeal, and vagal lobes, are present in the caudal zone of the medulla oblongata, in this order from rostral to caudal. This zone is hypertrophied in teleosts with well-developed taste systems, including goldfish, carp, and catfish, and the primary gustatory centers form distinct lobes. In other species such as the Japanese eel, the primary gustatory centers are not distinct, but rather form a continuous column of primary gustatory centers. In the caudal-most medulla oblongata, the commissural nucleus of Cajal, which receives sensory information from the viscera through the vagal nerve, is situated caudally adjacent to the vagal lobe. In Japanese eels, the area postrema, which lies on the commissural nucleus of Cajal and also receives viscerosensory information, is involved in the suppression of water intake in seawater by atrial natriuretic peptide (Tsukada et al. 2007).

9.2.1.6 Spinal Cord

The medulla oblongata gradually shifts to the spinal cord, without a prominent landmark. Although the spinal cord appears as a long, continuous rod-like structure, it is composed of a number of segments, each of which gives rise to a corresponding spinal nerve. The spinal nerve contains 2 components: sensory and motor. Sensory axons running through the spinal nerve continue into a nerve root called the dorsal root, which enters the spinal cord dorsolaterally. Motor axons originating from motor neurons in the spinal cord emerge as the ventral root from the ventrolateral aspect of the spinal cord. Corresponding to this positional relationship, sensory neurons receiving inputs from the sensory fibers of the dorsal root are situated in the dorsal part of the spinal cord (dorsal horn), whereas motor neurons giving rise to motor fibers to muscles are present in the ventral part of the spinal cord (ventral horn).

9.2.2 Peripheral Nervous System

The peripheral nervous system comprises cranial nerves and spinal nerves connected to the brain and spinal cord, respectively. Axons running through peripheral nerves can be categorized into: (1) sensory (afferent) fibers that convey sensory information to the central nervous system and (2) motor (efferent) fibers that send commands to the periphery. Efferent fibers either directly innervate the striated muscles or indirectly innervate the cardiac muscles, and smooth muscles in the walls of blood vessels, viscera, and glands (autonomic fibers). Autonomic fibers are either sympathetic (originating from the spinal cord) or parasympathetic (originating mostly from the brain). Efferent fibers that innervate sensory organs are also present.

9.2.2.1 Cranial Nerves

There are 12 pairs of cranial nerves in teleosts (Fig. 9.5: Japanese eel), most of which are named by number. The rostral-most cranial nerve is the olfactory nerve (cranial nerve I), which carries olfactory information from the olfactory epithelium to the olfactory bulb. Olfactory nerve fibers are axons of olfactory receptor cells in the olfactory epithelium. The olfactory nerve of the Japanese eel is thick, reflecting a large olfactory epithelium, and is composed of 4–5 rootlets. The optic nerve (cranial nerve II) sends visual information to the optic tectum in the mesencephalon and to several other centers in the diencephalon. The optic nerve also contains a minor component of efferent fibers to the eye (retina), which control visual sensory processing in the retina. The eyes of Japanese eels are small; hence, the optic nerve is rather thin. The oculomotor nerve (cranial nerve III) is a motor nerve with fibers originating from the oculomotor nucleus in the mesencephalic tegmentum that innervate 4 extraocular muscles (superior rectus, inferior rectus, medial rectus, and inferior oblique) to rotate the eyeball. As a minor component, the oculomotor nerve involves parasympathetic fibers that control the lens muscle to accommodate the eye. The trochlear nerve (cranial nerve IV) innervates the superior oblique muscle among the 6 extraocular muscles.

The trigeminal nerve (cranial nerve V) mediates skin sensations (touch, etc.) from the facial region, mainly to the pons (principal trigeminal nucleus) and medulla oblongata (descending trigeminal and medial funicular nuclei). The trigeminal nerve is divided into 3 major branches (ophthalmic, maxillary, and mandibular nerves), hence its name. The mandibular nerve also carries efferent fibers from the trigeminal motor nucleus, which innervates the mandibular and opercular muscles. The trigeminal nerve of the Japanese eel is quite thick. The abducens nerve (cranial nerve VI)

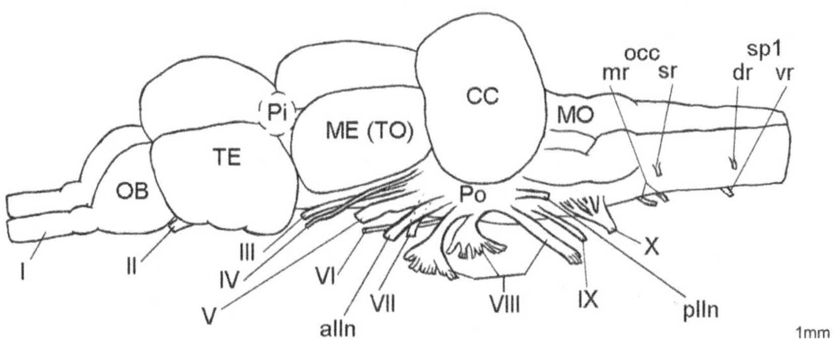

Fig. 9.5 Peripheral nerves of Japanese eel (lateral view). Rostral is to the left. *alln* anterior lateral line nerve, *CC* corpus cerebelli, *dr* dorsal root, *ME* mesencephalon, *MO* medulla oblongata, *mr* motor root, *OB* olfactory bulb, *occ* occipital nerve, *Pi* pineal organ, *Po* pons, *plln* posterior lateral line nerve, *sp1* first spinal nerve, *sr* sensory root, *TE* telencephalon, *TO* optic tectum, *vr* ventral root, *I* olfactory nerve, *II* optic nerve, *III* oculomotor nerve, *IV* trochlear nerve, *V* trigeminal nerve, *VI* abducens nerve, *VII* facial nerve, *VIII* octaval nerve, *IX* glossopharyngeal nerve, *X* vagal nerve. (Modified from Yamamoto and Hagio (2019a))

innervates the lateral rectus of the extraocular muscles and laterally rotates the eyeball. The facial nerve (cranial nerve VII) mediates gustatory information from taste buds in the anterior part of the oral cavity and from those on the body surface in species with such external taste buds. The target of the gustatory fibers in the facial nerve is the facial lobe of the medulla oblongata. The facial nerve also contains efferent fibers to the opercular muscles, which originate from the facial nucleus in the caudal medulla oblongata.

The octaval nerve (cranial nerve VIII) carries auditory and equilibrium senses from the otolith organs and semicircular canals to the octaval nuclei, within the octavolateral area of the medulla oblongata. The lateral line nerve sends the lateral line sense to the lateral line zone within the octavolateral area of the medulla oblongata. In the Japanese eel, a species without electrosense, the lateral line nerve only carries mechanosensory lateral line information (flow and vibration of water) from neuromasts. Both the octaval and lateral line nerves contain a minor component of efferent fibers (octavolateral efferent fibers) to the inner ear endorgans and lateral line neuromasts, which modulate the sensitivity of these receptor organs.

The glossopharyngeal nerve (cranial nerve IX) sends gustatory information from the posterior part of the oral cavity and anterior part of the pharynx to the glossopharyngeal lobe. Efferent fibers to the gill muscles are also included in the cranial nerve. The vagal nerve (cranial nerve X) mediates gustatory information from the caudal part of the pharynx, including the pharyngeal jaws, to the vagal lobe of the caudal medulla oblongata. This nerve also carries general visceral information from the thoracic and abdominal viscera to the commissural nucleus of Cajal. The vagal nerve also contains efferent fibers to the gill muscles and parasympathetic fibers that control the heart, stomach, and intestine to control visceral functions.

The occipital nerve originates from the transitional zone between the medulla oblongata and spinal cord and carries sensory information from the pectoral fin, and controls its movements. This is not homologous to the occipital nerve in humans, which is a branch of the spinal nerve and contains motor fibers alone. The occipital nerve of teleosts is sometimes regarded as the rostral-most component of the spinal nerve and is also called the spino-occipital nerve.

The last but not the least, a peculiar cranial nerve, or the terminal nerve (also called cranial nerve zero), is present. Fibers of this nerve originate from neurons located along the olfactory nerve and olfactory bulb, and have been reported to modulate olfactory and visual sensory processing and control the motivation for reproductive behavior (Yamamoto et al. 1997). A more recent study has suggested the involvement of the terminal nerve in the avoidance of CO_2 (Koide et al. 2018). Terminal nerve neurons that produce GnRH have also been identified in the Japanese eel (Nozaki et al. 1985).

9.2.2.2 Spinal Nerve

The dorsal and ventral roots from the spinal segment come together to form the spinal nerve. The spinal nerve gives rise to branches that course in different directions. The ventral root also contains sympathetic fibers, which is a common feature to other vertebrates.

References

Borg B, Eksröm P, van Veen T (1983) The parapinal organ of teleosts. Acta Zool 64:211–218. https://doi.org/10.1111/j.1463-6395.1983.tb00802.x

Bosch TJ, Roberts BL (2001) The relationship of brain stem systems to their targets in the spinal cord of eel, *Anguilla anguilla*. Brain Behav Evol 57:106–116. https://doi.org/10.1159/000047230

Caprio J, Shimohara M, Marui T, Harada S, Kiyohara S (2014) Marine teleost locates prey through pH sensing. Science 344:1154–1156. https://doi.org/10.1126/science.1252697

Hashimoto Y, Kounosu S, Fuseya S, Nose K (1968) Substances in Japanese littleneck clam extract attracting eels – I. Survey of effective substances by omission test. Nippon Suisan Gakkaishi 34: 78–83. https://doi.org/10.2331/suisan.34.78; in Japanese

Koide T, Yabuki Y, Yoshihara Y (2018) Terminal nerve GnRH3 neurons mediate slow avoidance of carbon dioxide in larval zebrafish. Cell Rep 22:1115–1123. https://doi.org/10.1016/j.celrep.2018.01.019

Kuroki M, Kawai M, Jónsson B, Aoyama J, Millar MJ, Noakes DLG, Tsukamoto K (2008) Inshore migration and otolith microstructure/microchemistry of anguillid glass eels recruited to Iceland. Environ Biol Fish 83:309–325. https://doi.org/10.1007/s10641-008-9341-y

Mukuda T, Ando M (2003) Brain atlas of the Japanese eel: comparison to other fishes. Mem Fac Integrated Arts and Sci, Hiroshima Univ, Ser IV 29:1–25. https://ir.lib.hiroshima-u.ac.jp/159/files/40918

Nishi T, Archdale MV, Kawamura G (2018) Behavioural evidence for the use of geomagnetic cue in Japanese glass eel *Anguilla japonica* orientation. Ichthyol Res 65:161–164. https://doi.org/10.1007/s10228-017-0587-2

Nozaki M, Fujita I, Saito N, Tsukahara T, Kobayashi H, Ueda K, Oshima K (1985) Distribution of LHRH-like immunoreactivity in the brain of the Japanese eel (*Anguilla japonica*) with special reference to the nervus terminalis. Zool Sci 2:537–547

Silver WL (1982) Electrophysiological responses from the peripheral olfactory system of the American eel, *Anguilla rostrata*. J Comp Physiol 148:379–388. https://doi.org/10.1007/BF00679022

Sudo R, Tosaka R, Ijiri S, Adachi S, Aoyama J, Tsukamoto K (2012) 11-ketotestosterone synchronously induces oocyte development and silvering-related changes in the Japanese eel, *Anguilla japonica*. Zool Sci 29:254–259. https://doi.org/10.2108/zsj.29.254

Tsukada T, Nobat S, Hyodo S, Takei Y (2007) Area postrema, a brain circuventricular organ, is the site of antidipsogenic action of circulating atrial natriuretic peptide in eels. J Exp Biol 210:3970–3978. https://doi.org/10.1242/jeb.010645

Uematsu K, Kohbara J, Yamamoto N (2013) Senses. In: Aida K, Kaneko T (eds) Basic of fish physiology – revised and Enlarged Edition. Kouseisha Kouseikaku, Tokyo, pp 65–102

Wood P, Partridge JC, Grip WJD (1992) Rod visual pigment changes in the elver of the eel *Anguilla anguilla* L. measured by microspectrophotometry. J Fish Biol 41:601–611. https://doi.org/10.1111/j.1095-8649.1992.tb02686.x

Yamamoto N, Hagio H (2019a) Brain and nerves. In: Tsukamoto K (ed) Science of eels. Asakura Publishing, Tokyo, pp 58–62; in Japanese

Yamamoto N, Hagio H (2019b) Senses. In: Tsukamoto K (ed) Science of eels. Asakura Publishing, Tokyo, pp 72–77; in Japanese

Yamamoto N, Ito H (2002) Visual pathways in bony fishes. In: Uematsu K, Oka Y, Ito H (eds) Fish neuroscience - frontier of fish neuroscience. Koseisha Kouseikaku, Tokyo, pp 122–136; in Japanese

Yamamoto N, Oka Y, Kawashima S (1997) Lesions of gonadotropin-releasing hormone-immunoreactive terminal nerve cells: effects on the reproductive behavior of male dwarf gouramis. Neuroendocrinology 65:403–412. https://doi.org/10.1159/000127203

Yamamoto N, Ishikawa Y, Yoshimoto M, Xue H-G, Bahaxar N, Sawai N, Yang C-Y, Ozawa H, Ito H (2007) A new interpretation on the homology of the teleostean telencephalon based on hodology and a new eversion model. Brain Behav Evol 69:96–104. https://doi.org/10.1159/000095198

Yoshii K, Kamo N, Kurihara K, Kobatake Y (1979) Gustatory responses to eel palatine receptors to amino acids and carboxylic acids. J Gen Physiol 74:301–317. https://doi.org/10.1085/jgp.74.3.301

Chapter 10
Digestion and Absorption

Soichi Watanabe

Contents

Digestion and absorption are essential for multicellular organisms which depend on exogenous nutrition. Digestive and absorptive organs have various shapes in different animals, from jellyfish and worms to humans, but play similar roles. This chapter describes the mechanisms and characteristics of digestion and absorption in eels.

10.1 Gross Anatomy of Eel Digestive Organs

The main digestive organs in fish are the teeth, esophagus, stomach, intestine, pancreas, gall bladder, and rectum, which are all present in eels after metamorphosis from leptocephalus (Fig. 10.1a). The anterior-most part of the digestive tract is the esophagus, which is located behind the liver. The length of the esophagus of the Japanese eel is relatively long compared to those of other fish species (Fig. 10.1b, c). This anatomical feature enables the stomach to be placed after the liver, and is advantageous to maximize the stomach capacity with the limited space in the abdominal cavity of eels.

The esophagus connects to the stomach at the cardia, which is the boundary between the organs. The stomach of the Japanese eel is a Y-shaped organ with very near inlet (cardia) and outlet (pylorus) areas and a digestive stomach area

S. Watanabe (✉)
Graduate School of Agricultural and Life Sciences, The University of Tokyo, Tokyo, Japan
e-mail: watanabe@marine.fs.a.u-tokyo.ac.jp

K. Tsukamoto et al. (eds.), *Eel Science*, Fisheries Science Series,
https://doi.org/10.1007/978-981-99-5692-0_10

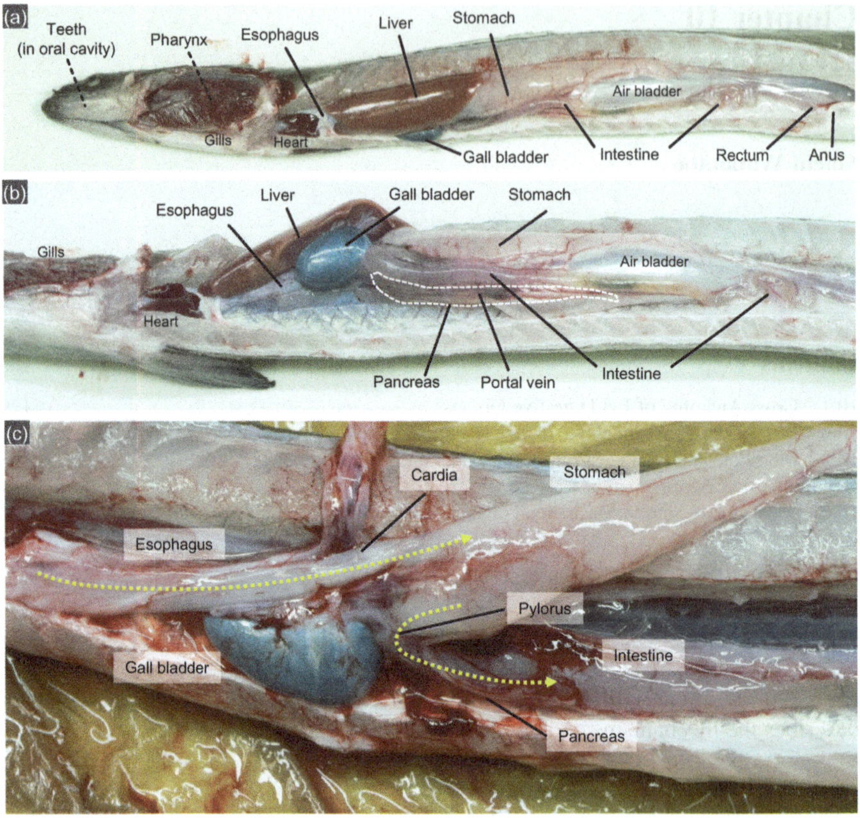

Fig. 10.1 Gross anatomy of digestive organs of Japanese eel. (**a**) Lateral view, (**b**) ventral view, and (**c**) magnified view of digestive organs. Yellow dotted arrows indicate the flow of the ingested feed

(Fig. 10.1c). A Y-shaped stomach can be found in many fish species; in the eel, this morphological feature allows the pylorus to be positioned as anterior as possible, which is beneficial to ensure sufficient length of the intestine in the short abdominal cavity of eels. The stomach of the eel possesses a very flexible wall that maximizes the amount of feed intake at one time.

The stomach connects to the intestine at the pylorus. The eel has no pyloric caeca, which is a digestive organ located in the anterior-most part of the intestine in many fish species. The intestine of Japanese eel can be divided into 2 sections: anterior and posterior intestines. The anterior intestine of the eel has a thicker wall than the posterior intestine, and thus appears opaque (Fig. 10.1b). The anterior intestine of the eel runs straight to the posterior intestine, and minor curvatures can be found in the posterior intestine, which is also likely related to the arrangement of organs in the abdominal cavity (Fig. 10.1a, b). Intestinal morphology is strongly related to the feeding habits of fish species. In general, the length of the intestine is shorter in carnivorous fish species and increases as they become omnivorous and herbivorous. This tendency has also been observed in mammals; however, the intestines of fish are much shorter than those of mammals. For example, the intestine lengths of cats, dogs, and cows are 4, 6, and 20 times longer than their body lengths, respectively. Even tilapia, which exhibits omnivorous to herbivorous behavior, is about 3–4 times longer than its body length. The intestinal length of a Japanese eel is about one-fourth of its body length.

The gallbladder and pancreas are adjacent to the anterior intestine, and provide bile and pancreatic juice to the intestinal lumen, respectively. Fasted eels usually have large and dark-colored gall bladders due to bile accumulation (Fig. 10.1b). Notably, the eel has an independent pancreas on the surface of the anterior part of the intestine (Fig. 10.1b), whereas many other fish species have a hepatopancreas, where pancreatic cells are distributed in the liver. The portal vein, connecting the intestine and liver, runs inside the pancreas, and is a beneficial morphological feature allowing pancreatic cells to directly monitor the nutrient absorption status. The posterior-most part of the digestive tract is the rectum, which is mildly thicker than the posterior intestine (Fig. 10.1a).

Digestive organs in eel leptocephali can be observed externally because of their translucent body. The digestive tract runs straight to the anus along the ventral edge of the body (Miller 2009). Usually, these organs are not fully developed at the larval stage in eels and an evident stomach is absent. The liver and gall bladder are located at the marginal area of the esophagus and intestine in eel larvae, and the gall bladder of eel leptocephali is colorless owing to a lack of bilirubin, a metabolite of hemoglobin. The pancreas is located in the same area as the liver and gall bladder; however, the pancreas is difficult to identify externally at the larval stage of the eel (Kurokawa et al. 2002).

10.2 Morphology and Functions of Eel Teeth and Esophagus

Teeth also play an important role in food intake. The morphology of the teeth varies according to the feeding habits of the fish species. Carnivorous fish species have sharp and well-developed teeth to prevent prey from escaping, whereas herbivorous fish species possess well-developed pharyngeal teeth located just before the esophagus to grind the ingested feed. Some carnivorous and omnivorous fish also have pharyngeal teeth to crush their prey shells. As in these examples, the morphology of eel teeth also varies depending on their feeding habits. For example, the pike eel and moray eel, which primarily eat fish, have well-developed sharp teeth. In contrast, Japanese eels, which mainly consume small fish and aquatic invertebrates, have small conical teeth. The Japanese eel has relatively small teeth and no evident pharyngeal teeth, indicating that the eel uses its teeth to hold prey. Interestingly, the morphology of the teeth of the eel leptocephali is completely different from that

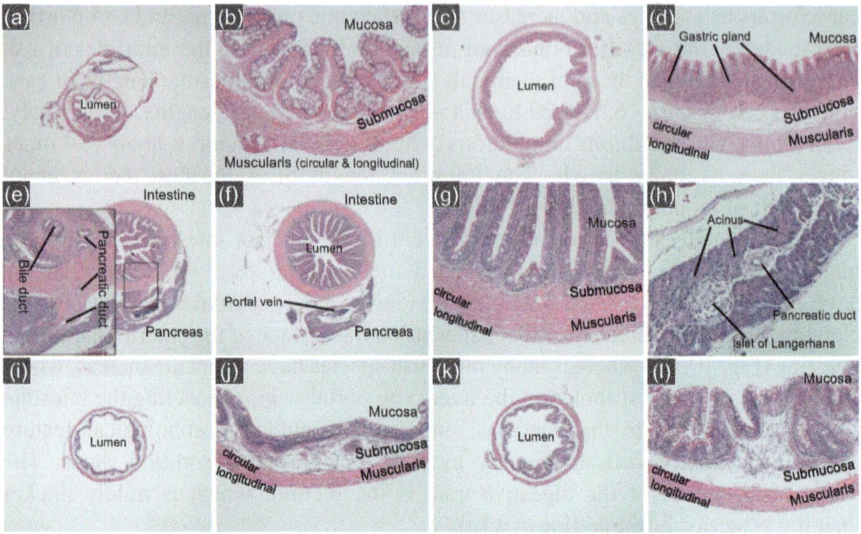

Fig. 10.2 Histology of digestive organs of Japanese eel. (**a** and **b**) Esophagus, (**c** and **d**) stomach and outlets of bile and pancreatic ducts, (**e**) anterior-most part of the intestine, (**f** and **g**) anterior intestine, (**f** and **h**) pancreas, (**i** and **j**) posterior intestine and (**k** and **l**) rectum

of the adult form. They have sparse, sharp, and long teeth projecting forward, which is a common morphology in leptocephali of many eel species (Miller 2009). It has been shown that Japanese eels feed mainly on organic suspended materials called marine snow during the leptocephalus stage (Miller et al. 2013); however, how these characteristics of teeth are advantageous for eel larvae in feeding on such materials remains controversial.

The esophagus is the anterior-most part of the digestive tract connecting the pharynx to the gastrointestinal tract. The main function of the esophagus is as a passageway for ingested feed. Many fish species have a very short esophagus; eel species, however, have a relatively long esophagus that play an important role in osmoregulation under hyperosmotic conditions by desalting the ingested environmental water (Hirano and Mayer-Gostan 1976). The esophageal mucosa is a mucus cell-rich epithelium coated with mucus to prevent the esophagus from being damaged by the shells and bones/spines of prey (Fig. 10.2a, b). The gastrointestinal tract passes ingested products using peristalsis with controlled contraction and relaxation of the circular and longitudinal muscular layers. These two muscular layers are orthogonal to each other, but their running direction is oblique to the longitudinal axis of the esophagus. Therefore, the boundary of these layers is not evident in the transverse section of the esophagus (Fig. 10.2b). The leptocephali of eels often have an extremely long esophagus compared to other fish larvae. Therefore, peristalsis of the esophagus is more important than that in other fish species for the proper delivery of ingested food to the digestive and absorptive organs.

10.3 Morphology of Eel Gastrointestinal Tract and Pancreas

The stomach of the Japanese eel consists of the mucosa, submucosa, muscularis, and serosa (Fig. 10.2d). The gastric gland, which secretes gastric acid and digestive enzymes into the mucosa, is a cellular complex responsible for gastric digestion. The outermost part of the mucosa is covered by mucus cells. Secreted mucus is vital for protecting mucosal cells from harsh luminal conditions. Leaky submucosa connecting the mucosal and muscular layers has been observed, which is important for the flexible structure of the stomach wall (Fig. 10.2d). The muscularis of the eel stomach consists of thick inner circular and thin outer longitudinal smooth muscle layers (Fig. 10.2d). No evident stomach-like structure has been observed, and gene expression of gastric enzymes is lacking in eel leptocephalus, indicating that the Japanese eel does not have a functional stomach at the larval stage (Kurokawa et al. 2011).

The bile and pancreatic ducts open at the anterior-most part of the intestine (a region adjacent to the pyloric sphincter; Fig. 10.2e). The pancreas is a digestive enzyme-producing organ essential for complete intestinal digestion. There are two distinctive structures in the pancreas, the Islet of Langerhans and acinus (Fig. 10.2h). The islets of Langerhans contain two types of endocrine cells: Langerhans α and β cells, which produce glucagon and insulin, respectively. Acinus is the gland responsible for the production of various digestive enzymes and their precursors, including trypsinogen, which is secreted from acinar cells in an exocrine manner. Digestive enzymes are transported to the intestine via the pancreatic duct and are secreted into the lumen of the intestine.

As mentioned in Sect. 10.1, the intestine of Japanese eels can be divided into anterior and posterior parts. The anterior intestine has well-developed folding mucosa, which contributes to maximizing the surface area of the absorptive epithelia (Fig. 10.2f). In contrast, the folding structure of the mucosal epithelia in the posterior intestine is not as developed as that in the anterior intestine (Fig. 10.2i). In addition, the circular muscularis in the anterior intestine is thicker than that in the posterior intestine (Fig. 10.2g, j). The histological features of the rectum are similar to those of the posterior intestine (Fig. 10.2j, k); however, these two parts of the digestive tract are clearly divided by the sphincter structure (Kim et al. 2008). The histology of the intestine of eels at the leptocephalus stage differs from that of the adult form. For example, European eel larvae intestine consists of a single absorptive epithelial layer with no evident folding structure, and there is no distinctive submucosa or muscularis (Knutsen et al. 2021). Furthermore, the intestines of Japanese eel larvae do not possess peristaltic ability because of the absence of intestinal muscular layers during the leptocephalus stage. It is most likely that peristaltic movement at the esophagus is the major driving force of digesta movement in the digestive tract of eel larvae. Notably, highly ciliated cells in the larval intestine have also been reported, suggesting that the transport mechanism of digesta in the eel larval intestine is completely different from that in the adult eel.

Another morphological feature of the intestine is the brash-border structure of the intestinal epithelial cells at the apical surface. This structure can be observed both in adult and larval eels, and is also common to many animals. This nanometer-scale structure contributes to maximizing nutrient absorption by increasing the area in contact with nutrients in the intestinal tract as much as possible.

10.4 Digestive Mechanism of Eel

In many fish, including eels, food passes through the esophagus to the stomach. The stomach participates in the digestion and temporal storage of ingested feed. The digestive mechanism in the fish stomach is similar to that of mammals in that it secretes hydrochloric acid to make the gastric lumen acidic and digest proteins through the action of pepsin, a digestive endoprotease. Gastric acid and pepsinogen, precursors of pepsin, are secreted by the gastric gland. Pepsinogen secreted from the gastric gland is autonomously degraded into its active form, pepsin, under acidic conditions. Pepsin usually shows the highest proteolytic activity under acidic conditions and exhibits a broader specificity than other digestive endoproteases. The stomach also secretes other digestive enzymes. Chitinase has been identified in the stomach of the Japanese eel, and its optimal pH is 4–5 (Kono et al. 1990). Therefore, eels might utilize chitin, a polymer chain of N-acetylglucosamine, as a nutrient. Gastric lipase, a digestive enzyme for triacylglycerol, is known to be present in some teleost species. In European eels, lipase activity is detected in stomach tissue; however, there is no information on gastric lipase molecules (Caruso et al. 2008). Fish species, such as medaka, saury, carp, and pufferfish, do not have a functional stomach and are therefore called stomach-less fish. These fish species do not undergo digestion in the stomach but rather complete digestion and absorption in the intestine. In general, fish stomachs do not develop during the larval stages. As previously mentioned, this is also true for eels. In the case of the Japanese eel, the expression of the pepsinogen gene, a precursor of pepsin, has not been detected during the leptocephalus stage (Kurokawa et al. 2011). However, the leptocephalus stage of eels is much longer than the larval stage of other fish species (~100 days or more). This is unique because they are able to continue to grow for an extended period without gastric digestive function.

The intestine is an essential organ for digestion and absorption in eels. However, almost all enzymes responsible for digestion in the intestine are not produced in intestinal epithelial cells. Most enzymes that are important for intestinal digestion are supplied by the pancreas.

The pancreas secretes pancreatic juice into the anterior-most part of the intestine via the pancreatic duct, which contains trypsin, chymotrypsin, and carboxypeptidase for protein digestion; lipase and phospholipase for lipid digestion; RNase and DNase for nucleic acid digestion; and amylase for carbohydrate digestion. The secretion control of pancreatic enzymes in eels is thought to be the same as that reported in mammals. Cholecystokinin (CCK), a stimulator of pancreatic enzymes (amylase, trypsin, and lipase) and bile secretion in mammals, is found in eel species, and the CCK gene is expressed in the intestine of Japanese eels both in the larval and adult phases (Kurokawa et al. 2004; Politis et al. 2018). Interestingly, secretin, a stimulator of pancreatic juice secretion in mammals, has not yet been found in the genome information of teleost species (Tam et al. 2014), suggesting that they possess an alternate pancreatic juice secretion control mechanism.

The major digestive proteolytic enzymes, trypsin and chymotrypsin, are secreted from pancreatic acinar cells in the form of their inactive precursors, trypsinogen and chymotrypsinogen, to prevent proteolytic damage to the pancreas. Trypsinogen undergoes cleavage by enterokinase into active trypsin in the intestine, and chymotrypsinogen undergoes cleavage by trypsin into active chymotrypsin. The cleavage site specificities of trypsin and chymotrypsin in eel are thought to be the same as those in mammals because trypsin and chymotrypsin purified from Japanese eel can cleave substrates for mammalian trypsin and chymotrypsin, respectively (Murashita et al. 2013). Generally, trypsin hydrolyzes peptide bonds at the carboxylic side of arginine and lysine residues, which have basic side chains. Chymotrypsin cleaves the carboxylic side of aromatic amino acid residues. Trypsin and chymotrypsin cleave the fragmented proteins using gastric digestion into shorter peptides and oligopeptides. The obtained peptides are further cleaved into free amino acids by carboxypeptidase and aminopeptidase.

Two types of pancreatic lipase, triglyceride lipase and bile salt-activated lipase, have been reported in Japanese eels (Hsu et al. 2015). Triglyceride lipase in the Japanese eel is highly similar to colipase-dependent lipase in gilthead seabream, suggesting that it requires colipases for its lipolytic activity. The main substrate of lipase is triacylglycerol, which consists of 3 fatty acid chains bound to a glycerol molecule with an ester bond. Lipase hydrolyzes the ester bonds in triacylglycerol. Bile salt-activated lipase in fish species can also hydrolyze ester bonds in water-soluble carboxylic esters (Tocher 2003). Pancreatic amylase, a starch-digesting enzyme (Murashita et al. 2013), has also been reported in eel species (Kurokawa et al. 2002). According to its characteristics, Japanese eel amylase can be categorized as an α-amylase, which digests glycogen and starch.

Pancreatic juice is alkaline because it contains high levels of bicarbonate, which contributes to alkalizing acidic digestive products from the stomach. The optimal pH range of the major pancreatic digestive enzymes (trypsin, chymotrypsin, lipase, and amylase) of Japanese eels is 7–9 (Murashita et al. 2013). Therefore, alkalization of the digestive chyme in the anterior-most part of the intestine is crucial to maximize the digestion efficiency in the intestinal lumen.

Digestive enzymes produced in the pancreas begin expression early on in the leptocephalus stage (5–8 days post-hatching) (Kurokawa et al. 2002: Murashita et al. 2013; Ahn et al. 2013), suggesting that these enzymes are responsible for food digestion in the larval stage when the stomach does not exist.

To digest lipids effectively, it is necessary for large oil droplets to be dispersed into smaller droplets; bile salts derived from cholesterol play this role. Bile components are produced in the liver and transported to the gallbladder through the hepatic ducts. Bile from the gall bladder is secreted into the anterior-most part of the intestine in the same manner as pancreatic juice. As previously mentioned in this section, bile secretion is thought to be governed by intestinal CCK in the eel species. Bile is alkaline because it contains a large amount of bicarbonate and contributes to alkalization of the luminal environment in the intestine. Bile salts are important for the activation of bile salt-activated lipases (Tocher 2003). The secreted bile salts are most likely to be reabsorbed in the intestine and transported through the portal venous system to the liver for recycling.

The intestine secretes digestive enzymes for proteins/peptides, including enterokinase, aminopeptidase, and dipeptidase. Enterokinase is a highly specific endoproteinase responsible for the activation of trypsin. Digestive enzymes break down proteins into oligopeptides and free amino acids, carbohydrates into oligosaccharides and monosaccharides, and lipids into fatty acids and glycerol, which are then absorbed in the intestine.

10.5 Nutrient Absorption Mechanism of Eels

Nutrient absorption is a fundamental function in maintaining the vital activities of all animals. The intestine is an essential organ for nutrient absorption in animals, with very few exceptions. The mechanism of nutrient absorption in fish intestines has been elucidated in recent decades, and there are two major nutrient absorption pathways: transporter-driven and pinocytosis-mediated pathways. Some nutrients are absorbed by passive diffusion; however, most nutrients, including amino acids and carbohydrates, cannot diffuse through cellular membranes.

Pinocytosis is a form of phagocytosis in which cell membranes intrude into the cell and the extracellular substances contained in the space are taken up as vesicles. Pinocytotic vesicles have been reported in the rectum of Japanese eel larvae (Kurokawa et al. 1996). The pinocytic pathway is thought to be important for macromolecule uptake into the intestinal absorptive epithelial cells, at least at the larval stage of Japanese eel. In other teleost species, oral administration of gonadotropin-releasing hormone analogs can accelerate oocyte growth (Amezawa et al. 2020), suggesting that the pinocytosis-driven macromolecule transporting pathway that runs across the intestinal epithelia is still active in adult fish. However, the report also implies that the macromolecule transporting system is not involved in nutrient absorption because the administered peptide can reach the blood circulation without lysosomal digestion in enterocytes. Pinocytosis is not a selective transport system and transports all substances in the intestinal lumen, including necessary, unnecessary, and even harmful substances, in a non-specific manner, which is problematic. Therefore, nutrient absorption via transporters specific to each nutrient is now considered to be important. In mammals, nutrient absorption mechanisms via specific transporters have been identified. These mechanisms, based on transporter molecules, enable efficient absorption of nutrients by minimizing the transfer of unnecessary toxic substances into the body.

The transporter-based mechanism of nutrient absorption has been studied in fish, and it has become clear that transporters that selectively take up nutrients are expressed in the intestine. It has been reported that at least small peptides and free amino acids, which are produced as a result of protein digestion, are absorbed into the intestine by transporters. In eel species, the oligopeptide transporter PepT1 has been identified in Japanese eels, and is specifically expressed only in the intestine (Ahn et al. 2013). PepT1 protein is localized at the apical membrane of intestinal absorptive epithelial cells, indicating that this molecule is involved in oligopeptide transport from the intestinal lumen into cells. Furthermore, the onset of PepT1 gene expression in Japanese eels synchronizes with the onset of trypsinogen gene expression during the leptocephalus stage of artificial seedlings—5–6 days after hatching. This timing coincides with the current initial feeding schedule at 6 days post-hatching, indicating that after the onset of feeding, trypsin and PepT1 of eel larvae can digest ingested proteins and absorb the resulting oligopeptides, respectively. Gene expression levels of PepT1 and trypsinogen are temporally upregulated in feed-deprived Japanese eel leptocephali (Ahn et al. 2013). This response to malnutrition suggests that digestion and nutrient absorption activities are regulated according to the nutritional conditions in the body, even at the very early life stage. Intestinal expression of amino acid transporters other than PepT1 in eel species remain unclear; however, Na^+-dependent and-independent transports of several amino acids have been reported in brush border membrane vesicles (BBMV) prepared from the intestine of European eels (Storelli et al. 1989). This study indicates that there are functional transport mechanisms specific to amino acids in the intestines of eel species.

The major absorptive pathway of lipids (fatty acids, cholesterols, monoacylglycerols, etc.) is likely to passively diffuse through the cell membrane; however, the involvement of receptor-mediated fatty acid transport and glycerol transporters cannot be ruled out. Passive diffusional absorption of lipids requires high concentrations of luminal lipids because the driving force of the transport is the concentration gradient of the lipids between the intestinal lumen and inside the enterocyte. Receptor-mediated fatty acid absorption in the intestine is thought to be able to transport fatty acids against the concentration gradient and is important for complete lipid absorption under low luminal lipid concentrations in mammalian species (Ko et al. 2020).

The absorption of carbohydrates has not received much attention because there are few carbohydrate-rich feeds in aquatic environments, and eel species are regarded as highly carnivorous species. However, glycosaminoglycan, an amino-sugar-containing polysaccharide, is one of the major components of eel leptocephali. In addition, amylase gene expression results suggest that eels can consume carbohydrates as nutrients. Feed formulation studies have indicated that eel species can accept starch as a nutrient to some extent. Furthermore, sodium-dependent glucose transport in intestinal BBMV of European eels has been reported (Storelli et al. 1986), suggesting that carbohydrate absorption in the intestine of eel species is mediated by the sodium-glucose cotransporter, which is a molecule responsible for intestinal glucose transport in the intestines of mammalian species. Therefore, understanding the mechanisms of digestion and absorption of carbohydrates in eels is becoming increasingly important to develop sustainable aquaculture feeds with less protein and higher carbohydrates for this species.

References

Ahn H, Yamada Y, Okamura A, Tsukamoto K, Kaneko T, Watanabe S (2013) Intestinal expression of peptide transporter 1 (PEPT1) at different life stages of Japanese eel, *Anguilla japonica*. Comp Biochem Physiol B 166:157–164. https://doi.org/10.1016/j.cbpb.2013.08.005

Amezawa K, Ozaki Y, Yazawa R, Takeuchi Y, Kazeto Y, Yoshizaki G (2020) Oral administration of gonadotropin-releasing hormone analogue enhances previtellogenic oocyte growth of blue mackerel *Scomber australasicus*. Nippon Suisan Gakkaishi 86:83–90. https://doi.org/10.2331/suisan.19-00029; in Japanese with English abstract

Caruso G, Denaro MG, Genovese L (2008) Temporal changes in digestive enzyme activities in the gastrointestinal tract of European eel (*Anguilla anguilla*) (Linneo 1758) following feeding. Mar Freshw Behav Physiol 41:215–218. https://doi.org/10.1080/10236240802492931

Hirano T, Mayer-Gostan N (1976) Eel esophagus as an osmoregulatiory organ. Proc Natl Acad Sci U S A 73:1348–1350. https://doi.org/10.1073/pnas.73.4.1348

Hsu H-Y, Chen S-H, Cha Y-R, Tsukamoto K, Lin C-Y, Han Y-S (2015) De novo assembly of the whole transcriptome of the wild embryo, preleptocephalus, leptocephalus, and glass eel of *Anguilla japonica* and deciphering the digestive and absorptive capacities during early development. PLoS One 10:e0139105. https://doi.org/10.1371/journal.pone.0139105

Kim YK, Ideuchi H, Watanabe S, Park SI, Huh MD, Kaneko T (2008) Rectal water absorption in seawater-adapted Japanese eel *Anguilla japonica*. Comp Biochem Physiol A 151:533–541. https://doi.org/10.1016/j.cbpa.2008.07.016

Knutsen HR, Sørensen SR, Munk P, Bardal T, Kjørsvik E (2021) Digestive tract and the muscular pharynx/esophagus in wild leptocephalus larvae of European eel (*Anguilla anguilla*). Front Mar Sci 8:545217. https://doi.org/10.3389/fmars.2021.545217

Ko CW, Qu J, Black DD, Tso P (2020) Regulation of intestinal lipid metabolism: current concepts and relevance to disease. Nat Rev Gastroenterol Hepatol 17:169–183. https://doi.org/10.1038/s41575-019-0250-7

Kono M, Matsui T, Shimizu C, Koga D (1990) Purification and some properties of chitinase from the stomach of Japanese eel, *Anguilla japonica*. Agric Biol Chem 54:973–978. https://doi.org/10.1080/00021369.1990.10870054

Kurokawa T, Tanaka H, Kagawa H, Ohta H (1996) Absorption of protein molecules by the rectal cells in eel larvae *Anguilla japonica*. Fish Sci 62:832–833. https://doi.org/10.2331/fishsci.62.832

Kurokawa T, Suzuki T, Ohta H, Kagawa H, Tanaka H, Unuma T (2002) Expression of pancreatic enzyme genes during the early larval stage of Japanese eel *Anguilla japonica*. Fish Sci 68:736–744. https://doi.org/10.1046/j.1444-2906.2002.00487.x

Kurokawa T, Iinuma N, Unuma T, Tanaka H, Kagawa H, Ohta H, Suzuki T (2004) Development of endocrine system regulating exocrine pancreas and estimation of feeding and digestive ability in Japanese eel larvae. Aquaculture 234:513–525. https://doi.org/10.1016/j.aquaculture.2003.12.002

Kurokawa T, Koshio M, Kaiya H, Hashimoto H, Nomura K, Uji S, Awaji M, Gen K, Tanaka H (2011) Distribution of pepsinogen- and ghrelin-producing cells in the digestive tract of Japanese eel (*Anguilla japonica*) during metamorphosis and the adult. Gen Comp Endocrinol 173:475–482. https://doi.org/10.1016/j.ygcen.2011.07.008

Miller MJ (2009) Ecology of anguilliform leptocephali: remarkable transparent fish larvae of the ocean surface layer. Aqua-BioSci Monogr 2:1–94

Miller MJ, Chikaraishi Y, Ogawa NO, Yamada Y, Tsukamoto K, Ohkouchi N (2013) A low trophic position of Japanese eel larvae indicates feeding on marine snow. Biol Lett 9:20120826. https://doi.org/10.1098/rsbl.2012.0826

Murashita K, Furuita H, Matsunari H, Yamamoto T, Awaji M, Nomura K, Nagao J, Tanaka H (2013) Partial characterization and ontogenetic development of pancreatic digestive enzymes in Japanese eel *Anguilla japonica* larvae. Fish Physiol Biochem 39:895–905. https://doi.org/10.1007/s10695-012-9749-3

Politis SN, Sørensen SR, Mazurais D, Servili A, Zambonio-Lnfante JL, Miest JJ, Clemmesen CM, Tomkiewicz J, Butts IAE (2018) Molecular ontogeny of first-feeding European eel larvae. Front Physiol 9:1477. https://doi.org/10.3389/fphys.2018.01477

Storelli C, Vilella S, Cassano G (1986) Na-dependent D-glucose and L-alanine transport in eel intestinal brush border membrane vesicles. Am J Phys 251:R463–R469. https://doi.org/10.1152/ajpregu.1986.251.3.R463

Storelli C, Vilella S, Romano MP, Maffia M, Cassano G (1989) Brush-border amino acid transport mechanisms in carnivorous eel intestine. Am J Phys 257:R506–R510. https://doi.org/10.1152/ajpregu.1989.257.3.R506

Tam JKV, Lee LTO, Jin J, Chow BKC (2014) Secretin/secretin receptors. J Mol Endocrinol 52:T1–T14. https://doi.org/10.1530/JME-13-0259

Tocher DR (2003) Metabolism and functions of lipids and fatty acids in teleost fish. Rev Fish Sci 11:10–184

Chapter 11
Osmoregulation

Toyoji Kaneko and Soichi Watanabe

Contents

Several teleost fish species exhibit spectacular migration between seawater (SW) and freshwater (FW) environments, including the Japanese eel *Anguilla japonica*. Eels are equipped with mechanisms for acclimation to both FW and SW, which are adjusted to forthcoming environments during their upstream and downstream migrations.

In general, teleost fish regulate blood ion concentrations and osmolality levels depending on the external environment. Blood plasma osmolality is maintained at ~1/3 of the SW osmolality in fish acclimated to both FW and SW. Because Na^+ and Cl^- are the major ions in the plasma, their regulation is critical for osmoregulation. Another important factor is the regulation of water balance, as water serves as a solvent for osmolytes. Teleost fish have developed superior osmoregulatory mechanisms to maintain blood ion and water balances in hypotonic (FW) and hypertonic (SW) environments. Osmoregulation in adult teleosts is primarily the result of integrated ion and water transport activities in the gills, kidney, and intestine (Marshall and Grosell 2006). FW teleosts face the risk of water load and salt loss through their permeable body surfaces, most of which are occupied by the gill epithelia. To remedy this, the kidney produces dilute urine to discharge excess water and the gill epithelia take up ions (Fig. 11.1a). In contrast, marine teleosts must deal with water loss and salt load. The loss of water is compensated for by drinking SW and absorbing water across the intestinal epithelia, while excess ions are actively excreted from the gills and kidney (Fig. 11.1b).

T. Kaneko (✉) · S. Watanabe
Graduate School of Agricultural and Life Sciences, The University of Tokyo, Tokyo, Japan
e-mail: kaneko31@marine.fs.a.u-tokyo.ac.jp

K. Tsukamoto et al. (eds.), *Eel Science*, Fisheries Science Series,
https://doi.org/10.1007/978-981-99-5692-0_11

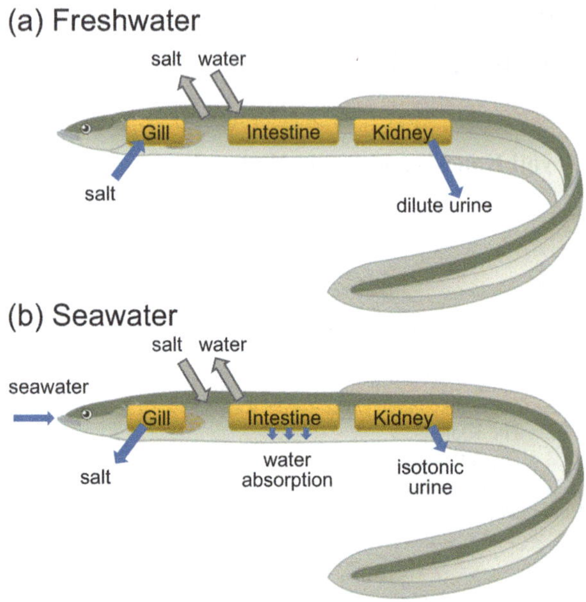

Fig. 11.1 Osmoregulatory mechanisms of teleosts in freshwater (**a**) and seawater (**b**) Gray and blue arrows indicate passive movements and active transports, respectively, of ion and water

In this chapter, to shed some light on eel migration from a physiological point of view, the osmoregulatory mechanism of Japanese eel is reviewed with focus on the ion and water transport activities of the gills, kidney, and intestine.

11.1 The Gills

In addition to its respiratory function, the gills are important for osmoregulation in teleosts. Keys and Willmer (1932) were the first to describe special secretory cells for Cl^- in the gills of the common eel *Anguilla vulgaris* (now known as European eel *Anguilla anguilla*) and called them "chloride-secreting cells." Thereafter, the terms "chloride-secreting cells," or "chloride cells," were used for those cells specializing in active ion transport in the gill epithelia and other tissues in fish; however, it has since been proven that the ion-transport function of those cells is not restricted to Cl^- secretion, and also involves bidirectional transport of various ions. Thus, it is more appropriate to refer to these cells as "ion-transporting cells" or "ionocytes." Gill ionocytes are the major site of ion absorption and secretion and are thus vital for acclimation to FW and SW, respectively.

The ionocytes in the gill epithelia are characterized by a rich population of mitochondria and an extensive tubular system continuous with the basolateral membrane, providing a large surface area for the placement of ion-transporting proteins, such as Na^+/K^+-ATPase (NKA) (Hirose et al. 2003; McCormick 1995). NKA generates ionic and electrical gradients that are used for ion secretion and ion uptake in SW and FW, respectively. The extensive distribution of NKA in ionocytes enables the immunocytochemical detection of ionocytes with an NKA-specific antibody.

11.1.1 Freshwater- and Seawater-Type Ionocytes in Cultured Eel

Distinct FW- and SW-type ionocytes have been reported in the gills of Japanese eels. Sasai et al. (1998a) examined the osmoregulatory ability and general morphology of gill ionocytes in FW-cultured eels and those acclimated to SW for 2 weeks. Although plasma osmolality was slightly but significantly higher in SW fish than in FW fish, it stayed within a physiological range, indicating that the SW eel prepared by transfer of FW eel to SW were well acclimated to the SW environment. Histological observations revealed that NKA-immunoreactive ionocytes were extensively distributed in gill epithelia. According to their location and shape, ionocytes were classified into 2 types: the first was round or columnar in shape and located at the base of lamellae in the gill filament (filament ionocytes); the other was flat and located in the gill lamella (lamellar ionocytes) (Fig. 11.2). Although the 2 distinct types of gill ionocytes were detectable in both FW and SW eels, there were apparent differences in their numbers and sizes between both groups. Filament ionocytes in SW eels were more abundant, larger, and stained more intensely than in FW eels. This finding indicates the activation of filament ionocytes in SW, suggesting that filament ionocytes are SW-type cells responsible for secretion of excess salt. In contrast to filament ionocytes, no apparent difference was observed in lamellar ionocytes between FW and SW eels. Lamellar ionocytes are generally considered FW-type cells responsible for ion uptake in hypotonic environments. Under natural conditions, mature eels migrate downstream to the sea for spawning. Because the

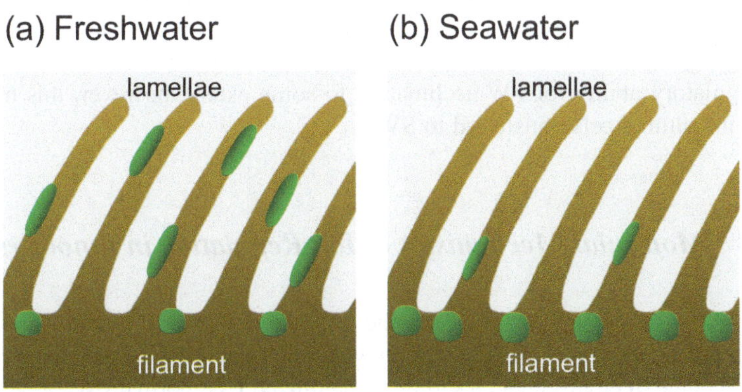

Fig. 11.2 Filament (seawater-type) and lamellar (freshwater-type) ionocytes in the gills of Japanese eel acclimated to freshwater (**a**) and seawater (**b**)

cultured eels used in their study were immature, they would not be fully ready for downstream migration despite their potential capability for SW acclimation. Activation of filament ionocytes could be interpreted as an adaptive response to unexpected exposure to SW, whereas the presence of lamellar ionocytes in the SW-transferred eel may imply the retention of FW adaptability.

11.1.2 Ionocytes in Yellow and Silver Eels

Based on the occurrence of FW-type (lamellar) and SW-type (filament) ionocytes in cultured eels, Sasai et al. (1998a) investigated the morphological modification of gill ionocytes in eels caught in wild stocks, focusing on the difference between yellow and silver eels. In this study, 4 groups of Japanese eels captured in natural habitats were used: small FW yellow eel (body weight: 100–192 g), large FW yellow eel (body weight: 443–536 g), FW silver eel (body weight: 576 g), and brackish water (BW) silver eel (body weight: 300 g). In both small and large yellow eels, ionocytes were detected in both the filament and lamellar epithelia of the gills. There was no significant difference in the number of filament ionocytes between small and large yellow eels, whereas lamellar ionocytes were fewer in the large yellow eels than those in the small yellow eels. Filament ionocytes were the dominant cell type in the FW and BW silver eels. Lamellar ionocytes, frequently observed in yellow eels, were markedly low in both silver eel groups. Filament ionocytes in silver eels were significantly larger than in yellow eels. Moreover, those ionocytes were larger in BW silver eels than in FW silver eels.

The development of filament ionocytes in silver eels agrees with the results obtained in cultured eels, suggesting the importance of filament ionocytes in SW acclimation (Fig. 11.2b). In contrast, lamellar ionocytes were more abundant in the yellow eels than in the silver eels, implying a significant role of lamellar ionocytes in FW acclimation (Fig. 11.2a). As eels grow in FW environments and begin gonadal development under natural conditions, they may develop hypo-osmoregulatory ability to lower blood osmolality for SW acclimation as preadaptation for downstream migration to the sea. At the same time, those eels reduce hyper-osmoregulatory ability for FW acclimation to some extent; however, this may not be true in cultured eels transferred to SW.

11.1.3 Molecular Mechanisms of Ion Regulation in Ionocytes

Seo et al. (2009, 2013) investigated the morphological changes and molecular mechanisms of ion regulation in the ionocytes of cultured Japanese eels acclimated to a wide range of environmental salinities: deionized FW, normal FW, 30%-diluted SW, and normal SW. Immunohistochemical observations revealed that filament ionocytes were most developed in normal SW, whereas lamellar ionocytes were most observed in deionized FW, confirming that filaments and lamellar ionocytes are responsible for ion secretion and absorption, respectively (Seo et al. 2009). Notably, the cystic fibrosis transmembrane conductance regulator (CFTR) was located in the apical region of ionocytes in eels in diluted and normal SWs but not in those in deionized and normal FWs. Because CFTR is an apical Cl^- channel specific for SW-type cells, the results clearly indicate that ionocytes developed in diluted and normal SWs function as ion-secreting sites. To further investigate the molecular mechanisms of ion regulation in ionocytes, researchers measured the mRNA expression of 2 ion-transporting proteins, Na^+/H^+ exchanger-3 (NHE3) and $Na^+/K^+/2Cl^-$ cotransporter-1α (NKCC1α) (Seo et al. 2013). Both ion transporters are important in the molecular mechanisms of ionocytes in Mozambique tilapia *Oreochromis mossambicus* (Hiroi et al. 2008). The gene expression of NHE3 was higher in deionized FW and normal SW than that in normal FW and diluted SW. NKCC1 expression increases with increasing environmental salinity. Immunofluorescence staining showed that the apical NHE3 immunoreaction was stronger in deionized FW and normal SW than that in the other groups. Basolateral NKCC1a immunoreactivity was most intense in normal SW. These results indicate that apical NHE3 is involved in ion uptake in eels acclimated to hypotonic environments and that basolateral NKCC1α is important for acclimation to hypertonic environments. The relatively high expression of NHE3 in normal SW further indicates a possible role for NHE3 in acid-base regulation in the gills of SW-acclimated eels, as seen in Mozambique tilapia (Watanabe et al. 2008). Although the molecular mechanisms of ion transport in eel ionocytes have not been fully elucidated, these findings suggest that the Japanese eel and Mozambique tilapia may share basic molecular mechanisms.

11.2 The Kidney

In teleosts, the kidney (body kidney) functions as an excretory and osmoregulatory organ. The functional unit of the kidney is the nephron, which consists of a renal corpuscle (Bowman's capsule and glomerulus), the first (PT-I) and second (PT-II) segments of the proximal tubule (PT) and distal tubule (DT), followed by a collecting duct (CD). Kidneys are generally composed of numerous nephrons and infilling lymphoid tissues. The kidneys of FW fish produce large volumes of diluted urine by filtering large volumes of blood in the glomeruli and reabsorbing ions from the filtrate in the renal tubules. Conversely, the kidneys of SW fish excrete excess divalent ions but produce relatively low volumes of isotonic urine to minimize water loss. The major functional significance of renal tubules in osmoregulation lies in the reabsorption and secretion of monovalent and divalent ions, respectively, to maintain blood ion balance.

11.2.1 Ion Transports in the Renal Tubules

Like ionocytes, active ion transport in the renal tubules involves electrochemical gradients created by NKA, which are mainly distributed in the basolateral membrane of epithelial cells in the renal tubules. In the Japanese eel, 3-dimensional analyses have shown that the renal tubules drift throughout lymphoid tissues in a random pattern and subsequently merge into the CD, which is aligned linearly. Among the various segments of the renal tubules, the DT and CD cells show more intense NKA immunoreaction in FW eels than in SW-acclimated eels, confirming that the DT and CD segments are important for FW acclimation (Teranishi and Kaneko 2010). These segments are responsible for the reabsorption of monovalent ions, such as Na^+ and Cl^- from primitive urine in the kidneys of FW fish (Fig. 11.3a). In contrast, the PT-II segment in marine teleosts, including SW-acclimated eel, is thought to secrete divalent ions (Ca^{2+} and Mg^{2+}) for acclimation to hypertonic environments (Fig. 11.3b).

11.2.2 Molecular Mechanisms of Monovalent-Ion Transports

To clarify the molecular mechanisms of monovalent-ion transport in the PT and CD of Japanese eel, Teranishi et al. (2013) examined the gene expression and immunocytochemical distributions of cation-chloride cotransporters involved in renal osmoregulation (NKCC2α and Na^+/Cl^- cotransporter (NCCα)). Duplicate pairs of cation-chloride cotransporters have been identified in some teleosts (Cutler and Cramb 2008; Hiroi et al. 2008; Teranishi et al. 2013). Both Japanese and

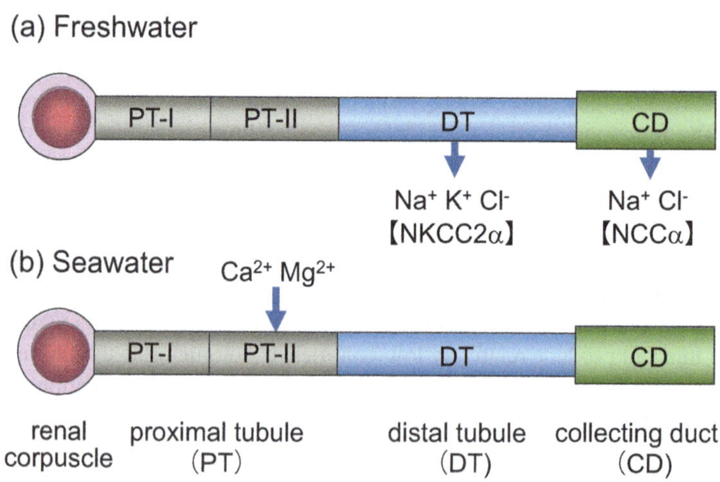

Fig. 11.3 The structure and osmoregulatory functions of nephrons in the kidney of Japanese eel acclimated to freshwater (**a**) and seawater (**b**)

European eels possess 2 NKCC2 isoforms (NKCC2α and NKCC2β) and 2 NCC isoforms (NCCα and NCCβ). Expressions of NKCC2α and NCCα mRNAs are limited to the renal tissue, whereas NKCC2β and NCCβ expressions primarily occur in the intestine (Cutler and Cramb 2008; Watanabe et al. 2011; Teranishi et al. 2013). In the kidneys of the Japanese eel, NCCα expression increases as the environmental salinity decreases, whereas NKCC2α expression is not significantly different between FW and SW (Teranishi et al. 2013). Interestingly, immunocytochemical observations have revealed that the apical membranes of epithelial cells in the DT and CD produce more intense immunoreactions of NKCC2α and NCCα, respectively, in FW eels than those in SW eels.

Preferential localizations of apical NKCC2α and basolateral NKA within DT cells and apical NCCα and basolateral NKA within CD cells indicate that NKCC2α and NCCα reabsorb Na^+ and Cl^- from primitive urine in eels acclimated to hypotonic environments (Fig. 11.3a). The expression patterns of NKCC2α and NCCα in the anterior and posterior epithelial cells, respectively, along the renal tubule and CD, may allow increased efficiency of monovalent ion reabsorption in the Japanese eel kidney. The concentration of K^+ in primitive urine likely decreases as it travels along the renal tubules and CD. NCCα could possibly reabsorb Na^+ and Cl^- more efficiently under K^+-deficient conditions in the posterior segment than NKCC2α, as NKCC2α requires K^+ for Na^+ and Cl^- reabsorption.

11.3 The Intestine

Marine teleosts compensate for osmotic water loss by ingesting large amounts of SW which is subsequently absorbed in the intestine, whereas drinking in FW teleosts is considered almost negligible. The osmotic gradient created between intestinal fluid and blood is the primary mechanism behind intestinal water absorption in SW fish. Prior to intestinal water absorption, ions are removed from the ingested SW (1000–1050 mOsm/kg) and osmolality reduces to near-isotonic (about 300 mOsm/kg) or sub-isotonic levels during its passage along the gastrointestinal tract (Fig. 11.4). In SW-acclimated eels, ingested SW is first desalted in the ion-permeable esophagus, and again (to a lesser extent) in the stomach, by passive diffusion of ions into the blood (Hirano and Mayer-Gostan 1976). Subsequently, the osmolality of ingested SW is further decreased by the active absorption of monovalent ions (Na^+, Cl^-, and K^+) and the removal of divalent ions (Ca^{2+} and Mg^{2+}) through the formation of Ca and Mg precipitates in the intestine. Finally, water absorption occurs in the posterior portion of the intestinal tract (the rectum) through aquaporin 1 (AQP1), a water-selective channel located in the apical membrane of intestinal epithelial cells.

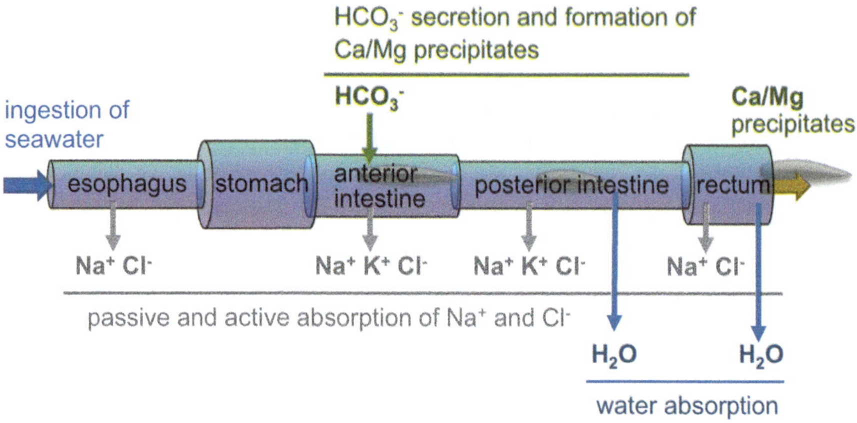

Fig. 11.4 Mechanisms of intestinal water absorption in Japanese eel acclimated to seawater

11.3.1 Active Absorption of Monovalent Ions

The current accepted model for NaCl absorption in intestinal epithelial cells involves the cooperation of ion-transporting proteins, including NKA, NKCC, and NCC (Grosell 2011). Electrochemical gradients created by NKA located in the basolateral membrane of intestinal epithelial cells expedite electroneutral ion uptake into cells across the apical membrane through NKCC and/or NCC. Duplicate pairs of NKCC2 and NCC isoforms have been identified in European and Japanese eels (Cutler and Cramb 2008; Watanabe et al. 2011). In both eel types, NKCC2β and NCCβ are mainly expressed in the intestinal tract, whereas the expression of NKCC2β and NCCβ is specific to the kidney.

The intestine of eels is roughly divided into 3 anatomically and physiologically distinct regions: anterior intestine, posterior intestine, and rectum (see Chap. 10). Quantitative PCR analysis shows that in SW-acclimated Japanese eels, NKCC2β is predominantly expressed in the anterior and posterior intestines, whereas NCCβ expression is significantly high in the rectum. Immunohistochemical observations further clarified that NKCC2β localizes in the apical surface of epithelial cells in the anterior and posterior intestines, whereas NCCβ is likely distributed in the rectum (Watanabe et al. 2011). These findings suggest that NKCC2β and NCCβ in Japanese eels are functional cation-Cl^- cotransporters involved in ion transport in the intestinal tract. Similar to the kidney, preferential localizations of NKCC2β in the anterior and posterior intestines and NCCβ in the rectum are likely to enable the efficient absorption of Na^+ and Cl^- along the intestinal tract. NKCC2β absorbs monovalent ions more effectively than NCCβ but cannot function properly under K^+-deficient conditions in the rectal fluid (Kim et al. 2008), whereas NCCβ in the rectum does not require K^+ for NaCl absorption.

11.3.2 Formation of Ca and Mg Precipitates

Marine teleosts produce white feces, which is often referred to as Ca precipitates mainly composed of calcium and magnesium carbonates. In the Japanese eel, Ca precipitates were not found in the intestines of fish acclimated to FW. Furthermore, the volume of Ca precipitates increased with increasing environmental salinity, indicating that Ca precipitate formation is closely related to SW acclimation or hypo-osmoregulation (Mekuchi et al. 2010). The crystal structure of the Japanese-eel Ca precipitate is amorphically similar to Mg-calcite, which contains Ca and Mg at a ratio of 7:2. In addition to the gastrointestinal absorption of monovalent ions, the formation and excretion of Ca precipitates contributes to a further reduction in the osmolality of ingested SW by eliminating Ca^{2+} and Mg^{2+} from the ingested SW.

Ca precipitates are formed in the intestine by the chemical reaction of Ca^{2+} and Mg^{2+} derived from ingested SW with HCO_3^-, which is secreted not only from the intestine (Grosell et al. 2009) but also from the pancreas (Mekuchi et al. 2013). In Japanese eels, when the pancreas was surgically removed, the amount of Ca precipitates decreased compared with that in control fish, indicating that HCO_3^- secretion from the pancreas contributed substantially to the formation of Ca precipitates in the intestine (Mekuchi et al. 2013).

11.3.3 Water Absorption in the Rectum

Following the absorption of Na^+ and Cl^- along the gastrointestinal tract and the formation of Ca precipitates in the intestine, the osmolality of ingested SW is reduced to a near-isotonic or sub-isotonic level, which is favorable for osmotic water absorption in the posterior part of the intestinal tract.

Aquaporins (AQPs) is a family of integral membrane proteins that function as water channels. Among them, AQP1 has been identified as a water-entry site in the intestinal epithelium of Japanese eels (Aoki et al. 2003). AQP1 is predominantly expressed in the intestine, and its expression levels are higher in SW eels than in FW eels. Using intestinal sacs filled with balanced salt solution (BSS) and incubated in isotonic BSS, Aoki et al. (2003) reported higher water permeability of the intestinal epithelium in SW eels than in FW eels. The higher water permeability is attributed to the higher expression of AQP1 in the intestinal epithelium of the SW eel. Immunohistochemical studies further revealed intense AQP1 immunoreactivity in the apical surface of columnar epithelial cells in the SW eel; however, the intensity of the AQP1 immunoreaction differed in different parts of the intestine (Aoki et al. 2003; Kim et al. 2008). Distinct immunoreactivity for AQP1 was observed in the apical region of epithelial cells in the rectum, whereas apical labeling of AQP1 in epithelial cells was minimal in the posterior intestine and even less in the anterior intestine, indicating that the rectum is the central site of intestinal water absorption.

The rectum is spatially separated from the posterior intestine and the anus by a valve structure and a sphincter, respectively. Such structures allow the rectum to swell as intestinal fluid flows into it, and a concomitant increase in hydrostatic pressure may provide an additional force for rectal water absorption. Although water absorption concentrates the rectal fluid, additional ion absorption driven by NKA and mediated by NCCβ restores the osmotic gradient favorable to water absorption. Simultaneously, the inflow of intestinal fluid into the rectum maintains high hydrostatic pressure. Maintenance of lowered rectal fluid osmolality and high hydrostatic pressure allows for continuous water absorption, which is likely to be maintained for a prolonged period in the rectum until it is evacuated through the anus.

11.4 Osmoregulation in Japanese Eel Embryos and Larvae

11.4.1 Ionocytes in Embryos and Hatched Larvae

During the early life stages, although osmoregulatory organs in adult fish have not yet developed or are not fully functional, extrabranchial ionocytes have been found in the epithelia covering the body and yolk sac (Sasai et al. 1998b). In the late embryonic stage, a large number of NKA-immunoreactive ionocytes were detected in the yolk sac membrane, which occupies a large proportion of the body surface. However, minimal ionocytes were observed in the epithelium that covered the developing embryo. On the day of hatching, ionocytes were extensively distributed in the epithelium of the anterior half of the body, which included the yolk sac membrane. In the larvae (pre-leptocephali) at 6 days post hatching (DPH), ionocytes in the body surface increased in number and were arranged along the muscle segments. The ionocytes in the yolk sac membrane and integument of the body surface are important for ion secretion in hypertonic environments in the absence of gill ionocytes.

11.4.2 Osmoregulation in Leptocephali and Glass Eels

Japanese eels are equipped with mechanisms to regulate ion and water balances during the pre-leptocephalus and leptocephalus stages, when gill and gill ionocytes are not yet developed. Lee et al. (2013) reported that the body fluid osmolality of leptocephali (total length: 8–49 mm) ranged from 360 to 540 mOsm/kg, which was marginally higher than that in adult eels but significantly lower than that of SW. Similar to pre-leptocephali, NKA-immunoreactive ionocytes were distributed over the entire body surface of the leptocephali. Using a fluorescent sodium indicator and chloride tests, they successfully localized Na^+- and Cl^--secreting sites at the apical region of the ionocytes. These findings clearly indicate that cutaneous ionocytes secrete excess Na^+ and Cl^- to maintain body fluid osmolality at a lower level than that in ambient SW.

In addition to functional ionocytes, the involvement of drinking in osmoregulation has been implicated in eel leptocephali. To examine the onset of drinking during the early life stages of the Japanese eel, Ahn et al. (2015) sought to detect ingested SW in the digestive tract using fluorescent dextran as an inert marker. Drinking was observed as early as 0 DPH, but ingested water remained in the forepart of the digestive tract. At 2 DPH, ingested water traveled along the digestive tract to the anus. To estimate water absorption in the intestine, Lee et al. (2013) exposed leptocephali to seawater containing dextran labeled with Alexa Fluor 488 and measured the fluorescence intensity along the digestive tract. They showed that water absorption predominantly occurred in the rectum, and was barely absorbed in the stomach and intestine. Furthermore, mRNA expression of NKCC2β and NCCβ, key ion transporters for intestinal ion absorption prior to water absorption, were detected mainly in the intestine and rectum, respectively, in eel larvae at 13 DPH, which coincides with the results observed in adult eels.

These findings indicate that eel leptocephali are equipped with mechanisms to regulate ion and water balances, exhibiting a hypo-osmoregulatory ability comparable to that of adult eels. It is highly likely that the kidney contributes to osmoregulation by secreting divalent ions in eel leptocephali; however, direct evidence of this is not yet available.

11.4.3 Development of the Gills and Gill Ionocytes

The major site of ionocytes shifts from the body surface to the gills as leptocephali gills further grow and develop. Sasai et al. (2007) reported that ionocytes first appeared in developing gill filaments in the mid-larval stage of leptocephalus. The number of ionocytes gradually increased as the fish grew to the late stage of leptocephalus. Gill lamellae developed from the gill filaments in glass eels immediately after metamorphosis, and a rich population of ionocytes was observed in the gill filaments. Furthermore, ionocytes were extensively distributed in the gill filaments of glass eels collected from coastal areas. These findings indicate that gill and gill ionocytes develop slowly during the extensive larval stage, followed by rapid differentiation over a short period of metamorphosis. Once eel leptocephali develop into a glass eel after metamorphosis, the osmoregulatory mechanism shifts from the larval type to the adult type.

References

Ahn H, Lee KM, Inokuchi M, Watanabe S, Okamura A, Tsukamoto K, Kaneko T (2015) Observations on initial water ingestion and ion absorption in the digestive tract of Japanese eel larvae. Fish Sci 81:283–290. https://doi.org/10.1007/s12562-014-0841-8

Aoki M, Kaneko T, Katoh F, Hasegawa S, Tsutsui N, Aida K (2003) Intestinal water absorption through aquaporin 1 expressed in the apical membrane of mucosal epithelial cells in seawater-adapted Japanese eel. J Exp Biol 206:3495–3505. https://doi.org/10.1242/jeb.00579

Cutler CP, Cramb G (2008) Differential expression of absorptive cation-chloride cotransporters in the intestinal and renal tissues of the European eel (*Anguilla anguilla*). Comp Biochem Physiol B 149:63–73. https://doi.org/10.1016/j.cbpb.2007.08.007

Grosell M (2011) Intestinal anion exchange in marine teleosts is involved in osmoregulation and contributes to the oceanic inorganic carbon cycle. Acta Physiol 202:421–434. https://doi.org/10.1111/j.1748-1716.2010.02241.x

Grosell M, Mager EM, Williams C, Taylor JR (2009) High rates of HCO_3^- secretion and Cl^- absorption against adverse gradients in the marine teleost intestine: the involvement of an electrogenic anion exchanger and H^+-pump metabolon? J Exp Biol 212:1684–1696. https://doi.org/10.1242/jeb.027730

Hirano T, Mayer-Gostan N (1976) Eel esophagus as an osmoregulatory organ. Proc Natl Acad Sci U S A 73:1348–1350. https://doi.org/10.1073/pnas.73.4.1348

Hiroi J, Yasumasu S, McCormick SD, Hwang PP, Kaneko T (2008) Evidence for an apical Na-Cl cotransporter involved in ion uptake in a teleost fish. J Exp Biol 211:2584–2599. https://doi.org/10.1242/jeb.018663

Hirose S, Kaneko T, Naito N, Takei Y (2003) Molecular biology of major components of chloride cells. Comp Biochem Physiol B 136:593–620. https://doi.org/10.1016/S1096-4959(03)00287-2

Keys AB, Willmer EN (1932) "Chloride secreting cells" in the gills of fish with special reference to the common eel. J Physiol 76:368–378. https://doi.org/10.1113/jphysiol.1932.sp002932

Kim YK, Ideuchi H, Watanabe S, Park SI, Huh MD, Kaneko T (2008) Rectal water absorption in seawater-adapted Japanese eel *Anguilla japonica*. Comp Biochem Physiol A 151:533–541. https://doi.org/10.1016/j.cbpa.2008.07.016

Lee KM, Yamada Y, Okamura A, Tsukamoto K, Kaneko T (2013) Hyposmoregulatory ability and ion- and water-regulatory mechanisms during the leptocephalus stages of Japanese eel *Anguilla japonica*. Fish Sci 79:77–86. https://doi.org/10.1007/s12562-012-0576-3

Marshall WS, Grosell M (2006) Ion transport, osmoregulation, and acid-base balance. In: Evans DH, Claiborne JB (eds) The physiology of fish 3. CRC Press, Boca Raton, pp 177–230

McCormick SD (1995) Hormonal control of gill Na^+, K^+-ATPase and chloride cell function. In: Wood CM, Shuttleworth TJ (eds) Cellular and molecular approaches to fish ionic regulation. Academic, New York, pp 285–315

Mekuchi M, Hatta T, Kaneko T (2010) Mg-calcite, a carbonate mineral, constitutes Ca precipitates produced as a byproduct of osmoregulation in the intestine of seawater-acclimated Japanese eel *Anguilla japonica*. Fish Sci 76:199–205. https://doi.org/10.1007/s12562-009-0199-5

Mekuchi M, Watanabe S, Kaneko T (2013) Bicarbonate secreted from the pancreas contributed to the formation of Ca precipitates in Japanese eel, *Anguilla japonica*. J Exp Zool A 319:53–62. https://doi.org/10.1002/jez.1774

Sasai S, Kaneko T, Hasegawa S, Tsukamoto K (1998a) Morphological alteration in two types of gill chloride cells in Japanese eels (*Anguilla japonica*) during catadromous migration. Can J Zool 76:1480–1487. https://doi.org/10.1139/z98-072

Sasai S, Kaneko T, Tsukamoto K (1998b) Extrabranchial chloride cells in early life stages of the Japanese eel, *Anguilla japonica*. Ichthyol Res 45:95–98. https://doi.org/10.1007/BF02678580

Sasai S, Katoh F, Kaneko T, Tsukamoto K (2007) Ontogenic change of gill chloride cells in leptocephalus and glass eel stages of the Japanese eel, *Anguilla japonica*. Mar Biol 150:487–492. https://doi.org/10.1007/s00227-006-0355-8

Seo MY, Lee KM, Kaneko T (2009) Morphological changes in gill mitochondria-rich cells in cultured Japanese eel *Anguilla japonica* acclimated to a wide range of environmental salinity. Fish Sci 75:1147–1156. https://doi.org/10.1007/s12562-009-0144-7

Seo MY, Mekuchi M, Teranishi K, Kaneko T (2013) Expression of ion transporters in gill mitochondrion-rich cells in Japanese eel acclimated to a wide range of environmental salinity. Comp Biochem Physiol A 166:323–332. https://doi.org/10.1016/j.cbpa.2013.07.004

Teranishi K, Kaneko T (2010) Spatial, cellular and intracellular localization of Na^+/K^+-ATPase in the sterically-disposed renal tubules of Japanese eel. J Histochem Cytochem 58:707–719. https://doi.org/10.1369/jhc.2010.955492

Teranishi K, Mekuchi M, Kaneko T (2013) Expressions of sodium/hydrogen exchanger 3 and cation-chloride cotransporters in the kidney of Japanese eel acclimated to a wide range of salinities. Comp Biochem Physiol A 164:333–343. https://doi.org/10.1016/j.cbpa.2012.11.011

Watanabe S, Niida M, Maruyama T, Kaneko T (2008) Na^+/H^+ exchanger isoform 3 expressed in apical membrane of gill mitochondrion-rich cells in Mozambique tilapia *Oreochromis mossambicus*. Fish Sci 74:813–821. https://doi.org/10.1111/j.1444-2906.2008.01593.x

Watanabe S, Mekuchi M, Ideuchi H, Kim YK, Kaneko T (2011) Electroneutral cation-Cl^- cotransporters NKCC2β and NCCβ expressed in the intestinal tract of Japanese eel *Anguilla japonica*. Comp Biochem Physiol A 159:427–435. https://doi.org/10.1016/j.cbpa.2011.04.009

Chapter 12
Reproduction

Shigeho Ijiri, Hajime Matsubara, Moemi Horiuchi, and Yukinori Kazeto

Contents

Eels do not mature under captive conditions. It is not uncommon for fish to begin and complete their final maturation under closed aquaculture conditions. In most cases, they proceed and successfully complete vitellogenesis and spermatogenesis. However, oocyte maturation and ovulation, and sperm maturation is not spontaneously triggered under aquaculture conditions, which is different from their natural spawning places. The difficulty of eel reproduction is that female and male eels will not proceed vitellogenesis and spermatogenesis, respectively. The Japanese eel undergoes vitellogenesis and spermatogenesis during long-distance migration toward the spawning site. Thus, because natural eels have not been captured during spawning migration in the ocean, the natural controls that promotes sexual

The original version of the chapter has been revised: Figure 12.6 has been corrected. A correction to this chapter can be found at https://doi.org/10.1007/978-981-99-5692-0_24

S. Ijiri (✉)
Graduate School of Fisheries Sciences, Hokkaido University, Hokkaido, Japan
e-mail: ijiri@fish.hokudai.ac.jp

H. Matsubara
Noto Center for Fisheries Science and Technology, Kanazawa University, Ishikawa, Japan

M. Horiuchi
IRAGO Institute Co., Ltd., Aichi, Japan

Y. Kazeto
Minamiizu Field Station, Fisheries Technology Institute, Japan Fisheries Research and Education Agency, Shizuoka, Japan

K. Tsukamoto et al. (eds.), *Eel Science*, Fisheries Science Series,
https://doi.org/10.1007/978-981-99-5692-0_12

maturation and final maturation are not known. Previous studies have provided the current known information on natural reproductive conditions by examining adult eels captured at their spawning ground, West Mariana Ridge; however all these eels had already completed ovulation or spawning (Tsukamoto et al. 2011; Shimizu et al. 2021). Therefore, reproductive physiological conditions can only be determined from artificially induced sexually maturing eels. In a previously published study, we described the control of ovarian vitellogenic growth with special focus on changes in the morphology of oocytes and steroid hormone production in the ovary of the Japanese eel (Adachi et al. 2003). In this chapter, we describe the latest insights, especially into natural sex differentiation and the mechanism of steroidogenic shift that promotes final oocyte maturation in the Japanese eel.

12.1 Gonadal Sex Differentiation

12.1.1 Morphological Gonadal Sex Differentiation

Most cultured eels differentiate into males, and although female eels are produced by oral estrogen administration, ovarian differentiation and development may not correspond to natural ovarian development. To understand natural gonadal sex differentiation in eels, we collected wild eels and investigated their gonadal sex differentiation status. A total of 189 eels ranging 20–40 cm in total length (TL) were collected from 10 rivers in Miyazaki and Oita prefectures. Age, which was determined by otolith analyses, was widely distributed between 2 and 8 years. Differentiated ovaries and testes were found as early as 3 and 4 years of age, respectively. Undifferentiated gonads were found until 7 years of age, indicating that the beginning of gonadal sex differentiation is not determined by aging. In relation to the growth assessed by TL, apparently differentiated ovaries and testes were observed after 32 and 30 cm, respectively, excluding 1 eel with an apparently differentiated ovary measuring 25 cm in TL. Undifferentiated gonads were not observed after 34 cm, suggesting that this size marked the completion of gonadal sex differentiation. Differentiating ovaries and testes emerged after 26 and 24 cm, respectively. Before the apparently differentiated ovary and testis emerged, early-stage differentiation of the ovary and testis could be distinguished by fine histological observation. These observations suggest that ovarian and testicular differentiation begin at 26–32 cm and 24–30 cm, respectively (Horiuchi et al. 2022).

12.1.2 Molecular Gonadal Sex Differentiation

Prior to morphological gonadal sex differentiation, it is well known that dimorphic gene expression pattern is observed between genetic male and female undifferentiated gonads (Ijiri et al. 2008). These genes are known to be related to sex differentiation. The period of the emergence of sexual dimorphic gene expression patterns in undifferentiated gonads before morphological gonadal differentiation begins is called the molecular sex differentiation period. In many fish species, the aromatase gene (*cyp19*) and a transcription factor involving *cyp19* transcription (*foxl2*) show higher expression in undifferentiated gonads of genetic females, and the resultant estrogen, which is produced by aromatase, promotes ovarian differentiation. Without estrogen production, testicle-specific genes such as gonadal-soma derived factor (*gsdf*) and/or anti-Müllerian hormone (*amh*) are expressed, and the gonad differentiates into the testis.

In wild Japanese eels, we investigated the gene expression patterns in the ovaries and testes shortly after morphological differentiation began. In addition, molecular sex differentiation was investigated in undifferentiated gonads. In differentiating ovaries, *cyp19* and *foxl2* typically showed higher expression than in differentiating testes, although the mRNA levels varied widely among individual eels. If ovaries were aligned in the order of detailed developmental stage according to morphological oocyte features, *cyp19* expression tended to be high at the earliest stage of differentiating ovaries consisting of oogonial cysts, early meiosis, and postpachytene oocytes. Furthermore, *cyp19* expression tended to decrease as oogonia growth progressed, suggesting that *cyp19* mRNA is only briefly expressed after triggering ovarian differentiation. This assumption contradicts the general understanding that *cyp19* expression lasts longer during ovarian differentiation and development than that during testicular differentiation and sexual maturation. Obtaining differentiating ovarian samples is difficult, probably due to the rapid differentiation period; however, molecular sex differentiation may occur during this limited period. Therefore, more gonadal samples within this narrow window should be investigated to understand the molecular control of sex differentiation in the eel.

Gsdf and *amh* mRNA levels in the early differentiated testis increased after reaching 33 cm in TL. In eels measuring <33 cm, gonads differentiated into testis did not show higher expression levels compared to ovaries. This suggests that *gsdf* and *amh* expression does not increase during the early stages of testis differentiation, further suggesting the existence of unknown genes that promote the early stage of testis differentiation.

In undifferentiated gonads, which are assumed to be found around the initial time of sex differentiation, distinct dimorphic gene expression patterns were not observed in 27 eels measuring from 30–33 cm in TL. The fact that molecular sex differentiation has not been detected in eel undifferentiated gonads may indicate that this window, in which gene expression shows distinct dimorphic patterns among undifferentiated gonads, is brief. Therefore, it would be difficult to obtain gonadal samples that may show dimorphic gene expression patterns. For further clarification,

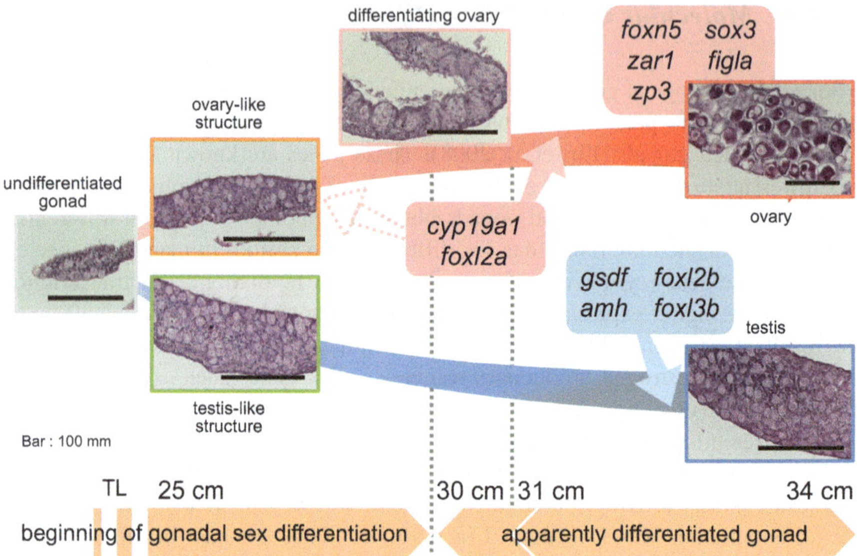

Fig. 12.1 Gonadal sex differentiation in the wild eel. Bar = 100 μm

collecting gonadal samples at the beginning of sex differentiation is necessary in future studies (Fig. 12.1; Horiuchi et al. 2022).

12.2 Pre-vitellogenic Development, Oil Droplet Accumulation

After ovarian differentiation begins, gonia, which are essentially equivalent to primordial germ cells, gradually multiply by mitotic division. During mitotic division, cluster of gonium form cysts in which the cluster of gonium is surrounded by somatic cells (Fig. 12.2). This morphological feature is common at the beginning of ovarian differentiation. Simultaneously, the epithelium of the gonad begins to invaginate, which differentiates into a future ovarian lamella. After this typical morphological differentiation was observed, the gonia were termed oogonia. After gradual mitotic proliferation, oogonia enter meiotic division to create oocytes (Horiuchi et al. 2022). The meiotic division is halted during the first division prophase. The morphological classification of this stage of oocytes is called the chromatin nucleus stage. This arrest of meiotic division is not re-initiated until oocyte maturation is triggered. As oocyte size gradually increases, oocytes are transformed to the next developmental stage, termed the perinucleolus stage. Morphological differentiation of oocytes has been described in detail in our previous report (Adachi et al. 2003). The perinucleolus oocyte proceeds to the next

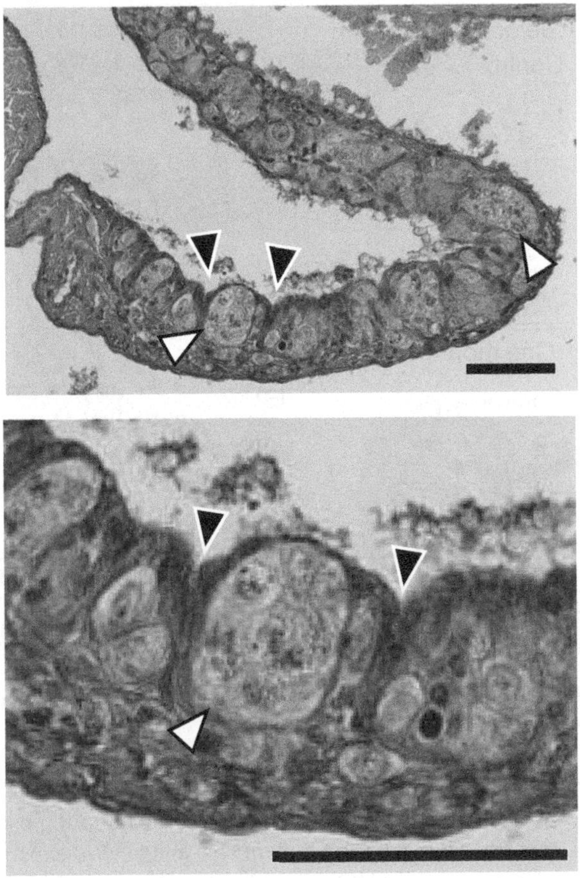

Fig. 12.2 Morphological indication at the beginning of ovarian differentiation. Lower panel represents high magnification of upper panel. Arrowheads, notches in the epithelium of the ovary. White arrowheads, cluster of active mitotic gonial cell division. Bar = 50 μm

developmental stage characterized by the accumulation of oil droplets in oocytes, which is called the oil droplet stage. Oil droplet accumulation occurs from summer to autumn in feminized eels (estradiol-administered eels) in the second year of the glass eel stage in captive conditions. During this period, oil droplet accumulation accelerates when the rearing water temperature gradually decreases in response to natural seasonal changes. Regarding changes in serum steroid hormone levels during this period, the 11-ketotestosterone (11KT) concentration tended to be high from October to February, in contrast to E2 levels that tended to be high in February. Because oil droplet accumulation tended to accelerate after September, 11KT was considered a candidate factor for oil droplet accumulation. To confirm this, we incorporated 11KT into silastic tubes that were subsequently implanted into the eel abdomen; concurrently, E2 implantation was conducted as a control. One month after implantation, 11KT implantation successfully facilitated oil droplet accumulation and increased oocyte diameter, in contrast to E2 implantation, which did not affect oocyte growth (Fig. 12.3; Matsubara 2003). We also demonstrated that 11KT facilitated VLDL incorporation into oocytes, which resulted in oil droplet

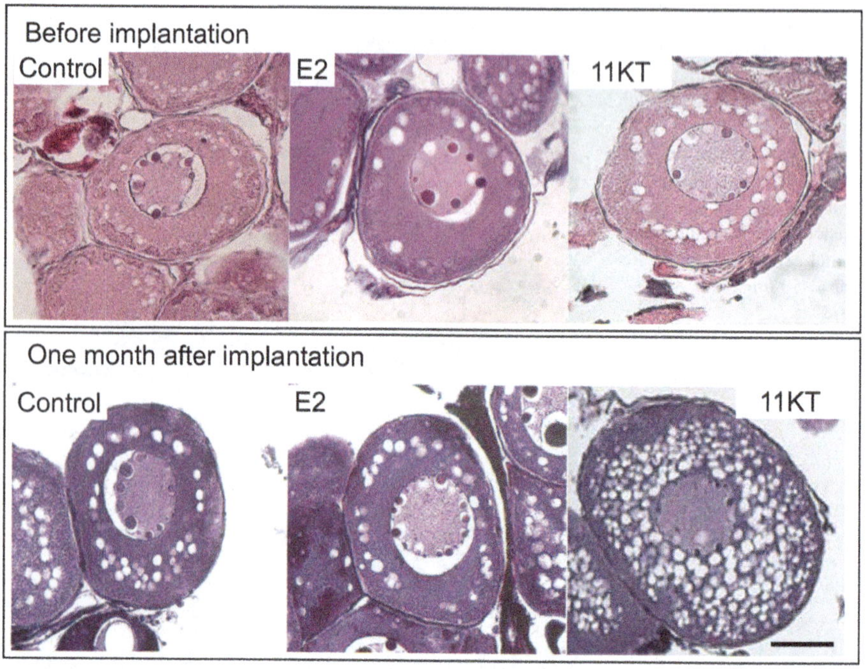

Fig. 12.3 Morphological changes in oocytes before and 1-month after received E2 or 11KT implantation. Bar = 100 μm

accumulation and individual oocyte growth if ovarian fragments were incubated with 11KT and purified eel VLDL was added to the incubation (Endo et al. 2011). The results obtained from both *in vivo* and *in vitro* experiments clearly demonstrated the pivotal role of 11KT in oil droplet accumulation in oocytes at the perinucleolus stage. Furthermore, 11KT was implanted into first-year eels from the glass eel stage to induce oil droplet accumulation at a very small size. After oil droplet accumulation proceeded, the eels were able to enter vitellogenic growth if they received gonadotropic hormone administration in captive conditions (Fig. 12.4).

12.3 Vitellogenic Oocyte Growth

After oil droplet accumulation in oocytes proceeds, eels become ready for development to the next stage toward sexual maturation. The developmental stage of oocytes of natural silver eels, which migrate downstream and prepare for spawning migration, is the beginning of the vitellogenic growth stage, in which small yolk vesicles can be detected in the peripheral area of oocytes ranging from 200 to 330 μm in diameter. Cultivated feminized eels also accumulate oil droplets, and some begin vitellogenesis spontaneously if rearing temperatures gradually decrease after

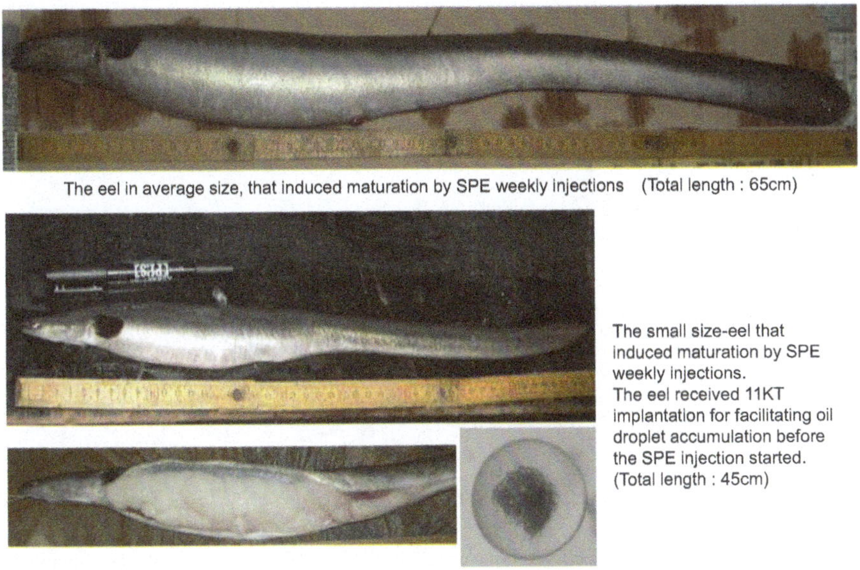

Fig. 12.4 Artificial maturation induced by weekly salmon pituitary extract in young eel (45 cm in total length) which induced oil droplet accumulation by 11KT implantation

summer (Ijiri et al. 1998; Chai et al. 2010). These observations suggested that eels begin sexual maturation without exogenous hormone treatment. However, neither silver eels nor feminized eels that begin vitellogenic growth undergo further vitellogenic growth beyond the early stage in captive conditions. These eels maintain very early vitellogenic oocytes during winter, and oocytes degenerate after spring under captive conditions. Gonadotropin administration is commonly used for artificial eel maturation to induce further vitellogenic development. Morphological oocyte development during artificial maturation has been previously reported in detail (Adachi et al. 2003); thus, in this section, we describe how vitellogenic growth proceeds in response to weekly injections of salmon pituitary extract (SPE) based on gene expression related to oocyte development. Weekly injection of SPE into eels whose oocytes accumulate oil droplets induces vitellogenic growth. Before SPE injection, the eel pituitary contained follicle-stimulating hormone (FSH); however, luteinizing hormone (LH) was not detected by immunohistochemistry (Fig. 12.5). After SPE injection, *fshβ* subunit mRNA expression rapidly decreased and maintained a very low expression level. In contrast, *lhβ* subunit rapidly increased after SPE injection and was maintained at a high level until oocyte maturation (Fig. 12.6). However, eel FSH and LH do not seem necessary for artificial maturation because hypophysectomy eels can complete vitellogenesis and oocyte maturation if SPE is injected weekly and DHP is injected at the migratory nucleus stage (unpublished data). SPE contains salmon LH while FSH content is very low; therefore, salmon LH contained in SPE mainly proceeds via eel vitellogenesis until the migratory nuclear stage.

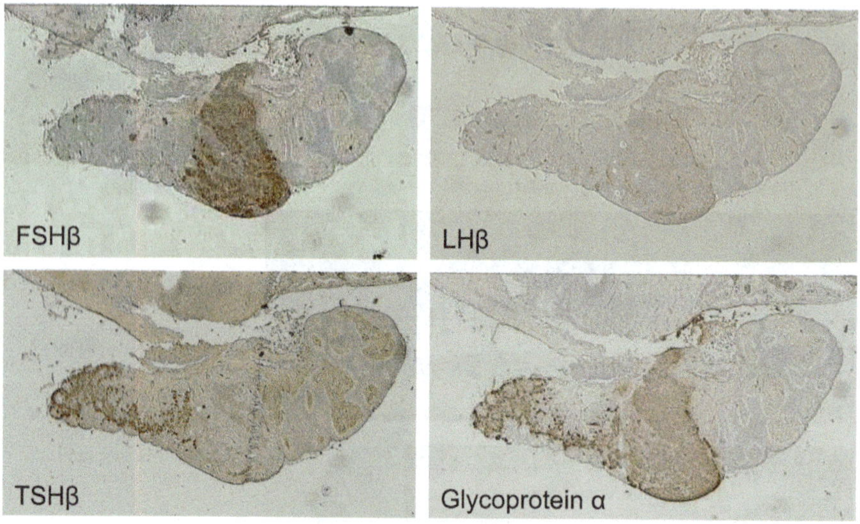

Fig. 12.5 Immunohistological observations of FSHβ, LHβ, TSHβ, and glycoprotein α subunits in eel pituitaries before salmon pituitary extract injection started

SPE was injected weekly from 8 to >15 weeks until the eels reached the migratory nucleus stage. During this period, serum estradiol-17β (E2) and 11KT levels increased linearly as the pulse mode was stimulated by every weekly SPE injection. However, DHP levels are consistently maintained at very low levels during oocyte development until the migratory nucleus stage before the exogenous DHP injection that induced oocyte maturation and ovulation (Adachi et al. 2003). The pivotal role of SPE injection in vitellogenic growth seems to be to stimulate E2 production in the ovary and simultaneously facilitating vitellogenin incorporation into oocytes. The mechanism of vitellogenin incorporation into oocytes remains unknown; however, evidence suggests that it is facilitated by gonadotropin; however, the synthesis of vitellogenin in the liver was facilitated by E2 stimulation. In fact, the expression of *vitellogenin 1* and *2* mRNA levels in the liver increased synchronously with serum E2 elevation during artificial maturation induced by SEP injections (Adachi et al. 2003). To understand how E2 production is induced by SPE injection at the molecular level, cDNAs encoding steroidogenic enzymes responsible for E2 production were isolated, and changes in expression during artificial maturation were investigated. *Cholesterol side-chain cleavage* (*P450scc*) and *steroid-17-α-hydroxylase/*$C_{17\text{–}20}$ *lyase* (*cyp17a1*) mRNA levels were elevated during SPE injections and maintained at high levels during late-vitellogenic and migratory nucleus stages in the ovary. Furthermore, *3β-hydroxysteroid dehydrogenase/Δ5-Δ4 isomerase* (*3β-HSD*) mRNA was expressed at a high level before SPE injection and did not show changes in expression during induced vitellogenic growth. *Cyp19a1* encoding aromatase mRNA increased after SPE injection, reached a peak at the mid-vitellogenic stage, and these high expression levels were maintained until the migratory nucleus stage (Adachi et al. 2003). Taken together, the data suggests that

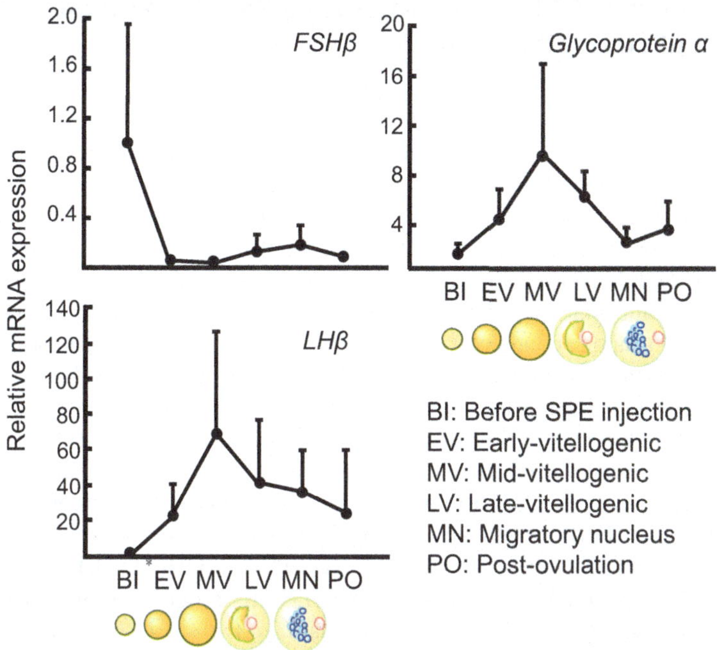

Fig. 12.6 Changes in mRNA levels of FSHβ, LHβ, and glycoprotein α in pituitaries during induced sexual maturation in the eel

weekly SPE injections stimulate the expression of almost all steroidogenic enzymes repeatedly, which facilitates incremental E2 production in the ovary as SPE injections proceed. SPE also likely facilitates the incorporation of vitellogenin into oocytes, and may also play a role in facilitating vitellogenin synthesis in the liver, although clear evidence has not been obtained. During artificial maturation, serum 11KT levels increase consistently. Oocytes continuously incorporate oil droplets inside during vitellogenic growth; therefore, consistent production of 11KT would facilitate the incorporation of oil droplets into oocytes during vitellogenic growth, as seen in pre-vitellogenic oocyte development.

After weekly SPE injections, the eel oocytes reached the migratory nucleus stage (prophase of the first meiotic division), and are still immature. Generally, to proceed with the oocyte at this stage to maturation, stimulation of maturation-inducing steroids (MIS) onto oocytes is required. In the next section, the control of MIS production in relation to LH surge is discussed.

12.4 Control of Maturation-Inducing Steroid Synthesis and Oocyte Maturation

After vitellogenic growth is completed, previously halted meiotic first division in oocytes is re-initiated by stimulation of the LH surge secreted from the pituitary until the second meiotic metaphase, which is called final oocyte maturation. After maturation, oocytes are stripped from follicular cells, which is called ovulation, and then become fertilizable eggs. During fertilization, the egg re-initiates and completes the meiotic second division and initiates ontogenetic development.

The environmental trigger causing the LH surge is unknown during natural eel maturation. In captive conditions, the LH surge does not occur spontaneously; thus, 17α, 20β-dihydroxy-4-pregnen-3-one (DHP) is routinely injected to obtain eel eggs. DHP was first identified in 1985 as a maturation-inducing steroid hormone in amago salmon (*Oncorhynchus masou ishikawae*) (Nagahama and Adachi 1985), and has since been confirmed in Japanese eels (Adachi et al. 2003). In series of *in vivo* and *in vitro* studies concerning DHP production in salmonids, it was demonstrated that LH stimulates DHP production through 20β-hydroxysteroid dehydrogenase (20β-HSD) which converts 17α-hydroxyprogesterone (17OHP) to DHP in granulosa cells, while LH simultaneously stimulates 17OHP production in thecal cells. Recently, we demonstrated for the first time that *17α-hydroxysteroid dehydrogenase type 12-like* (*hsd17b12L*) encodes 20β-HSD, which is responsible for DHP production in granulosa cells during oocyte maturation in *O. masou masou* (Ijiri et al. 2017). The ortholog gene *hsd17b12L* in Nile tilapia (*Oreochromis niloticus*) and amur sturgeon (*Acipenser schrenckii*) were also shown to be responsible for DHP production (Aranyakanont et al. 2020; Hasegawa et al. 2022a, b). We also isolated orthologous genes and investigated their roles in DHP production in the Japanese eel.

The coding region of the Japanese eel *hsd17b12L* was inserted into the mammalian expression vector pSI harboring the SV40 promoter and was subsequently transfected into human embryonic kidney 293 T cells (HEK293T). Eel *hsd17b12L* transfected HEK293T cells in 24-well plate converted over 20% of exogenous 17OHP (100 ng/mL) to DHP, confirming that eel Hsd17b12L also possesses strong 20β-HSD activity against 17OHP. Unlike salmon, in which *hsd17b12L* mRNA expression was limited to the maturing ovarian follicle layer, eel *hsd17b12L* expression was ubiquitously detected in all examined tissues, such as the brain, pituitary, heart, muscle, liver, kidney, ovary, and testis; in contrast, DHP was not detected in the serum throughout all reproductive stages. Throughout vitellogenic growth, *hsd17b12L* expression in ovaries did not show any distinct changes. Before SPE injection at the late oil droplet stage of oocytes, *hsd17b12L* was detected and tended to increase until the early vitellogenic stage after SPE injection. *Hsd17b12L* expression was maintained until the late vitellogenic stage, and then slightly decreased until the migratory nucleus stage; however, the mRNA level was identical to that of the ovary before SPE injection. Despite injecting DHP after completion of

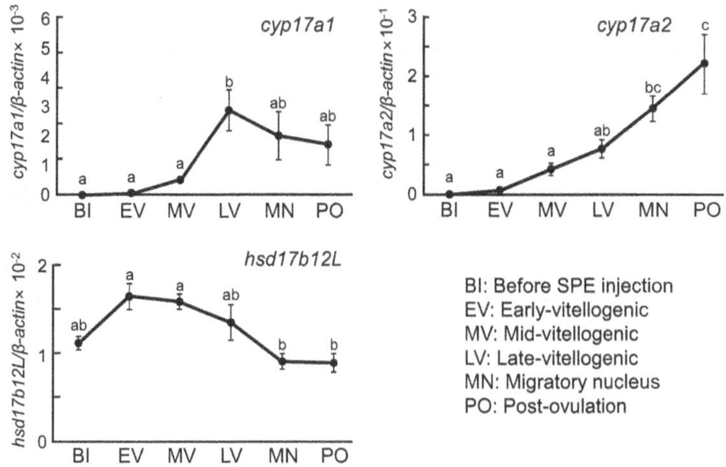

Fig. 12.7 Changes in mRNA levels of *cyp17a1*, *cyp17a2*, and *hsd17b12L* during induced sexual maturation in the eel

vitellogenesis, *hsd17b12L* levels remained at a constant level compared with the late vitellogenic ovary (Fig. 12.7).

The reason that DHP is not detected in eel serum, although all tissues show constant *hsd17b12L* mRNA expression, may be due to the absence of 17OHP production as the precursor of DHP. Progesterone (P4) produces 17OHP via 17-α-hydroxylase, or 17α-hydroxypregnenolone (17P5) via 3β-HSD. Many years ago, we identified Cyp17a1 in Japanese eel and demonstrated that this enzyme rapidly converts pregnenolone (P5) and P4 to dehydroepiandrosterone (DHEA) and androstenedione (A4), respectively. Intermediate metabolites, such as 17P5 or 17OHP, were only detected in a limited time span, when P5 or P4 were added to the incubation medium on COS7 cells transfected with eel *cyp17a1* (Kazeto et al. 2000). This indicates that Cyp17a1 produces DHEA or A4 without accumulating 17P5 or 17OHP. Currently, the mechanisms of 17OHP production are completely unknown, although salmon thecal cells produce substantial amounts of 17OHP during oocyte maturation (Young et al. 1986); this was explained several years ago by the discovery of a novel enzyme, Cyp17a2. Cyp17a2 was first discovered in medaka, followed by Nile tilapia. In contrast to Cyp17a1, which possesses both 17α-hydroxylase and C_{17-20} lyase activities, Cyp17a2 possesses 17α-hydroxylase activity, but lacks C_{17-20} lyase activity, resulting in the production of 17P5 or 17OHP as terminal metabolites. In medaka and tilapia, *cyp17a1* expression decreased instead of increasing *cyp17a2* at the time of oocyte maturation, consequently resulting in 17OHP production (Zhou et al. 2007). In these species, *hsd17b12L* expression was also elevated and 17OHP was converted to DHP, which triggered oocyte maturation.

Based on these new discoveries, we isolated eel *cyp17a2* from Japanese eels. HEK293T cells transfected with pSI-inserted eel *cyp17a2* coding region converted

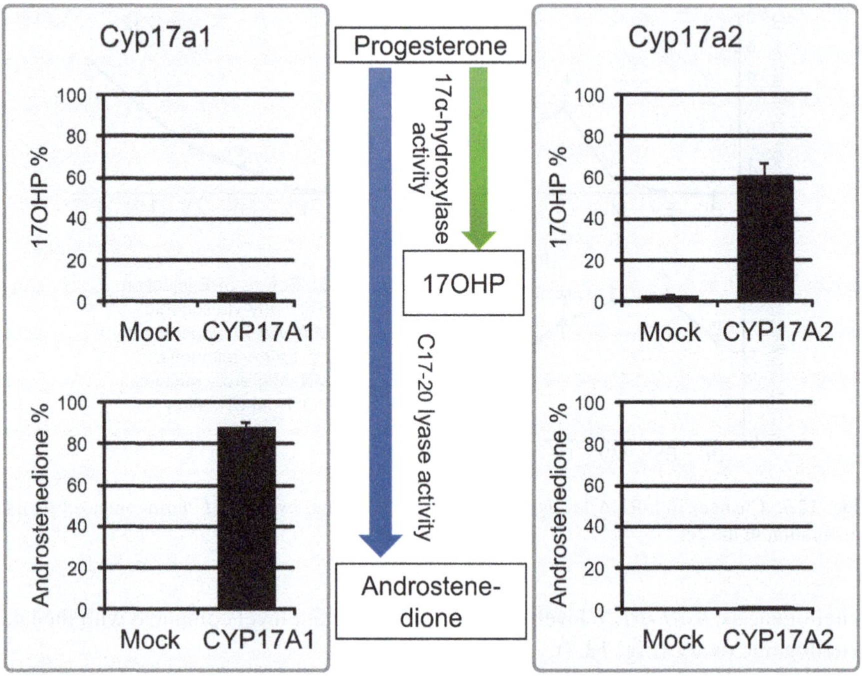

Fig. 12.8 Characterization of enzymatic activities of eel Cyp17a1 and Cyp17a2 that express in HEK293T cells

P4 efficiently to 17OHP; however, it was not further converted to A4, confirming the lack of C_{17-20} lyase activity, similar to medaka and tilapia Cyp17a2. Simultaneously, eel *cyp17a1* completely converted P4 to A4, which confirmed its strong C_{17-20} lyase activity (Fig. 12.8). In eel ovaries, *cyp17a1* expression gradually increased after SPE injection began, reached a peak at the late vitellogenic stage, and was maintained at a high level even after DHP-induced oocyte maturation and ovulation. *Cyp17a2* expression linearly increased in the ovaries after SPE injection until DHP-induced oocyte maturation and ovulation were completed (Fig. 12.7). The expression patterns of *cyp17a1* and *cyp17a2* were very different from those of medaka and tilapia. This is likely because DHP-injected oocyte maturation did not reflect the natural steroidogenic changes that would naturally occur at the beginning of oocyte maturation in the eel.

A decade ago, Dr. Hirohiko Kagawa developed a novel method to induce eel oocyte maturation without injecting DHP. Kagawa demonstrated that a high dose (tenfold) SPE injection (300 mg/kg BW) instead of DHP injection in eels whose oocytes reached the migratory nucleus stage (after vitellogenesis completion) could induce oocyte maturation and ovulation. This procedure implies that regular SPE injection (30 mg/kg BW) can induce vitellogenic growth; however, this injection dose is not sufficient to induce oocyte maturation and ovulation. High-dose SPE injections may mimic the natural LH surge to stimulate oocyte maturation and

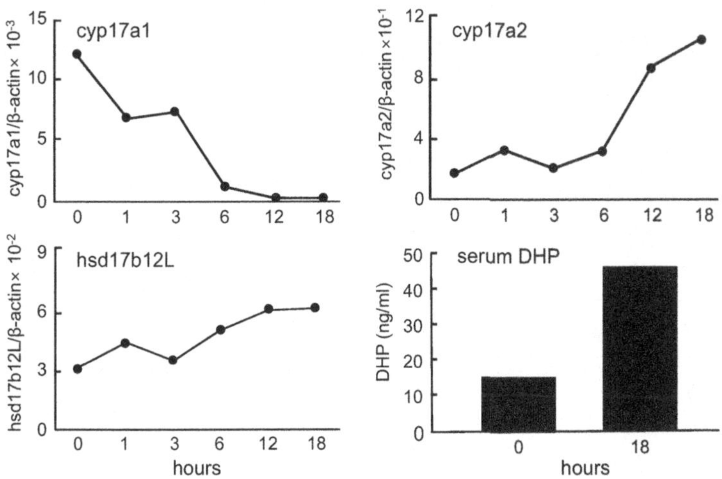

Fig. 12.9 Changes in serum DHP concentrations and mRNA levels of *cyp17a1*, *cyp17a2*, and *hsd17b12L* in the ovary that underwent induced oocyte maturation and ovulation by injection with high-dose salmon pituitary extract (300 mg/kg-BW)

ovulation in eels. This indicates that it is possible to investigate natural steroidogenic shift to induce oocyte maturation in eels if we could obtain biological samples of eels that induced oocyte maturation and ovulation by high-dose SPE injection rather than DHP-induced oocyte maturation.

The steroidogenic shift during oocyte maturation and ovulation in eels induced by high-dose SPE injection was investigated. Eels whose oocytes reached the migratory nucleus stage, over 800 μm in diameter, received a high-dose SPE injection. In a representative case of eels in which oocyte maturation and ovulation were induced, ovulation was completed 18 h after injection. One hour after injection, *cyp17a1* mRNA expression rapidly decreased, reached the lowest at 8 h, and consequently became undetectable at the time of ovulation at 18 h. *Cyp17a2* expression was maintained before the high-dose SPE injection, and then increased linearly from 6 to 18 h. In contrast to these distinct changes in expression, *hsd17b12L* expression showed a gradual increase after injection until 18 h after ovulation was complete. In relation to these gene expression changes, the serum DHP concentration increased from 15 ng/mL to 50 ng/mL during induced oocyte maturation and ovulation (Fig. 12.9). In correlation with the changes in *cyp17a1* and *cyp17a2* mRNA expression, the expression of proteins observed by immunohistochemistry also showed drastic changes. Before injection, Cyp17a1 was detected in the thecal cells; however, immunohistochemically positive cells were not detected in the ovary after ovulation. Cyp17a2 positive cells were also detected in the thecal cells before the injection, and clusters of positive cells were widely observed in post-ovulatory follicular cells after ovulation (Fig. 12.10). These findings show for the first time that high-dose SPE injection mimics the physiological changes induced by the expected natural LH surge that occurs at the time of oocyte maturation. Furthermore, high-dose SPE

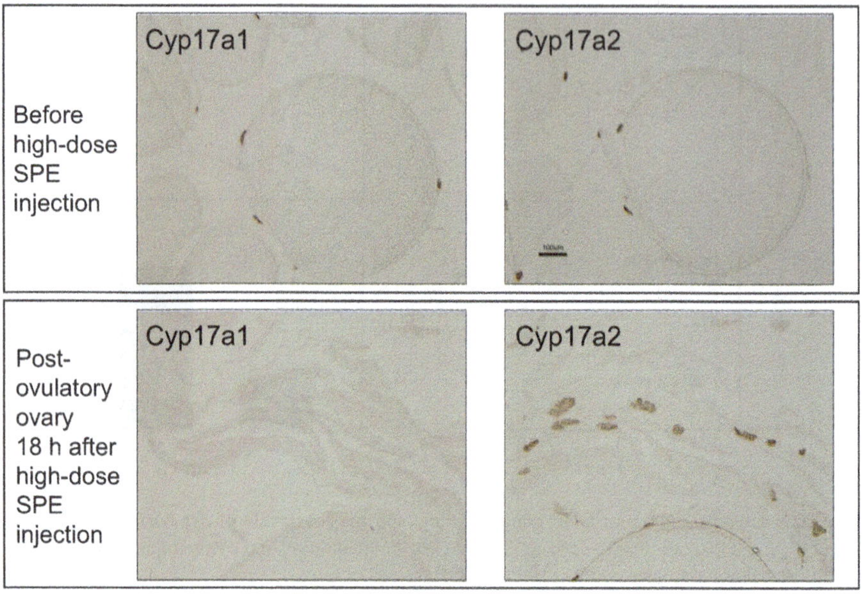

Fig. 12.10 Changes in immunohistochemical expression of Cyp17a1 and Cyp17a2 in the ovary that underwent induced oocyte maturation and ovulation by injection with high-dose salmon pituitary extract (300 mg/kg in body weight)

injection stimulates downregulation of *cyp17a1* expression at the transcriptional level, together with a linear increase in *cyp17a2* expression. These gene expression changes likely enabled the production of 17OHP because of the disappearance of $C_{17–20}$ lyase activity in the ovary. 17OHP enables the production of DHP via Hsd17b12L, which is constantly expressed in the eel ovary. A distinct difference in the mechanism behind DHP production at final maturation in salmon and tilapia is that eel DHP production seems to be regulated by complete downregulation of *cyp17a1* gene expression rather than rapid upregulation of *hsd17b12L* (Fig. 12.11). These observations revealed, for the first time, the mechanism of DHP production stimulated by LH surge to induce oocyte maturation and ovulation in the eel. However, the molecular mechanisms that control the transcription of these key enzymes remain unknown. A complete understanding could be achieved if the action of LH surge is linked with transcriptional regulation of genes encoding steroidogenic enzymes at the molecular level.

12.5 Perspectives

Most cultured eels differentiate into males. Because of the limited availability of naturally differentiating females other than E2 induced females, investigations of morphological and molecular gonadal differentiation are not commonly conducted.

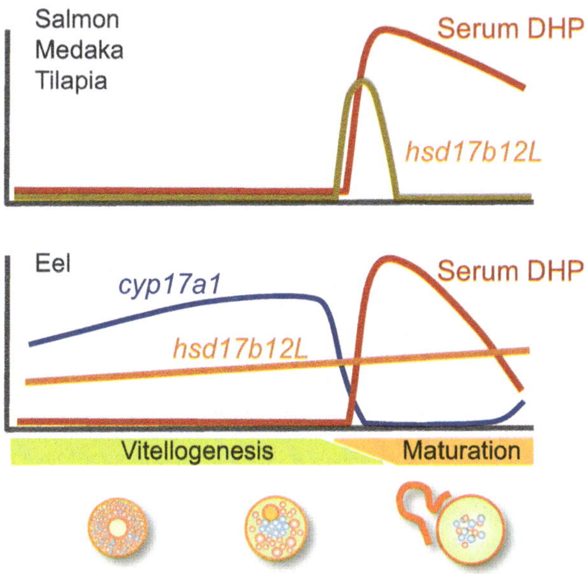

Fig. 12.11 Control of DHP production by *cyp17a1* and *hsd17b12L* expressional changes during oocyte maturation and ovulation in salmon and eel

Our investigation of wild eels reveals a part of natural eel gonadal sex differentiation. Natural morphological gonadal differentiation is clearly understood; however, molecular gonadal differentiation is not. The collection of wild eels is strictly regulated; therefore, wild eels are limited during the period of sex differentiation. The fact that male and female eels exist equally in the wild river indicates that cultivation conditions that lead to 50% female ratio must exist and should be disclosed. These rearing conditions would enable an in-depth investigation of the molecular control of gonadal sex differentiation. After ovarian differentiation, the overall process of oocyte development until oil droplet accumulation is largely understood. After oil droplet accumulation proceeded, the endocrine control to promote vitellogenic growth induced by SPE injection is also largely understood. However, the natural control of vitellogenic growth, which is probably stimulated by eel FSH action, should be investigated in the future. The biggest mystery is that FSH is not secreted under aquaculture conditions, which is the main reason why captive eels do not undergo vitellogenic growth. Unfortunately, this cannot be clarified until the mechanism for the control of synthesis and secretion of FSH is elucidated, which is also currently unknown throughout all fish species. After completion of vitellogenesis, an understanding of the initiation mechanisms of oocyte maturation was also largely developed in our recent study.

Throughout eel gonadal differentiation until completion of maturation, a shift in steroid hormone production, such as E2, 11KT, and DHP, plays a pivotal role in oocyte development. To further understand the gene expression levels involved in steroid production, the regulation of transcription governing steroid production is necessary to achieve a complete understanding of the total molecular controls of eel gonadal development preceding egg production.

References

Adachi S, Ijiri S, Kazeto Y, Yamauchi K (2003) Oogenesis in the Japanese eel. In: Aida K, Tsukamoto K, Yamauchi K (eds) Eel biology. Springer, Tokyo, pp 301–317

Aranyakanont C, Ijiri S, Hasegawa Y, Adachi S (2020) 17β-Hydroxysteroid dehydrogenase type 12 is responsible for maturation-inducing steroid synthesis during oocyte maturation in Nile tilapia. Gen Comp Endocrinol 290:113399. https://doi.org/10.1016/j.ygcen.2020.113399

Chai Y, Tosaka R, Abe T, Sago K, Sago Y, Hatanaka E, Ijiri S, Adachi S (2010) The relationship between the developmental stage of oocytes in various seasons and the quality of the egg obtained by artificial maturation in the feminized Japanese eel *Anguilla japonica*. Aquaculture Sci 58:269–278. https://doi.org/10.11233/aquaculturesci.58.269; in Japanese with English abstract

Endo T, Todo T, Lokman PM, Kudo H, Ijiri S, Adachi S, Yamauchi K (2011) Androgens and very low density lipoprotein are essential for the growth of previtellogenic oocytes from Japanese eel, *Anguilla japonica*, in vitro. Biol Reprod 84:816–825. https://doi.org/10.1095/biolreprod.110.087163

Hasegawa Y, Ijiri S, Surugaya R, Sakai R, Adachi S (2022a) 17β-hydroxysteroid dehydrogenase type 12-like is associated with maturation-inducing steroid synthesis during induced oocyte maturation and ovulation in sturgeons. Aquaculture 564:737238. https://doi.org/10.1016/j.aquaculture.2021.737238

Hasegawa Y, Surugaya R, Adachi S, Ijiri S (2022b) Regulation of 17α-hydroxyprogesterone production during induced oocyte maturation and ovulation in Amur sturgeon (*Acipenser schrenckii*). J Mar Sci Eng 10:86. https://doi.org/10.3390/jmse10010086

Horiuchi M, Hagihara S, Kume M, Chushi D, Hasegawa Y, Itakura H, Yamashita Y, Adachi S, Ijiri S (2022) Morphological and molecular gonadal sex differentiation in the wild Japanese eel *Anguilla japonica*. Cell 11:1554. https://doi.org/10.3390/cells11091554

Ijiri S, Kayaba T, Takeda N, Tachiki H, Adachi S, Yamauchi K (1998) Pretreatment reproductive stage and oocyte development induced by salmon pituitary homogenate in the Japanese eel *Anguilla japonica*. Fish Sci 64:531–537. https://doi.org/10.2331/fishsci.64.531

Ijiri S, Kaneko H, Kobayashi T, Wang DS, Sakai F, Paul-Prasanth B, Nakamura M, Nagahama Y (2008) Sexual dimorphic expression of genes in gonads during early differentiation of a teleost fish, the Nile tilapia *Oreochromis niloticus*. Biol Reprod 78:333–341. https://doi.org/10.1095/biolreprod.107.064246

Ijiri S, Shibata Y, Takezawa N, Kazeto Y, Takatsuka N, Kato E, Hagihara S, Ozaki Y, Adachi S, Yamauchi K, Nagahama Y (2017) 17β-HSD type 12-like is responsible for maturation-inducing hormone synthesis during oocyte maturation in masu salmon. Endocrinology 158:627–639. https://doi.org/10.1210/en.2016-1349

Kazeto Y, Ijiri S, Todo T, Adachi S, Yamauchi K (2000) Molecular cloning and characterization of Japanese eel ovarian P450c17 (CYP17) cDNA. Gen Comp Endocrinol 118:123–133. https://doi.org/10.1006/gcen.1999.7449

Matsubara H (2003) Studies on ovarian steroidogenesis and artificial control of oocyte growth in eel. PhD thesis, 159 pp

Nagahama Y, Adachi S (1985) Identification of maturation-inducing steroid in a teleost, the amago salmon (*Oncorhynchus rhodurus*). Dev Biol 109:428–435. https://doi.org/10.1016/0012-1606(85)90469-5

Shimizu A, Ijiri S, Izumi H, Gen K, Kurogi H, Hashimoto H, Tanaka H, Jinbo T, Saito H, Chow S (2021) Histological evidence of multiple spawning in wild female Japanese eel *Anguilla japonica*. Zool Stud 60:61. https://doi.org/10.6620/zs.2021.60-61

Tsukamoto K, Chow S, Otake T, Kurogi H, Mochioka N, Miller MJ, Aoyama J, Kimura S, Watanabe S, Yoshinaga T, Shinoda A, Kuroki M, Oya M, Watanabe T, Hata K, Ijiri S, Kazeto Y, Nomura K, Tanaka H (2011) Oceanic spawning ecology of freshwater eels in the western North Pacific. Nat Commun 2:179. https://doi.org/10.1038/ncomms1174

Young G, Adachi S, Nagahama Y (1986) Role of ovarian thecal and granulosa layers in gonadotropin-induced synthesis of a salmonid maturation-inducing substance (17α, 20-β-dihydroxy-4-pregnen-3-one). Dev Biol 118:1–8. https://doi.org/10.1016/0012-1606(86)90067-9

Zhou LY, Wang DS, Kobayashi T, Yano A, Paul-Prasanth B, Suzuki A, Sakai F, Nagahama Y (2007) A novel type of P450c17 lacking the lyase activity is responsible for C21-steroid biosynthesis in the fish ovary and head kidney. Endocrinology 148:4282–4291. https://doi.org/10.1210/en.2007-0487

Chapter 13
Metamorphosis and Silvering

Seishi Hagihara and Ryusuke Sudo

Contents

Anguillid eels undergo 2 major external changes during their life cycle: metamorphosis, which occurs when they shift their lifestyle from planktonic life in the ocean to sedentary life in inland water, and silvering, which occurs when they begin their spawning migration to the oceanic spawning area. Both metamorphosis and silvering occur synchronously with lifestyle and habitat changes, and are periods of great importance in the ecology of eels. This chapter provides an overview of metamorphosis and silvering with respect to external and internal changes, and their physiological regulation.

13.1 Metamorphosis

13.1.1 Metamorphosis in Fish

Metamorphosis in fish generally refers to a marked change from the planktonic larval form to a juvenile form that resembles that of an adult fish. It is well known that flatfish (Pleuronectiformes), which have the most asymmetrical bodies of all vertebrates, exhibit metamorphosis. During the larval stage, they have symmetrical

S. Hagihara (✉)
Nanae Freshwater Station, Field Science Center for Northern Biosphere, Hokkaido University, Hokkaido, Japan
e-mail: shagihara@fsc.hokudai.ac.jp

R. Sudo
Minamiizu Field Station, Fisheries Technology Institute, Japan Fisheries Research and Education Agency, Shizuoka, Japan

K. Tsukamoto et al. (eds.), *Eel Science*, Fisheries Science Series,
https://doi.org/10.1007/978-981-99-5692-0_13

bodies similar to those of many other fish, allowing them to maintain symmetry and upright swimming position. Toward the end of the larval stage, 1 eye migrates across the top of the head to the contralateral side of the head, resulting in both eyes located on the same side of the head. Pigmentation of the body also becomes asymmetrical: the ocular side becomes dark with intense pigmentation and the blind side becomes white. Parallel to these morphological changes, their lifestyle passes from free-swimming pelagic plankton feeders to sedentary carnivores, lying on the bottom with both eyes facing up. Anguilliformes are also representative of fish that exhibit drastic morphological changes during the transition from larvae to juveniles. Their larvae, called leptocephali, are morphologically unique. Their body is laterally compressed, leaf-like in appearance, and transparent, and their head is small in relation to body size. During metamorphosis, they transform into a cylindrically shaped pigmented body typical of eels. In association with the transformation of their body shape, their habitat changes from pelagic to benthic in many anguilliform species. For these fish, the period of metamorphosis is often a crucial ecological transition period marking when they join a new habitat and change their mode of life.

13.1.2 External Changes During Metamorphosis and Staging

Since metamorphosing anguillid eels are rarely caught in the ocean, little is known about their metamorphosis in nature. However, with the establishment of artificial seedling production in the Japanese eel *Anguilla japonica*, it is possible to study the metamorphosis process from larvae to juveniles in detail in the laboratory. Under laboratory conditions, fully grown leptocephali reach 50–60 mm in length before commencing metamorphosis (Tanaka et al. 2003), while the size at which metamorphosis begins is considered to be more than 60 mm in the wild (Fukuda et al. 2018).

When metamorphosis begins, the forward-protruding teeth disappear, the face becomes rounded, and the position of the anus and the anterior edges of the dorsal and anal fins move forward (Fig. 13.1). Subsequently, the body length decreases, and the body depth drastically reduces. Concomitantly, a black pigment appears in the caudal region and extends along the lateral line to the head, the gill develop, and eye diameter is reduced. Following these changes, the eels metamorphose into glass eels in an eel-like form. Glass eels become elvers when they become pigmented and then become yellow eels after depositing guanine on the intra-abdominal membrane, causing their belly to turn a yellowish color.

From the onset of metamorphosis to the glass eel stage, the ratio of pre-anal length and body depth to total length (PAL/TL and BD/TL) is often used as a criterion for external metamorphic changes during staging. In many Japanese eel studies, larvae with a PAL/TL $\leq$ 70% are classified as in the metamorphosis stage, whereas juveniles with a sufficiently shrunken body depth (BD/TL $\geq$ 7 or 10%) are classified as the glass eel stage (Mochioka 2003; Kuroki et al. 2010). From the glass eel stage to the yellow eel stage (including the elver stage) pigments of the body

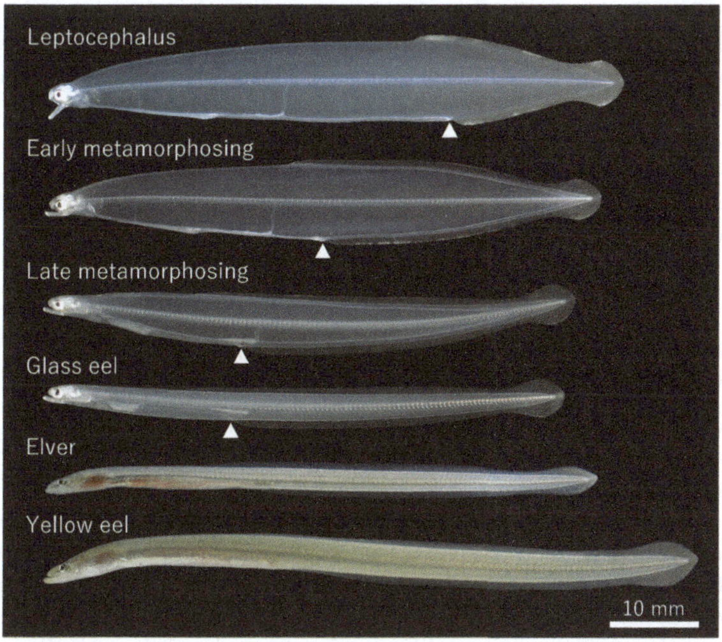

Fig. 13.1 The developmental stages of artificially read Japanese eel from leptocephalus stage to yellow eel stage. Arrow heads show the position of the anus, which moves anteriorly as metamorphosis advances. Bar = 10 mm. Photographed by Rui Hatakeyama

surface and guanine deposition on the intra-abdominal membrane have been used for staging, as described by Fukuda et al. (2013).

13.1.3 Internal Histological Changes During Metamorphosis

Hatakeyama et al. (2021) conducted a detailed histological observation of organogenesis from the leptocephalus to the yellow eel stage. The results showed that during early metamorphosis from the leptocephalus to the glass eel stage, the formation of gills and lateral muscles progressed conspicuously and obvious regression in the esophageal muscle and pancreas occurred concomitantly with a drastic body shrinkage from leaf-like to eel-like. This body shrinkage is linked to a decrease in hyaluronan, which is the main component of the highly hydrated extracellular matrix of leptocephali (Okamura et al. 2018; Kawakami 2002). The formation of the lateral line canals advanced continuously until the yellow eel stage. At the glass eel stage, cone photoreceptor cells appeared, and the formation of the esophageal, stomach, and intestinal muscles was initiated. Differentiation of the gastric glands began after 1 week in glass-eel form. Erythrocyte density increased continuously in glass eels and elvers, and the morphological features of cone cells and olfactory epidermal cells became clearer with stage progression. In the elderly, the swimbladder initiated inflation, the stomach fully expanded, the rectal longitudinal fold changed to a circle, and swimbladder gas glands appeared. Most organ structures formed during the yellow eel stage. In general, completion of metamorphosis is often considered to be up to the glass eel stage. However, as described above, organogenesis in the internal body continues to progress even after the glass eel stage, and thus some studies have defined metamorphosis as the process leading up to the yellow eel stage.

13.1.4 Thyroidal Regulation of Eel Metamorphosis

As is well demonstrated in amphibian metamorphosis, the thyroid axis has been shown to play a major role in the control of flatfish metamorphosis. Under the control of pituitary thyrotropin (thyroid-stimulating hormone [TSH]), thyroid glands produce thyroid hormones (TH; thyroxine [T4]; triiodothyronine [T3]), which act on target organs via specific receptors. During metamorphosis, TSH stimulates the thyroid gland to increase TH in the Japanese flounder *Paralichthys olivaceus*. TH has been shown to cause morphological changes (i.e., left-right asymmetries and shortening dorsal-fin extensions), changes in various internal organs (i.e., gastric gland formation, red blood cells, and adult type of muscle), and behavioral changes (i.e., floating to bottoming) from larvae to juveniles in all aspects considered in *in vivo* and *in vitro* experiments (Inui and Miwa 2012).

The thyroid axis also plays an important role in the regulation of metamorphosis from leptocephalus to juveniles, although endocrinological studies on anguillid eels are limited. Thyroid follicles first appear in larvae 12.2 mm in total length and the number of follicles increases as they grow (Yamano et al. 2007). During metamorphosis, the epithelial-cell height of follicles increases, and the inner colloid of thyroid follicles shows positive immunoreactivity for T4 (Sudo et al. 2014). Histological observations of wild Polynesian shortfin eel *A. obscura* and Indian bicolor eel *A. bicolor bicolor* revealed that TSH-producing cells in the pituitary appear just prior to the initiation of metamorphosis (Ozaki et al. 2000). These histological observations indicate that the development and activation of the thyroid gland are related to metamorphosis in anguillid eels.

Two studies examined TH levels during metamorphosis in Japanese eels (Yamano et al. 2007; Sudo et al. 2014). Yamano et al. (2007) demonstrated that T4 levels continued to increase during metamorphosis; however, T3 reached a maximum level at the late metamorphosis stage and subsequently declined toward the end of metamorphosis. Sudo et al. (2014) reported that both T4 and T3 levels showed moderate but significant increases in the leptocephalus to glass eel stage (Fig. 13.2). Although the pattern of variation differed between the 2 studies, it was confirmed that TH levels increased during metamorphosis in Japanese eels. As for TSH, a small peak in TSHβ mRNA levels was observed; however, they were highest

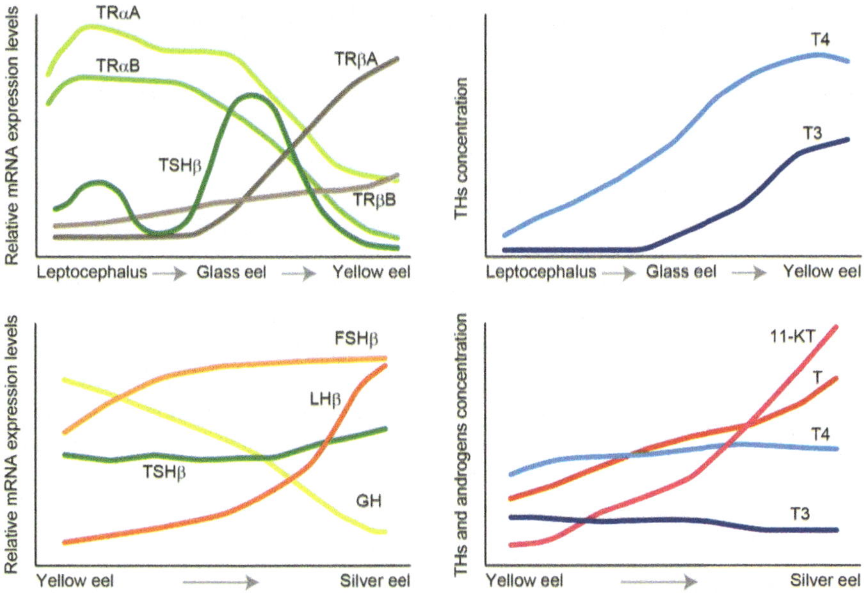

Fig. 13.2 Diagrammatic profiles in relative mRNA expression and concentration of endocrine factors during metamorphosis (upper) and silvering (lower) of Japanese eels

in the glass eel stage, and then decreased markedly in the elver stage (Sudo et al. 2014).

At present, studies administering TH for anguillid eels leptocephali are scarce; however, T4 administration experiments have been conducted in Whitespotted conger *Conger myriaster,* which is a related species of anguillid eels that metamorphoses from leptocephalus. It has been reported that exposure to seawater with T4 at 100 nM can induce metamorphosis in conger eels, while exposure to seawater with thiourea at 5 mM can delay the metamorphosis process (Yamano 2012). These treatments also induced the same effects on Japanese eel leptocephali (Sudo et al. unpublished data).

In higher vertebrates, it has been well known that the pleiotropic effects of THs are mediated through TH receptors (TRs), which are members of the nuclear hormone family. Accordingly, it is assumed that eel metamorphosis is triggered by modification of gene expression by TRs. In Japanese eels, 4 types of TRs (TRαA, TRαB, TRβA, and TRβB) have been identified, and their mRNA expression patterns have been measured during metamorphosis (Kawakami et al. 2013). The mRNA expression levels of TRαA and TRαB increased from the onset of metamorphosis, peaked in the middle stage of metamorphosis, and decreased after the glass eel stage. In contrast, the mRNA expression of TRβA and TRβB increased from the middle stage of metamorphosis to the glass eel stage and maintained high expression levels at the elver stage. The distinctly different mRNA expression patterns between TRs and TRs suggest that the roles of TH are regulated by TRs during metamorphosis.

13.1.5 The Effect of Starvation on Eel Metamorphosis

Although the exogenous factors involved in triggering the metamorphosis of anguillid eels remain unclear, it has been demonstrated in the laboratory that starvation triggers metamorphosis (Okamura et al. 2012). Fully grown larvae with a mean total length of >55 mm were divided into unfed and fed groups and reared for 2 weeks. As a result, more larvae in the unfed group began metamorphosis than in the fed group, suggesting that starvation triggers metamorphosis. In amphibians, starvation induces metamorphosis onset. Stress during starvation stimulates corticotropin-releasing hormone (CRH) secretion from the hypothalamus, leading to pituitary TSH secretion. Subsequently, the T4 concentration in the body increases and allows metamorphosis from tadpoles to frogs. In addition, CRH is also associated with the secretion of cortisol, which is involved in organogenesis during metamorphosis with TH. Further studies are required to understand the relationship between starvation and stress, and the role of cortisol in eel metamorphosis.

13.2 Silvering

13.2.1 Secondary Sexual Characteristics in Fish

Secondary sexual characteristics (SSCs) are features that appear during sexual maturity in animals and, unlike sex organs (primary sexual characteristics), are not a direct part of the reproductive system. SSCs have evolved in many animal species, including fish, with respect to fitness during reproduction. In fish, SSCs are generally significant in males (e.g., tubercles, fin length, and coloration), work in interindividual competition and sexual selection, and are under the control of an androgen, 11- ketotestosterone (11KT) (Borg et al. 1993; Ogino et al. 2016). However, some fish species allow females to develop SSCs, including weaponry and ornamentation (Houde 2001). In addition, several fish species, such as the genera *Tribolodon* and *Anguilla*, show very similar body color changes between sexes at the onset of sexual maturity (Nakamura 1969; Okamura et al. 2007).

13.2.2 Silvering and Related Morphological Changes

Anguillid eel species grow as yellow eels in terrestrial waters for several years to several decades (Hagihara et al. 2018a), and then transform into a silver eel with a metallic luster on its body prior to spawning migration to specific spawning grounds at sea. More precisely, their bodies darken, becoming metallic and dark brown rather than silvery; this color change is known as silvering. Silvering is accompanied by a variety of morphological and physiological changes such as enlargement of the eyes and pectoral fins, development of the swim bladder and heart, and degeneration of the digestive tract (Aoyama and Miller 2003; Durif et al. 2005).

The silvering and related changes are thought to be adaptations to the open ocean environments, which are drastically different from their typical habitats in freshwater or coastal areas. The changes in body coloration may reduce predation. The increase in eye diameter is considered an adaptation for swimming in the low-light environments of the epipelagic and mesopelagic zones. The change in the spectral sensitivity of the retina during silvering also indicates adaptation to different light environments. Elongation of the pectoral fins may contribute to improved swimming ability. The development of the swim bladder improves buoyancy control, which may play a significant role in their vertical diel migration during oceanic migration. Heart development is also thought to play a role in improving circulatory function during long-distance swimming. The degeneration of the digestive tract of silver eels may be related to both the facilitation of osmoregulation and the halting of redundant systems; however, the latter may be more significant because silvering-related digestive tract degeneration is also observed in sea eels that spend their entire lives in the sea.

Silvering-related changes have long been studied in temperate eel species, including the Japanese eel (Matsui 1972; Okamura et al. 2007) and European eel *A. anguilla* (Pankhurst 1982; Durif et al. 2005); however, it has recently become clear that similar changes occur in tropical eel species, such as the Celebes eel *A. celebesensis*, Indo-Pacific eel *A. marmorata,* and bicolor eel *A. bicolor* (Hagihara et al. 2012, 2020a, 2022). Hence, it is highly probable that these changes are general characteristics of anguillid eel species, regardless of the latitude of their growth habitat or their migration distance.

13.2.3 Silvering Stages

The original meaning of silvering is, as the word "silvering" implies, a change in body color; however, various criteria for silvering have been proposed using morphological and/or physiological characteristics. For example, Pankhurst (1982) considered European eels with an eye index higher than 6.5 to be silver eels. Acou et al. (2005) used a combination of multiple morphological characteristics (melanization of lateral line pores, color contrast between the dorsal and ventral skin, and eye index) to distinguish between silver and yellow European eels. Durif et al. (2005) conducted a multivariate analysis using the morphological and physiological indices of European eels and classified the silvering process into multiple stages.

Okamura et al. (2007) classified the silvering stages of Japanese eels into Y1, Y2, S1, and S2 based on their pectoral fins and body coloration in order of progression: Y1, without a metallic hue at the base of pectoral fins; Y2, with a metallic hue at the base of the pectoral fins but without melanization at the tip of pectoral fins; S1, with complete melanization at the tip of pectoral fins, but without a fully black or dark brown pigmented belly; and S2, with a black or dark brown belly. The silvering

Fig. 13.3 The silvering stages of (**a**) *Anguilla japonica*, (**b**) *Anguilla celebesensis*, (**c**) *Anguilla marmorata*, and (**d**) *Anguilla bicolor*, classified according to the silvering index of Okamura et al. (2007): Y1, Y2, S1, and S2. (Images were modified from Okamura et al. (2007) and Hagihara et al. (2012, 2020a)). Bar = 50 mm

index is now used in many ecological studies of Japanese eels because it can determine the silvering stages without dissection or in-depth analysis; therefore, it is highly convenient for field studies. Furthermore, the silvering index has some validity for other anguillid species such as *A. celebesensis*, *A. marmorata* and *A. bicolor* (Fig. 13.3; Hagihara et al. 2012, 2020a, 2022).

13.2.4 Seasonality of Silvering

Silvering and downstream migration occur mainly from autumn to winter in temperate anguillid species (Todd 1981a; Okamura et al. 2002; Durif and Elie 2008; Sudo et al. 2017). Treatment with decreasing water temperature increases the blood levels of 11KT, which is involved in silvering, gonadal development, and downstream migration (Sudo et al. 2011a; Sudo and Tsukamoto 2013). Thus, decreasing water temperature in autumn is thought to be 1 of the main factors regulating the seasonality of silvering and downstream migration of temperate eel species.

Tropical areas have less seasonality in terms of day length and water temperature than temperate areas, so it would be interesting to know whether seasonality exists in the silvering and downstream migration of tropical eels; however, the seasonality of silvering in tropical eels remains unknown. Regarding the seasonality of gonadal development in the bicolor eel *A. bicolor* and Bengal eel *A. bengalensis*, it has been reported that relatively high GSI are observed in individuals for almost the entire year (Arai and Abdul Kadir 2017). In contrast, even in areas with minimal seasonal variation in day length and water temperature, there was seasonality of downstream migration of tropical eels associated with the rainy and dry season changes that typically characterize tropical climates. In the Poso Lake system on Sulawesi Island, Indonesia, most *A. celebesensis* begin downstream migration during the early to middle rainy season, while *A. marmorata* migrates almost year-round, with a peak from the late rainy to middle dry season (Hagihara et al. 2018b). The physiological and ecological mechanisms underlying silvering and downstream migration in tropical eels and environmental stimuli are important in order to understand the evolution of downstream migration in anguillid eels.

13.2.5 Silvering and Gonadal Development

The gonads of yellow eels remain relatively immature for a long period. Spawning migration begins after puberty (onset of sexual maturation) and silvering. The gonads of silver eels of anguillid species are distinctly at more advanced developmental stages than those of yellow eels. Many previous studies have reported that temperate eels (American eel, Japanese eel, and Australian eel *A. australis*) start their downstream migrations at the oil droplet or early vitellogenic stage (Todd 1981b; Cottrill et al. 2001; Sudo et al. 2011b), and European eels start migrating at the oil droplet stage (Palstra et al. 2011). The subsequent oogenesis process is thought to advance during oceanic spawning migration after downstream migration. After arrival at the spawning area, oocyte maturation and ovulation occur, leading to spawning events.

The developmental status of the gonads of silver eels varies among species, which may be associated with the migration scale of each. Two New Zealand eels, the New Zealand longfin eel *A. dieffenbachii* and *A. australis*, showed a difference in average GSI values at the beginning of their spawning migration, with the females of the former species having a GSI of 8.1 and the latter 3.5 (Todd 1981b). This difference in the level of sexual maturation may indicate that *A. dieffenbachii* migrates a shorter distance to spawn than *A. australis*. Comparing European and Japanese eels, the ovarian development of silver eels is relatively advanced in Japanese eels, which migrate shorter distances to their spawning areas (Japanese eel: ~3000 km; European eel: ~6000 km). In addition, *A. celebesensis* begin their spawning migration with remarkably developed gonads (GSI: 3.3–11.4;

Fig. 13.4 Typical examples of ventral dissection of female migrating (silver-phase) eels, *Anguilla celebesensis* (upper) and *Anguilla japonica* (lower). Arrowed lines: ovaries

> 60% of individuals are at the midvitellogenic stage) (Hagihara et al. 2012, 2020b; Fig. 13.4); notably, in the case of Tomini Bay, this species could likely migrate the shortest distance within the genus *Anguilla*. It is probable that anguillid eel species have adapted their gonadal development levels at the start of migration during their evolutionary change from primitive small-scale to derivative large-scale migrations, depending on the time required for spawning migration and the migratory environment (e.g., water temperature), to allow them to reach the spawning area at the appropriate maturity conditions for spawning.

13.2.6 Silvering and Behavioral Change

With the silvering of the eels, a drastic change in their behavior has been observed. Silver-phase Japanese eels exhibit higher locomotor activity and reduced negative phototactic behavior during their spawning migration season compared to that of yellow eels (Sudo and Tsukamoto 2015). This increase in locomotor activity of eels is similar to the migratory restlessness in migratory birds, in which migratory birds exhibit high locomotor activity during their migratory season, which is thought to be indicative of the urge to migrate. Therefore, the high locomotor activity in silver eels is probably restless, reflecting the internal motivation of eels to start spawning migration. In addition, the administration of 11KT induced high locomotor activity levels in non-migrating yellow eels, suggesting that this hormone may be directly involved in motivating silver eels for spawning migration (Setiawan et al. 2012; Sudo and Tsukamoto 2015).

13.2.7 Physiological Mechanisms of Silvering

During silvering, the thyrotrophic and somatotropic axes are stable and/or inactivated (Sudo and Yada 2020). For example, the mRNA expression of Gh decreased and thyroid hormones (T4 and T3) remain stable during silvering (Fig. 13.2; Sudo et al. 2015). In contrast, the gonadotropic axis is activated during silvering. Both transcription of the Fshβ subunit (*fsh*β) and Lhβ subunit (*lh*β), and plasma levels of androgens, increase during silvering (Fig. 13.2; Sudo et al. 2011b). Among androgens, 11KT was significantly increased in both sexes; 11KT is required in eels not only for spermatogenesis in males but also for oil droplet accumulation during oogenesis in females (Endo et al. 2011). The fact that artificially decreased water temperature treatment increases blood 11KT levels in yellow eels and that 11KT administration to yellow eels induces oogenesis and silvering-related changes (early stage oocyte growth, enlargement of eyes, degeneration of the digestive tract, and development of the swim bladder) suggests that the decreasing water temperature in autumn stimulates 11KT production and results in silvering (Sudo et al. 2011a, 2012; Sudo and Tsukamoto 2013). Furthermore, 11KT administration to yellow eels induced an increase in their locomotor activity, which could be considered migratory restlessness, suggesting that 11KT is also closely related to increasing the drive for spawning migration (Sudo and Tsukamoto 2015).

As described above, silvering is accompanied by morphological and functional changes for spawning migration, gonadal development for spawning, and an increase in the drive for spawning migration, which seem to be common traits in anguillid eel species. Notably, 11KT is involved in all of these processes, and is thus considered to be an extremely important hormone for the silvering and spawning migration of eels. However, the endocrinological regulatory mechanism of 11KT elevation that occurs during silvering in autumn remains poorly understood. Further research to clarify the physiological and ecological mechanisms of 11KT elevation related to silvering will greatly enhance our understanding of the silvering and migration of anguillid eels.

References

Acou A, Boury P, Laffaille P, Crivelli AJ, Feunteun E (2005) Towards a standardized characterization of the potentially migrating silver European eel (*Anguilla anguilla* L.). Archiv fur Hydrobiologie 164:237–255. https://doi.org/10.1127/0003-9136/2005/0164-0237

Aoyama J, Miller MJ (2003) The silver eel. In: Tsukamoto K, Yamauchi K (eds) Aida K. Eel biology, Springer-Verlag, pp 107–117

Arai T, Abdul Kadir S (2017) Opportunistic spawning of tropical anguillid eels *Anguilla bicolor bicolor* and *A. bengalensis bengalensis*. Sci Rep 7:41649. https://doi.org/10.1038/srep41649

Borg B, Antonopoulou E, Andersson E, Carlberg T, Mayer I (1993) Effectiveness of several androgens in stimulation kidney hypertrophy, a secondary sexual character, in castrated male three-spined sticklebacks, *Gasterosteus aculeatus*. Can J Zool 71:2327–2329. https://doi.org/10.1139/z93-326

Cottrill RA, McKinley RS, van der Kraak G, Dutil JD, Reid KB, McGrath KJ (2001) Plasma non-esterified fatty acid profiles and 17β-oestradiol levels of juvenile immature and maturing adult American eels in the St Lawrence River. J Fish Biol 59:364–379. https://doi.org/10.1111/j.1095-8649.2001.tb00136.x

Durif CMF, Elie P (2008) Predicting downstream migration of silver eels in a large river catchment based on commercial fishery data. Fish Manag Ecol 15:127–137. https://doi.org/10.1111/j.1365-2400.2008.00593.x

Durif C, Dufour S, Elie P (2005) The silvering process of *Anguilla anguilla*: a new classification from the yellow resident to the silver migrating stage. J Fish Biol 66:1025–1043. https://doi.org/10.1111/j.0022-1112.2005.00662.x

Endo T, Todo T, Lokman PM, Kudo H, Ijiri S, Adachi S, Yamauchi K (2011) Androgens and very low density lipoprotein are essential for the growth of previtellogenic oocytes from Japanese eel, *Anguilla japonica*, in vitro. Biol Reprod 84:816–825. https://doi.org/10.1095/biolreprod.110.087163

Fukuda N, Miller MJ, Aoyama J, Shinoda A, Tsukamoto K (2013) Evaluation of the pigmentation stages and body proportions from the glass eel to yellow eel in *Anguilla japonica*. Fish Sci 79: 425–438. https://doi.org/10.1007/s12562-013-0621-x

Fukuda N, Kurogi H, Ambe D, Chow S, Yamamoto T, Yokouchi K, Shinoda A, Masuda Y, Sekino M, Saitoh K, Masujima M, Watanabe T, Mochioka N, Kuwada H (2018) Location, size and age at onset of metamorphosis in the Japanese eel *Anguilla japonica*. J Fish Biol 92:1342–1358. https://doi.org/10.1111/jfb.13590

Hagihara S, Aoyama J, Limbong D, Tsukamoto K (2012) Morphological and physiological changes of female tropical eels, *Anguilla celebesensis* and *Anguilla marmorata*, in relation to downstream migration. J Fish Biol 81:408–426. https://doi.org/10.1111/j.1095-8649.2012.03332.x

Hagihara S, Aoyama J, Limbong D, Tsukamoto K (2018a) Age and growth of migrating tropical eels, *Anguilla celebesensis* and *Anguilla marmorata*. J Fish Biol 92:1526–1544. https://doi.org/10.1111/jfb.13608

Hagihara S, Aoyama J, Limbong D, Tsukamoto K (2018b) Interspecific difference in downstream migratory season between two tropical eels, *Anguilla celebesensis* and *Anguilla marmorata*. J Fish Biol 93:729–732. https://doi.org/10.1111/jfb.13750

Hagihara S, Aoyama J, Limbong D, Tsukamoto K (2020a) Morphological, ecological and physiological characteristics of downstream migrating and non-migrating Pacific bicolor eels *Anguilla bicolor pacifica*. J Fish Biol 97:1842–1845. https://doi.org/10.1111/jfb.14528

Hagihara S, Aoyama J, Sudo R, Limbong D, Ijiri S, Adachi S, Tsukamoto K (2020b) Reproductive physiological characteristics of tropical Celebes eels *Anguilla celebesensis* in relation to downstream migration and ovarian development. J Fish Biol 96:558–569. https://doi.org/10.1111/jfb.14231

Hagihara S, Wakiya R, Maeda T, Kimura S (2022) Morphological and gonadal histological characteristics of the silver-phase male *Anguilla marmorata*. J Fish Biol 101:749–752. https://doi.org/10.1111/jfb.15139

Hatakeyama R, Sudo R, Yatabe T, Yamano K, Nomura K (2021) Developmental features of Japanese eels, *Anguilla japonica*, from the late leptocephalus to the yellow eel stages: an early metamorphosis to the eel-like form and a prolonged transition to the juvenile. J Fish Biol 100:454–473. https://doi.org/10.1111/jfb.14956

Houde AE (2001) Sex roles, ornaments, and evolutionary explanation. Proc Natl Acad Sci U S A 98:12857–12859. https://doi.org/10.1073/pnas.241503598

Inui Y, Miwa S (2012) Metamorphosis of flatfish (Pleuronectiformes). In: Dufour S, Rousseau K, Kapoor BG (eds) Metamorphosis in fish. CRC Press, New Hampshire, pp 107–153

Kawakami Y (2002) Metabolism of hyaluronic acid during early development of the Japanese eel, *Anguilla japonica*. Comp Biochem Physiol A 268:111203. https://doi.org/10.1016/j.cbpa.2022.111203

Kawakami Y, Nomura K, Ohta H, Tanaka H (2013) Characterization of thyroid hormone receptors during early development of the Japanese eel (*Anguilla japonica*). Gen Comp Endocrinol 194: 300–310. https://doi.org/10.1016/j.ygcen.2013.09.020

Kuroki M, Fukuda N, Yamada Y, Okamura A, Tsukamoto K (2010) Morphological changes and otolith growth during metamorphosis of Japanese eel leptocephali in captivity. Coast Mar Sci 34:31–38. https://doi.org/10.15083/00040668

Matsui I (1972) Mangaku. An eel science. Kouseisha-Kouseikaku, Tokyo; in Japanese

Mochioka N (2003) Leptocephali. In: Aida K, Tsukamoto K, Yamauchi K (eds) Eel biology. Springer, pp 51–60

Nakamura M (1969) Cyprinid fishes of Japan: studies on the life history of cyprinid fishes of Japan. Research Institute of Natural Resources, Tokyo; in Japanese

Ogino Y, Kuraku S, Ishibashi H, Miyakawa H, Sumiya E, Miyagawa S, Matsubara H, Yamada G, Baker ME, Iguchi T (2016) Neofunctionalization of androgen receptor by gain-of-function mutations in teleost fish lineage. Mol Biol Evol 33:228–244. https://doi.org/10.1093/molbev/msv218

Okamura A, Yamada Y, Tanaka S, Horie N, Utoh T, Mikawa N, Akazawa A, Oka HP (2002) Atmospheric depression as the final trigger for the seaward migration of the Japanese eel *Anguilla japonica*. Mar Ecol Prog Ser 234:281–288. https://doi.org/10.3354/meps234281

Okamura A, Yamada Y, Yokouchi K, Horie N, Mikawa N, Utoh T, Tanaka S, Tsukamoto K (2007) A silvering index for the Japanese eel *Anguilla japonica*. Environ Biol Fish 80:77–89. https://doi.org/10.1007/s10641-006-9121-5

Okamura A, Yamada Y, Mikawa N, Horie N, Tsukamoto K (2012) Effect of starvation, body size, and temperature on the onset of metamorphosis in Japanese eel (*Anguilla japonica*). Can J Zool 90:1378–1385. https://doi.org/10.1139/cjz-2012-0146

Okamura A, Sakamoto Y, Yamada Y, Tsukamoto K (2018) Accumulation of hyaluronan in read Japanese eel *Anguilla japonica* during early ontogeny. Aquaculture 497:220–225. https://doi.org/10.1016/j.aquaculture.2018.07.066

Ozaki Y, Okumura H, Kazeto Y, Ikeuchi T, Ijiri S, Nagae M, Adachi S, Yamauchi K (2000) Developmental changes in pituitary-thyroid axis, and formation of gonads in leptocephali and glass eels of *Anguilla* spp. Fish Sci 66:1115–1122. https://doi.org/10.1111/j.1444-2906.2005.00938.x

Palstra AP, Guerrero MA, De Laak G, Breteler JPGK, Van den Thillart GEEJM (2011) Temporal progression in migratory status and sexual maturation in European silver eels during downstream migration. Fish Physiol Biochem 37:285–296. https://doi.org/10.1007/s10695-011-9496-x

Pankhurst NW (1982) Relation of visual changes to the onset of sexual maturation in the European eel *Anguilla anguilla* (L.). J Fish Biol 21:127–140. https://doi.org/10.1111/j.1095-8649.1982.tb03994.x

Setiawan AN, Wylie MJ, Forbes EL, Lokman PM (2012) The effect of 11-ketotesotrerone on occupation of downstream location and seawater in the New Zealand shortfinned eel, *Anguilla australis*. Zool Sci 29:1–5. https://doi.org/10.2108/zsj.29.1

Sudo R, Tsukamoto K (2013) The onset mechanisms of the spawning migrations of anguillid eels. In: Ueda H, Tsukamoto K (eds) Physiology and ecology of fish migration. CRC Press, pp 56–80

Sudo R, Tsukamoto K (2015) Migratory restlessness and the role of androgen for increasing behavioral drive in the spawning migration of the Japanese eel. Sci Rep 5:17430. https://doi.org/10.1038/srep17430

Sudo R, Yada Y (2020) Anguillid eels as model species for understanding endocrinological influences on the onset of spawning migration of fishes. Biology 11:934. https://doi.org/10.3390/biology11060934

Sudo R, Tosaka R, Ijiri S, Adachi S, Suetake H, Suzuki Y, Horie N, Tanaka S, Aoyama J, Tsukamoto K (2011a) The effect of temperature decrease on oocyte development, sex steroids and gonadotropin β-subunit mRNA expression levels in female Japanese eels *Anguilla japonica*. Fish Sci 77:575–582. https://doi.org/10.1007/s12562-011-0358-3

Sudo R, Suetake H, Suzuki Y, Utoh T, Tanaka S, Aoyama J, Tsukamoto K (2011b) Dynamics of reproductive hormones during downstream migration in females of the Japanese eel, *Anguilla japonica*. Zool Sci 28:180–188. https://doi.org/10.2108/zsj.28.180

Sudo R, Tosaka R, Ijiri S, Adachi S, Aoyama J, Tsukamoto K (2012) 11-ketotestosterone synchronously induces oocyte development and silvering related changes in the Japanese eel, *Anguilla japonica*. Zool Sci 29:254–259. https://doi.org/10.2108/zsj.29.254

Sudo R, Okamura A, Kuroki M, Tsukamoto K (2014) Changes in the role of the thyroid axis during metamorphosis of the Japanese eel, *Anguilla japonica*. J Exp Zool A 321:357–364. https://doi.org/10.1002/jez.1861

Sudo R, Okamura A, Tsukamoto K (2015) Profiles of thyroid hormones and mRNA expression for thyroid stimulating hormone in Japanese eel downstream migration. Coast Mar Sci 38:1–7. https://doi.org/10.15083/00040614

Sudo R, Okamura A, Fukuda N, Miller MJ, Tsukamoto K (2017) Environmental factors affecting the onset of spawning migrations of Japanese eels (*Anguilla japonica*) in Mikawa Bay Japan. Environ Biol Fish 100:237–249. https://doi.org/10.1007/s10641-017-0575-4

Tanaka H, Kagawa H, Ohta H, Unuma T, Nomura K (2003) The first production of glass eel in captivity: fish reproductive physiology facilitates great progress in aquaculture. Fish Physiol Biochem 28:493–497. https://doi.org/10.1023/B:FISH.0000030638.56031.ed

Todd PR (1981a) Timing of periodicity of migrating New Zealand freshwater eels (*Anguilla* spp.). NZ J Mar Freshw Res 15:225–235. https://doi.org/10.1080/00288330.1981.9515915

Todd PR (1981b) Morphometric changes, gonad histology, and fecundity estimates in migrating New Zealand freshwater eels (*Anguilla* spp.). NZ J Mar Freshw Res 15:155–170. https://doi.org/10.1080/00288330.1981.9515908

Yamano K (2012) Metamorphosis of elopomorphs. In: Dufour S, Rousseau K, Kapoor BG (eds) Metamorphosis in fish. CRC Press, New Hampshire, pp 76–106

Yamano K, Nomura K, Tanaka H (2007) Development of thyroid gland and changes in thyroid hormone levels in leptocephali of Japanese eel (*Anguilla japonica*). Aquaculture 270:499–504. https://doi.org/10.1016/j.aquaculture.2007.05.033

Part IV
Applied Science

Chapter 14
Artificial Maturation

Shinji Adachi, Moemi Horiuchi, and Toshiomi Tanaka

Contents

The Japanese eel, *Anguilla japonica* is one of the most important species in the Japanese and East Asian aquaculture industries. Although the supply of Japanese eels to aquaculture farms is fully dependent on wild-caught glass eels, their stocks are rapidly declining. Therefore, it is necessary to establish artificial seedling production technology; however, Japanese eels do not mature in captivity. Because the pituitary gland of captive eels secretes gonadotropins, artificial hormone treatment is essential to induce sexual maturation (Ijiri et al. 2011). In this chapter, we review previous studies on artificial maturation in Japanese eels together with our results.

S. Adachi (✉)
Graduate School of Fisheries Sciences, Hokkaido University, Hokkaido, Japan
e-mail: s-adachi@fish.hokudai.ac.jp

M. Horiuchi
IRAGO Institute Co., Ltd., Aichi, Japan

T. Tanaka
Hamanako Branch, Shizuoka Prefectural Research Institute of Fishery and Ocean, Shizuoka, Japan

K. Tsukamoto et al. (eds.), *Eel Science*, Fisheries Science Series,
https://doi.org/10.1007/978-981-99-5692-0_14

14.1 Gonadal Stage of Broodstock Eels Before the Induction of Artificial Maturation

Attempts to induce the artificial maturation of the Japanese eel began in the 1960s. Yamamoto and Yamauchi (1974) were the first to successfully obtain fertilized eggs and hatched larvae of Japanese eels using hormone treatment. The subsequently obtained larvae reached a total length of 7 mm after 2 weeks of rearing (Yamauchi et al. 1976). At that time, both female and male wild silver eels were used as broodstocks for artificial maturation. The gonadal stage of silver eels involves early vitellogenesis in females and early spermatogenesis in males. However, because of difficulties in securing a stable supply of silver eels, cultivated eels were used in addition to silver eels. Since most cultivated eels are male, oral estradiol-17β (E2) administration enables stable production of feminized eels to avoid a male-biased sex ratio in cultivated eels in captivity (Tachiki et al. 1997; Ijiri et al. 1998).

It is empirically known that feminized eels are less likely to produce good-quality eggs as they age; thus, we investigated the seasonal changes in ovarian development in each age group. Two-, 3- and 6-year-old feminized eels (mean weight: 550 g, 920 g, 1220 g, respectively) were reared with feeding in freshwater at high temperatures (23–28 °C) and natural day length. The oocyte diameter in the 2-year-old fish was significantly smaller in May but gradually increased thereafter, and conversely tended to be larger in October than that in the others, maintaining a high value until April (Fig. 14.1a). Although oocytes in the early oil droplet stage were observed in May, many oocytes were in the late oil droplet stage after October and some reached

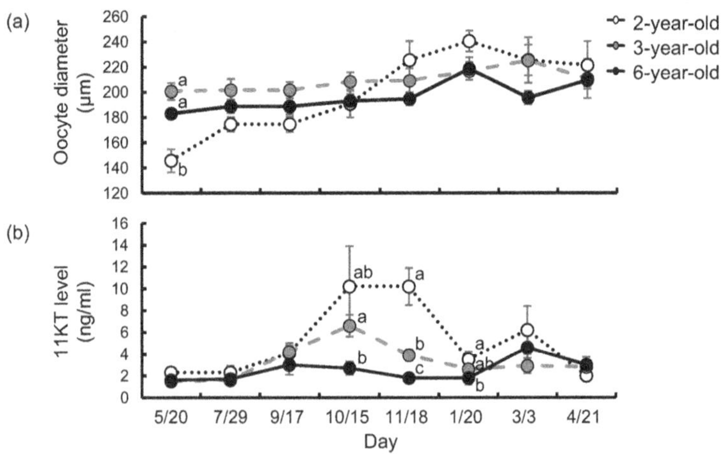

Fig. 14.1 Seasonal changes in (**a**) oocyte diameter (mean ± SE) and (**b**) serum 11KT levels (mean ± SE) in feminized Japanese eels of each age group under high temperature (23–28 °C) and natural day length. Statistically significant differences between groups are indicated by different letters

the early vitellogenic stage. However, oocytes at the oil droplet stage in 3-year-old fish regressed, except in some individuals in May. Lipidation of oocyte has been shown to be promoted by 11-ketotestosterone (11KT) (Endo et al. 2011). Serum 11KT levels were ~2 ng/mL in each group until July; however, that in 6-year-old fish barely increased from October to November, while that in 2-year-old fish was ~10 ng/mL (Fig. 14.1b). Furthermore, it was discovered that 2-year-old fish of sufficient size reach puberty even under high water-temperature conditions and that ovarian recrudescence is less likely to progress as they grow older. In males, hormone treatment is not required for sex control, but it is not clear which age of fish is suitable for artificial maturation and whether older fish produce insufficient results (females).

14.2 Method for the Induction of Artificial Maturation

At the start of artificial maturation, wild silver eels were injected with salmon pituitary homogenate (SPH) and commercially available human chorionic gonadotropin (hCG) in females and males, respectively. Weekly intramuscular injections of SPH can induce oocyte growth, and fully grown oocytes at the migratory nucleus stage can be obtained after 8–15 injections. However, final oocyte maturation (FOM; re-initiation of meiosis) and ovulation are not induced spontaneously in most cases. Nagahama and Adachi (1985) identified 17α, 20β-dihydroxy-4-pregnen-3-one (DHP) for the first time in vertebrates as a maturation-inducing steroid (MIS) in amago salmon. Following this finding, Yamauchi (1990) succeeded in inducing FOM and ovulation with a high probability by administering DHP to female silver eels that have ovaries containing fully grown oocytes at the migratory nucleus stage.

In the late 1990s, methods for artificial maturation of both females and males were improved, and a sufficient number of eggs were consistently obtained (Ohta and Izawa 1996; Ohta et al. 1997; Kagawa et al. 1997, 1998; Unuma et al. 2012). Briefly, after acclimation to seawater, feminized eels were intraperitoneally injected with salmon pituitary extract (SPE), but not SPH, every week. Female eels that gained 5–10% body weight were injected with SPE for priming 2 days after the day of weekly injection. An injection of DHP or its precursor (17-α-hydroxyprogesterone; 17αOHP), administered 24 h after a priming injection of SPE, successfully induced FOM and ovulation in almost all female eels with oocytes at the migratory nucleus stage. Typically, most fish ovulate 14–18 h after DHP injection. In the current protocols, vitellogenesis is induced at 15–20 °C, and FOM and ovulation at 20–23 °C. Therefore, in seedling production facilities, eels are usually kept at 15–20 °C during weekly SPE injections and at 20–23 °C from priming with SPE to egg collection. Cultivated male eels were injected intraperitoneally with human chorionic gonadotropin (hCG) weekly. Sperms were repeatedly obtained by stripping spermiating males, diluted 100 times with artificial seminal plasma, and used for artificial fertilization within 3 weeks. Instead of artificial insemination, fertilized eggs can be obtained by spontaneous spawning using

1 female and 3 males after DHP injection (Horie et al. 2008; Okamura et al. 2009). Furthermore, leptocephali were successfully produced (Tanaka et al. 2001) and subsequently metamorphosed into glass eels (Tanaka et al. 2003). The full life cycle of artificially produced Japanese eels was successfully completed in 2010 (Masuda et al. 2012).

14.3 Seasonal Effects on the Induction of Artificial Maturation

Despite these efforts, glass eels have never been commercially produced, as their survival rates from eggs to the leptocephali and glass eel stages are poor. One of the obstacles to mass production of glass eels is the unstable quality of eggs obtained from females induced by artificial maturation. In addition, female eels have less developed ovaries than silver eels, and better-quality eggs are obtained in autumn than in spring. Especially in spring, it is known that there are a considerable number of individuals who do not respond to SPE and hardly develop ovaries. Oocytes that develop from autumn to winter may regress owing to subsequent continuous rearing in captivity. It has also been suggested that artificial maturation in spring can be improved by controlling the environment (i.e., water temperature and salinity) during winter. Environmental manipulation may be effective in supplying broodstock with developed ovaries throughout the year to significantly improve egg quality.

Therefore, we examined winter rearing conditions to prevent the regression of oocytes. Female eels were reared in freshwater with feeding under natural day length at 26–28 °C water temperature. The experimental subjects consisted of 3 groups: a group reared in freshwater at 26–28 °C until early December while being fed, and acclimated to seawater at 15 °C within 1 week without feeding (Dec); a group reared in freshwater at 26–28 °C until February while being fed, and similarly acclimated to seawater at 15 °C (Feb); and a group reared in freshwater at 26–28 °C until March 28 while being fed (fresh). In this experiment, the water temperature was maintained at a higher level than that in the above experiment. Changes in the water temperature for 3 groups are shown in Fig. 14.2a. The fresh group was also acclimated to 15 °C seawater from March 28, and ovaries of all groups were biopsied on April 5 and 6 for histological observation and to measure oocyte diameter. Figure 14.2b shows the histology of the average-sized oocytes in each group. Regarding the stage of oocytes, in the 2 groups other than the fresh group, there were many individuals before and after the start of vitellogenesis; however, individual differences were significant, and some individuals had oocytes in the early vitellogenic stage, equivalent to silver eels (Fig. 14.2b, d). However, in the fresh group, many oocytes did not start vitellogenesis and some individuals showed regressive oocytes. After that, sexual maturation was artificially induced in all fish at 15 °C according to the standard method. The average hatching rates in the December, February, and fresh groups were ~ 60%,

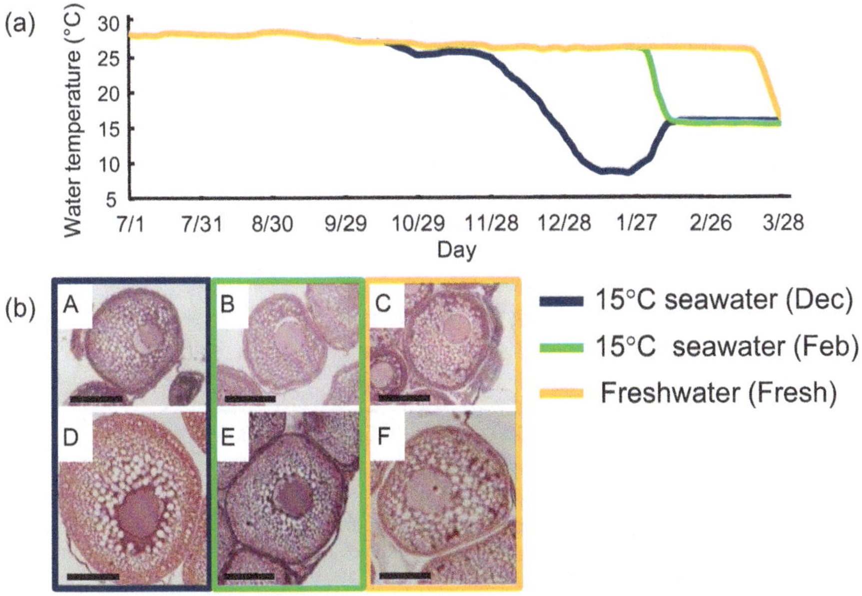

Fig. 14.2 (**a**) Changes in the rearing water temperature for 3 groups in feminized Japanese eels before artificial maturation, (**b**) Light micrographs of oocytes before artificial maturation in 3 groups: a group that was acclimated to 15 °C seawater from early December (Dec), a group similarly adapted to 15 °C from February (Feb), and a group reared in freshwater at 26–28 °C (Fresh). (a–c): Average-sized oocyte in each group. (d–f): The most developed oocyte in each group. Bar: 100 μm

40%, and 30%, respectively. Extremely good-quality eggs were obtained from many individuals in the Dec group.

Therefore, good results can be obtained in spring if the fish are reared under high water temperatures during winter. Figure 14.3 shows the relationship between oocyte diameter and serum 11KT levels in each group before artificial maturation, which tended to increase as the oocyte diameter increased. Typically, the larger the oocyte diameter before artificial maturation, the shorter the period until reaching full grown ovary; however, there were also significant individual differences within the group.

Thus, the results of artificial maturation in spring were nearly identical to those in autumn using feminized eels without lowering the water temperature in autumn and winter. It was reconfirmed that the decrease in water temperature and acclimation to seawater before spring promotes ovarian development and that a higher water temperature and freshwater environment suppresses ovarian development. In addition, it is clear that there are significant individual differences in ovarian development, and possibly in the expression of various endocrine factors.

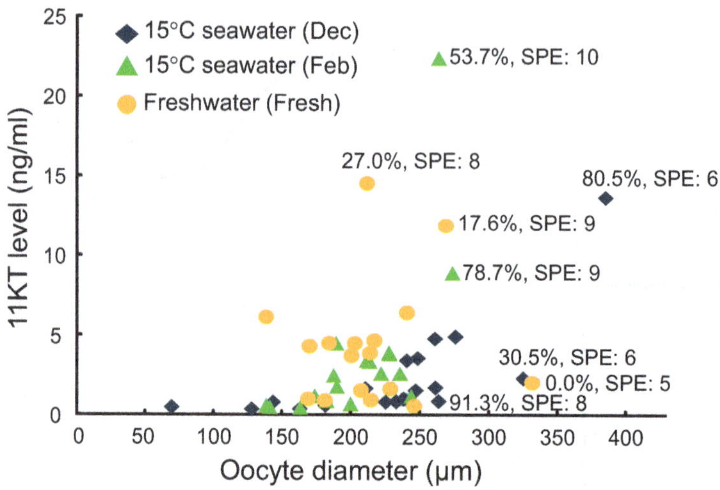

Fig. 14.3 The relationship between the oocyte diameter and the serum 11KT levels in each group before artificial maturation. Percentage (%) indicates hatchability. The number of SPEs indicates the number of injections until egg collection

14.4 Improvement of the Timing for the Induction of Ovulation Using the Fusion Stage of Lipid Droplets as a Biomarker

The main cause of poor egg quality seems to be the timing of FOM induction and ovulation. Therefore, we attempted to improve the timing of ovulation induction in detail using lipid droplets in oocytes as a biomarker (Tanaka et al. 2021). The lipid droplet fusion stage within oocytes was further classified into 10 stages, and it was found that stage 6 (average diameter of the 5 largest droplets: 90–110 μm) was the best for MIS (DHP or its precursor 17αOHP) administration (Unuma et al. 2011). Thus, injecting all females with MIS at stage 6 may reduce egg quality loss caused by inappropriately timed MIS injections. However, due to individual differences in the progression of oocyte maturation, it is difficult to predict when the ovarian stage in each female reaches the optimal stage. Therefore, to administer MIS to all females at the best time, females must be monitored daily for 24 h during the maturation-induction period; MIS should be injected on scheduled days and times in order to avoid daily monitoring. Therefore, we investigated the feasibility of optimizing the maturational status of female eels by rearing them at high and low temperatures on a fixed day and time. The results demonstrated that good-quality eggs can be obtained at a fixed day and time by either accelerating or slow down the progression of oocyte maturation in each female in order to optimize the oocyte stage through proper rearing at 20 °C or 15 °C (Tanaka et al. 2021).

14.5 Molecular Mechanism Underlying Egg-Quality Loss

Egg quality in eels, as well as in other fish, depends on various factors, including broodstock type (silver, feminized, or cultivated), environmental, and nutritional conditions. Furthermore, the period from ovulation to insemination strongly affects fertility and hatchability because eel eggs are rapidly overripened after ovulation (Ohta et al. 1996; Unuma et al. 2012). In addition, eel oocytes undergo intrafollicular overripeness before ovulation (Adachi 2000).

However, the molecular mechanism underlying the loss of egg quality is unknown; thus, it is necessary to clarify the molecular biological characteristics of good- and poor-quality eggs that are defined by fertility, hatchability, and survival rate. Recent studies focusing on maternal mRNA have been reported. Maternal mRNA accumulates during oogenesis and plays a role in early embryonic development and the transcriptional activity of zygotic genes (Wagner et al. 2004; Lee et al. 2013). The mRNA levels of some transcripts related to cell division are higher in good-quality eggs than those in poor-quality eggs (Izumi et al. 2016). In addition, the transcription of 5 genes (*igf2*, *npm2*, *phb2*, *pigf5*, and *cpt1*) may be upregulated in embryos after mid-blastula transition, leading to successful hatching (Rozenfeld et al. 2016). In RNA-seq analysis using good- and poor-quality eggs, insufficient expression levels of some genes affected egg quality, and multiple regression analysis showed that the mRNA levels of 5 genes (*gnpat*, *b4galntl*, *acsl6*, *rtkn*, and *trim24*) showed a strong correlation with hatchability (Izumi et al. 2019). However, there were no differences in the mRNA levels of many genes.

The localization of maternal mRNAs in eggs was investigated. The correct localization of maternal mRNA during oogenesis is essential for early development. In all investigated genes (*grip2*, *dazl*, *sybu*, *trim36*, *pou5f3*, and *npm2*), correct localization was observed in the oocytes from the perinucleolus to the migratory nucleus stage in artificially matured Japanese eels. In Japanese eel eggs, like other species, maternal mRNA was correctly localized in the cortex of the vegetal (*grip2*, *dazl*, *sybu*, and *trim36*) or animal pole (*pou5f3* and *npm2*). However, some ovulated eggs showed mRNA localization not only in the cortex (type A), but also in the more central cortex (type B). For all genes, eggs with high fertility and hatchability tended to have more type A genes, and eggs with low fertility and hatchability tended to have more type B genes (Fig. 14.4); however, this was not reflected in the target gene mRNA levels. In addition, many eggs with advanced oil-droplet coalescence (such as no. 13; Fig. 14.4), were observed in eggs with low fertility and hatchability. Since these eggs are not postovulatory overripes, it was suggested that the localization of oocytes, such as type B, is caused by preovulatory intrafollicular overripes. Our results suggest that the loss of quality of eggs from artificially matured Japanese eels is caused by abnormal mRNA localization, even if there is no difference in mRNA levels (Horiuchi et al. 2020).

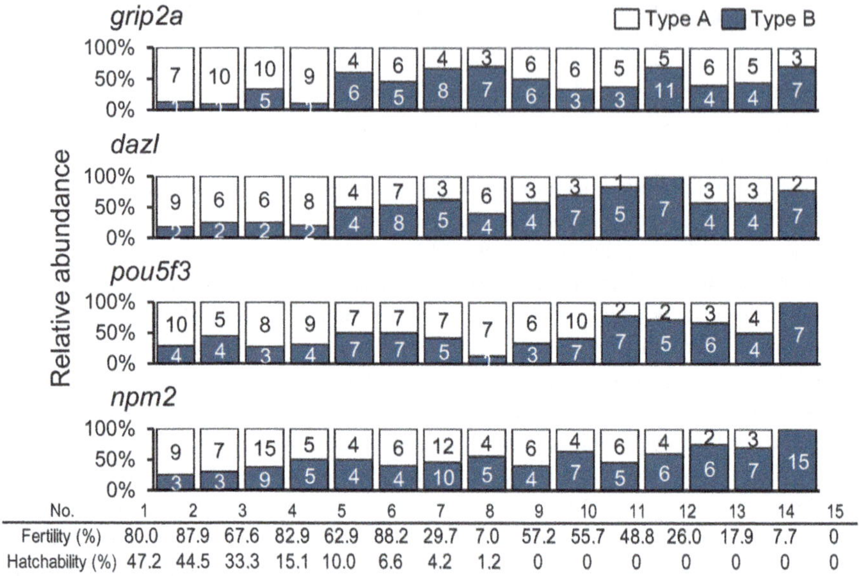

Fig. 14.4 Relationship between localization pattern of maternal mRNA and egg quality. Relative abundance of type A and type B in eggs from 15 artificially matured Japanese eels. Type A: mRNA localized in the cortex. Type B: mRNA distributed more centrally. Number in the column: number of eggs with type A or type B localization patterns. (Adapted from Horiuchi et al. 2020)

14.6 Conclusion

Currently, feminized eels are the best female broodstock for artificial maturation. It was found that 2-year-old fish of sufficient size attain puberty and are best suited for artificial maturation, whereas ovarian recrudescence is less likely to progress as they grow older. Seasonal changes in ovarian development were also observed. In an environment where rearing water temperature changes seasonally, it is best to induce artificial maturation in autumn. However, artificial maturation in spring was almost the same as that in autumn without lowering the water temperature in autumn and winter. It was reconfirmed that the decrease in water temperature and acclimation to seawater before spring promotes ovarian development and that a higher water temperature and freshwater environment suppress ovarian development.

The main cause of poor egg quality seems to be the timing of FOM induction and ovulation. Therefore, we attempted to improve the timing of ovulation induction in detail using lipid droplets in oocytes as a biomarker (Tanaka et al. 2021). The results indicated that proper rearing at 20 °C or 15 °C could either accelerate or slow down the progression of oocyte maturation in each female, resulting in better quality eggs.

Recently, recombinant gonadotropic hormones (GTH), including follicle stimulating hormone and luteinizing hormone, in the Japanese eel have been commercially produced and are more effective than conventional SPE or hCG injections for inducing oogenesis and spermatogenesis (Kazeto and Tanaka 2020). However, even if SPE is replaced with recombinant GTH, it is not always possible to obtain high-quality eggs. Repeated experience is important for obtaining good-quality eggs, as there are various techniques for artificial maturation to master, including handling and management of broodstock.

In addition, it is clear that there are large individual differences in ovarian development, and possibly in the expression of various endocrine factors. This is rather notable. The ease of raising larvae and the difference in growth can be judged based on appearance and behavior, but the developmental state of the gonads cannot be judged. Therefore, by accurately grasping ovarian and testicular development using various reproduction-related factors as indicators, and selective breeding of individuals with the most advanced gonad in the rearing environment, it may be possible to eliminate the induction of artificial maturation in the future.

Based on the results of this research, together with recent knowledge, it may be considered that the improvement of egg quality, which is one of the important issues in the development of artificial seedling production technology in Japanese eels, is nearly solved in terms of technology, although the molecular mechanism of egg quality loss remains unclear. Future research should focus on improving feed and facilities for larvae and reducing the costs of glass eel production.

References

Adachi S (2000) Artificial control of maturation and spawning, and egg quality. In: Matsuyama M, Adachi S, Kobayashi M (eds) Mechanism of gametogenesis in fish. Kaiyou, vol 32. Kaiyousyuppan, Tokyo, pp 120–126; in Japanese

Endo T, Todo T, Lokman PM, Kudo H, Ijiri S, Adachi S, Yamauchi K (2011) Androgens and very low density lipoprotein are essential for the growth of previtellogenic oocytes from Japanese eel, *Anguilla japonica*, in vitro. Biol Reprod 84:816–825. https://doi.org/10.1095/biolreprod.110.087163

Horie N, Utoh T, Mikawa N, Yamada Y, Okamura A, Tanaka S, Tsukamoto K (2008) Influence of artificial fertilization methods of the hormone-treated Japanese eel *Anguilla japonica* upon the quality of eggs and larvae (comparison between stripping-insemination and spontaneous spawning methods). Nippon Suisan Gakkaishi 74:26–35. https://doi.org/10.2331/suisan.74.26; in Japanese with English abstract

Horiuchi M, Izumi H, Lokman PM, Ijiri S, Adachi S (2020) Relationship between abundance and localization of maternal messenger RNA and egg quality in artificially matured Japanese eel *Anguilla japonica*. Fish Sci 86:43–56. https://doi.org/10.1007/s12562-019-01377-1

Ijiri S, Kayaba T, Takeda N, Tachiki H, Adachi S, Yamauchi K (1998) Pretreatment reproductive stage and oocyte development inducted by salmon pituitary homogenate in the Japanese eel *Anguilla japonica*. Fish Sci 64:531–537. https://doi.org/10.2331/fishsci.64.531

Ijiri S, Tsukamoto K, Chow S, Kurogi H, Adachi S, Tanka H (2011) Controlled reproduction in Japanese eel (*Anguilla japonica*) past and present. Aquaculture Europe 36:13–17

Izumi H, Gen K, Horiuchi M, Matsuya N, Ijiri S, Adachi S (2016) Quantitative comparisons of maternal transcripts related to cell division between good and poor quality eggs from artificially matured Japanese eel *Anguilla japonica*. Aquac Sci 64:29–42. https://doi.org/10.11233/aquaculturesci.64.29

Izumi H, Gen K, Lokman PM, Hagihara S, Horiuchi M, Tanaka T, Ijiri S, Adachi S (2019) Maternal transcripts in good and poor quality eggs from Japanese eel, *Anguilla japonica* —their identification by large-scale quantitative analysis. Mol Reprod Dev 86:1846–1864. https://doi.org/10.1002/mrd.23273

Kagawa H, Tanaka H, Ohta H, Okuzawa K, Iinuma N (1997) Induced ovulation by injection of 17,20β-dihydroxy-4-pregnen-3-one in the artificially matured Japanese eel, with special reference to ovulation time. Fish Sci 63:365–367. https://doi.org/10.2331/fishsci.63.365

Kagawa H, Iinuma N, Tanaka H, Ohta H, Okuzawa K (1998) Effects of rearing period in seawater on induced maturation in female Japanese eel *Anguilla japonica*. Fish Sci 64:77–82. https://doi.org/10.2331/fishsci.64.77

Kazeto Y, Tanaka T (2020) Study on artificial induction of maturation and ovulation in Japanese eel using the recombinant gonadotropins: application to the seed production. Nippon Suisan Gakkaishi 86:364–366. https://doi.org/10.2331/suisan.WA2754; in Japanese with English abstract

Lee MT, Bonneau AR, Takacs CM, Bazzini AA, DiVito KR, Fleming ES, Giraldez AJ (2013) Nanog, Pou5f1 and SoxB1 activate zygotic gene expression during the maternal-to-zygotic transition. Nature 503:360–364. https://doi.org/10.1038/nature12632

Masuda Y, Imaizumi H, Oda K, Hashimoto H, Usuki H, Teruya K (2012) Artificial completion of the Japanese eel, *Anguilla japonica*, life cycle: challenge to mass production. Bull Fish Res Agency 35:111–117. https://www.fra.affrc.go.jp/bulletin/bull/bull35/35-13.pdf. Accessed 6 June 2023

Nagahama Y, Adachi S (1985) Identification of a maturation-inducing steroid in a teleost, the amago salmon (*Oncorhynchus rhodurus*). Dev Biol 109:428–435. https://doi.org/10.1016/0012-1606(85)90469-5

Ohta H, Izawa T (1996) Diluent for cool storage of the Japanese eel (*Anguilla japonica*) spermatozoa. Aquaculture 142:107–118. https://doi.org/10.1016/0044-8486(95)01246-X

Ohta H, Kagawa H, Tanaka H, Okuzawa K, Hirose K (1996) Changes in fertilization and hatching rates with time after ovulation induced by 17, 20β-dihydroxy-4-pregnen-3-one in the Japanese eel, *Anguilla japonica*. Aquaculture 139:291–301. https://doi.org/10.1016/0044-8486(95)01167-6

Ohta H, Kagawa H, Tanaka H, Okuzawa K, Iinuma N, Hirose K (1997) Artificial induction of maturation and fertilization in Japanese eel, *Anguilla japonica*. Fish Physiol Biochem 17:163–169. https://doi.org/10.1023/A:1007720600588

Okamura A, Yamada Y, Mikawa N, Horie N, Utoh T, Kaneko T, Tanaka S, Tsukamoto K (2009) Growth and survival of eel leptocephali (*Anguilla japonica*) in low-salinity water. Aquaculture 296:367–372. https://doi.org/10.1016/j.aquaculture.2009.08.039

Rozenfeld C, Butts IAE, Tomkiewicz J, Zambonino-Infante JL, Mazurais D (2016) Abundance of specific mRNA transcripts impacts hatching success in European eel, *Anguilla anguilla* L. Comp Biochem Physiol 191:59–65. https://doi.org/10.1016/j.cbpa.2015.09.011

Tachiki H, Nakagawa T, Tamura K, Hirose K (1997) Effects of oral administration of estradiol-17β to young on gonadal sex and growth of Japanese eel *Anguilla japonica*. Suisanzousyoku 45:61–66. https://doi.org/10.11233/aquaculturesci1953.45.61; in Japanese

Tanaka H, Kagawa H, Ohta H (2001) Production of leptocephali of Japanese eel (*Anguilla japonica*) in captivity. Aquaculture 201:51–60. https://doi.org/10.1016/S0044-8486(01)00553-1

Tanaka H, Kagawa H, Ohta H, Unuma T, Nomura K (2003) The first production of glass eel in captivity: fish reproductive physiology facilitates great progress in aquaculture. Fish Physiol Biochem 28:493–497. https://doi.org/10.1023/B:FISH.0000030638.56031.ed

Tanaka T, Adachi S, Nomura K, Tanaka H, Unuma T (2021) Effects of rearing temperature manipulation on oocyte maturation progress in Japanese eel. Fish Sci 87:681–691. https://doi.org/10.1007/s12562-021-01531-8

Unuma T, Hasegawa N, Sawaguchi S, Tanaka T, Matsubara T, Nomura K, Tanaka H (2011) Fusion of lipid droplets in Japanese eel oocytes: stage classification and its use as a biomarker for induction of final oocyte maturation and ovulation. Aquaculture 322–323:142–148. https://doi.org/10.1016/j.aquaculture.2011.10.001

Unuma T, Sawaguchi S, Hasegawa N, Tsuda N, Tanaka T, Nomura K, Tanaka H (2012) Optimum temperature of rearing water during artificial induction of ovulation in Japanese eel. Aquaculture 358–359:216–223. https://doi.org/10.1016/j.aquaculture.2012.07.004

Wagner DS, Dosch R, Mintzer KA, Wiemelt AP, Mullins MC (2004) Maternal control of development at the midblastula transition and beyond: mutants from the zebrafish II. Dev Cell 6:781–790. https://doi.org/10.1016/j.devcel.2004.04.001

Yamamoto K, Yamauchi K (1974) Sexual maturation of Japanese eel and production of eel larvae in the aquarium. Nature 251:220–222. https://doi.org/10.1038/251220a0

Yamauchi K (1990) Studies of gonadal steroids involved in final gonadal maturation in the Japanese eel, *Anguilla japonica*. Int Revue Ges Hydrobiol 74:859–860. https://doi.org/10.1002/iroh.19900750630

Yamauchi K, Nakamura M, Takahashi H, Takano K (1976) Cultivation of larvae of Japanese eel. Nature 263:412. https://doi.org/10.1038/263412a0

Chapter 15
Larval Rearing

Akihiro Okamura, Yoshiaki Yamada, Noriyuki Horie, and Naomi Mikawa

Contents

Since the first hatched larvae (preleptocephali) of freshwater eel were obtained in the 1970s (Yamamoto and Yamauchi 1974), rearing techniques for eel larvae (leptocephali) have significantly improved through the efforts of several groups (Tanaka et al. 2006; Okamura et al. 2014; Tanaka 2015). Accumulation of knowledge from field surveys elucidating the natural habitat of eel leptocephali (Tsukamoto 1992; Tsukamoto et al. 2011) has also supported the development of rearing techniques. In the Japanese eel *Anguilla japonica*, Tanaka et al. (2003) were the first to successfully obtain captive-bred glass eels, and over thousands of glass eels per year have been produced in laboratories since then (Okamura et al. 2021). Efforts have also been made to obtain captive-bred larvae of other temperate eels (*A. anguilla*, *A. rostrata*, *A. dieffenbachii*, and *A. australis*) (Lokman and Young 2000; Oliveira and Hable 2010; Butts et al. 2016). However, the survival rate of leptocephali in captivity remains low, and the time required to begin metamorphosis is still over 200 days, probably because of insufficient nutrition in the artificial larval diet. Thus, the mass production of captive-bred glass eels on a commercial scale has not yet been achieved. In this section, we outline the recent progress in the rearing techniques of eel leptocephali, mainly based on the findings on *A. japonica*, and clarify the current remaining issues.

A. Okamura (✉) · Y. Yamada · N. Horie · N. Mikawa
IRAGO Institute Co., Ltd., Aichi, Japan
e-mail: aokamura@irago.co.jp

K. Tsukamoto et al. (eds.), *Eel Science*, Fisheries Science Series,
https://doi.org/10.1007/978-981-99-5692-0_15

15.1 Incubating Eggs and Preleptocephali

Fertilized eggs of *A. japonica* are kept in a cylindrical mesh basket floating in a polycarbonate tank (Fig. 15.1) in UV-exposed seawater flowing at a rate of 2–3 L/min at 25 °C and 35 practical salinity units (psu). Soon after hatching, preleptocephali (total length: 3–4 mm) are transferred from the mesh basket to a 180-L polycarbonate tank (Fig. 15.2) at a mean density of 500–1000 individuals/L. The tank are maintained in UV-exposed seawater flowing at a rate of 2–3 L/min at 25 °C and 35 psu until the formation of a functional mouth (5–6 days after hatching) (Okamura et al. 2007). In the water column, they typically drift upward, but respond to light and swim erratically despite the eyes being still not pigmented; therefore, they are preferably kept in the dark. The temperature and salinity of the rearing water are approximately equal to those seen in the natural habitat, where *A. japonica* spawns at a depth of around 160 m near seamounts at the southern end of the West Mariana Ridge (Tsukamoto et al. 2011).

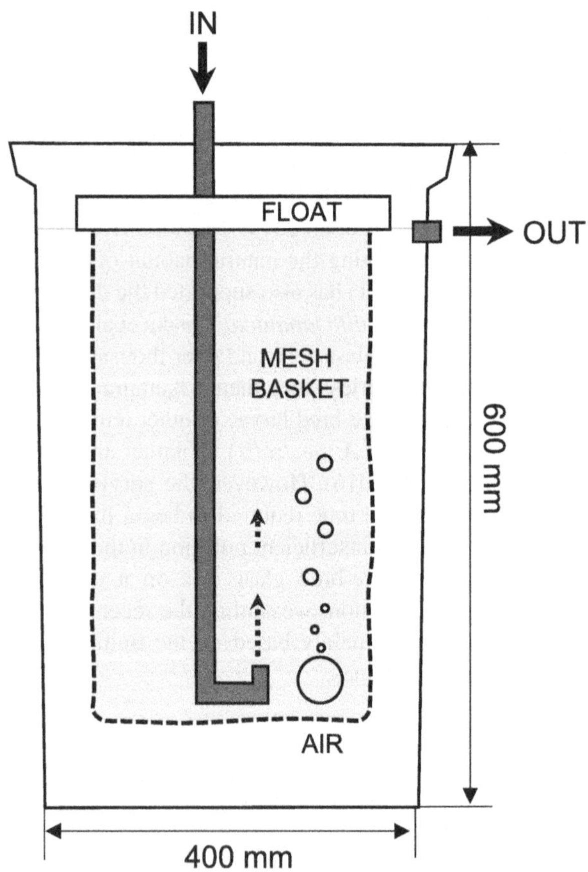

Fig. 15.1 An example of incubating system for fertilized eel eggs. Eggs are kept in the mesh basket (50 L) floating in a polycarbonate tank (100 L) until hatching. The mesh opening of the basket is 0.3 mm. Eggs are suspended in the water column with upwelling water and aeration

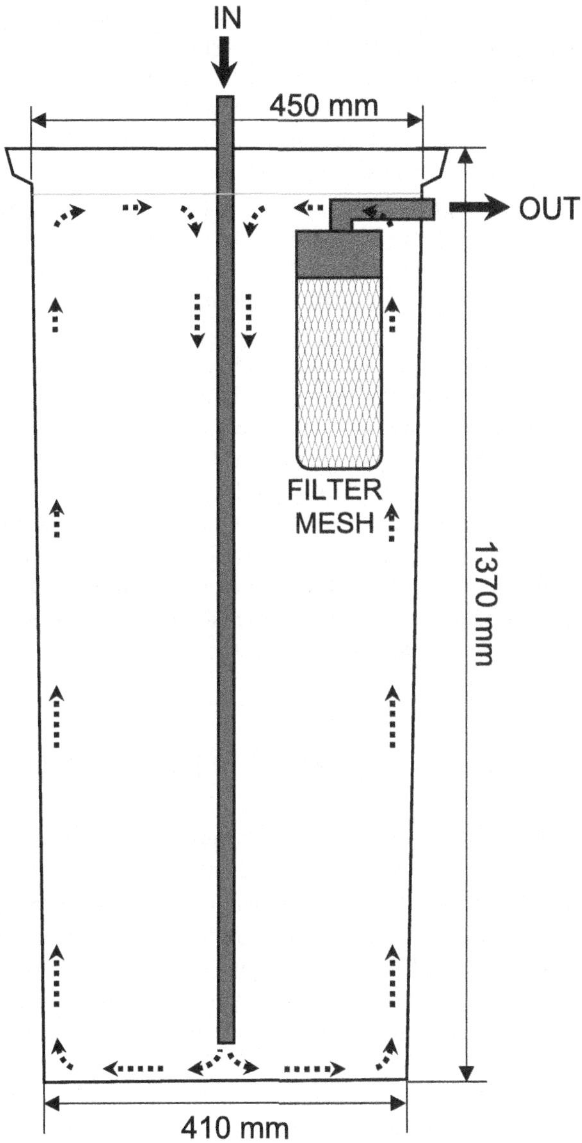

Fig. 15.2 Schematic drawing of a 180-L polycarbonate tank for rearing eel pre-leptocephali. Solid arrows show a water supply and an overflowing drain. Broken arrows show the directions of water flow. The mesh opening of the drain filter is 0.25 mm

During this period, embryonic and pre-larval development are greatly influenced by water temperature. In *A. japonica*, hatching occurred 24 hours after fertilization at 28 °C and 58 h at 19 °C and did not occur at 16 °C or 31 °C. The hatching and survival rates until the mouth formation were the highest at 25 °C (Okamura et al. 2007; Ahn et al. 2012).

Typically, temperatures less than 25 °C induce abnormal embryonic development, resulting in various morphological deformities in leptocephali (Okamura et al.

2007). However, favorable water temperatures may differ between species. In *A. anguilla*, the effects of rearing temperature on hatching and larval survival were assessed in the range of 16–24 °C. The most successful hatching was observed at 18–20 °C; however, all embryos quickly died at 24 °C (Politis et al. 2017). The survival rate of preleptocephali was the highest at 18 °C, which is considered to be an optimum condition for this species. While *A. japonica* eats 6 days after hatching, *Anguilla anguilla* larvae can begin feeding 12 days after hatching. Although there are a few examples in other species, embryos of *A. dieffenbachii* and its hybrids with *A. australis* developed and hatched at 18.2–22.7 °C (Lokman and Young 2000) and eggs and larvae of *A. rostrata* appeared to develop normally at 20 °C (Oliveira and Hable 2010).

Despite the fact that all eel larvae are marine organisms, they are highly tolerant of a wide range of salinity conditions. *A. japonica* leptocephali can tolerate low salinity conditions 5–10 days after hatching (5–20 psu), and the survival duration is prolonged at 5–10 psu without feeding (Chang et al. 2004). A gradual reduction in salinity (36–16 psu) during pre-larval development in *A. anguilla* increased their growth and survival rates (Politis et al. 2018). However, a negative effect of low salinity on embryogenesis has also been suggested; salinity below 33 psu produced morphological deformities in preleptocephali in *A. japonica* (Okamoto et al. 2009) and a reduction in salinity increased the occurrence of pericardial edema in *A. anguilla* (Politis et al. 2018).

15.2 Captive Environment for Leptocephali

During the larval migration of *A. japonica* in the Pacific, they perform daily vertical migrations from 10–100 m in depth (water temperature 26–28 °C) at night and 100–200 m (20–26 °C) during the day (Otake et al. 1998). In captivity, *A. japonica* leptocephali have been reared at 21–22 °C (Tanaka et al. 2001; Okamura et al. 2009a), relatively lower than nature. However, the lower temperature effectively prevents bacterial growth in the tanks, thereby enhancing larval survival rates and extending the rearing duration until metamorphosis. A recent rearing experiment of *A. japonica* leptocephali at 19–27 °C reported that they grew faster with increasing water temperature, and the best growth was achieved at 27 °C for 2 weeks after the start of feeding, consistent with the natural environment upper than 100 m in depth. However, the growth rate gradually decreased at higher temperatures after that, most likely because the increased basal metabolism at higher temperatures was insufficiently compensated for by poor nutrition in underdeveloped artificial diets (Kuroki et al. 2019).

Unlike embryonic development, leptocephali at the growth stage do not necessarily require the same salinity conditions as the natural habitat. Half-diluted seawater (17.5 psu) remarkably increases the survival rate of *A. japonica* leptocephali and accelerated its growth (Okamura et al. 2009b). The tissue osmolality of reared *A. japonica* leptocephali ranges from 360–540 mOsm per kg H_2O (Lee et al. 2013), which is equivalent to that of 50% or more diluted seawater, suggesting that the best survival and growth of leptocephali results from the saved energy due to the lower cost of osmoregulation under near-iso-osmotic conditions (Okamura et al. 2009b).

Leptocephali have strong negative phototaxis and continue to swim toward the bottom of the tank to avoid light even under a low light intensity of around 2 lx (Yamada et al. 2009). However, movement in the dark is markedly minimal. Thus, to reduce energy consumption, they should be maintained under dark conditions (0 lx), except for feeding time. While *A. japonica* leptocephali are being fed, a light (approximately 200 lx at the bottom of the tank) is placed above the tank (Okamura et al. 2009a).

15.3 Rearing Tanks

Several types of tanks for rearing leptocephali, such as hemispherical and horizontal cylindrical tanks, have been developed (Tanaka et al. 2006; Okamura et al. 2009a). Tanaka et al. (2001) initially used a 5 L round-bottomed hemisphere tank, which was originally designed for rearing the phyllosoma larvae of spiny lobsters (Matsuda and Takenouchi 2007), and later developed a cylindrical tank (~20 L) equipped with a long center strainer and a sweeping water flow for easy removal of uneaten-food remnants (Tanaka et al. 2006). Okamura et al. (2009a) used a planktonkreisel (Fig. 15.3a), which was originally developed by Greve (1975) for the maintenance of planktonic animals. The planktonkreisel can produce a stable vertical revolving

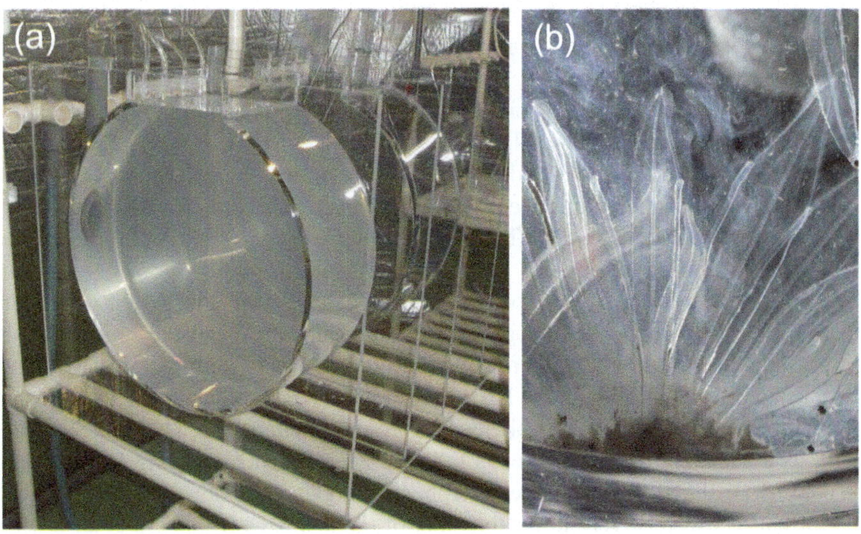

Fig. 15.3 (**a**) A planktonkreisel (19 L) developed for rearing eel leptocephali and (**b**) fully grown *Anguilla japonica* leptocephali fed with a slurry-type diet. The main component of the tank is made from clear acrylic resin and the outlet chamber from polyvinyl chloride

water flow through the inlet water, which has the following advantages: (a) Only about 20 s is required to remove uneaten food, because the fast-revolving current can easily remove the uneaten diet at the bottom of the tank, keep it suspended in the water column, and prevent it from adhering to the walls of the tank; (b) The water flow pulls leptocephali away from the wall of the tank, which is often contaminated by residual food, thereby preventing leptocephali from scratching and damaging their skin; (c) It can prevent the occurrence of lower jaw deformity, where the lower jaw becomes fixed in a downward position and the notochord behind the head bends from excessive swimming downward with the mouth pressed against the bottom of the tank. A strong and constant water flow, on the other hand, may cause notochordal abnormalities in leptocephali (Kuroki et al. 2016).

15.4 Artificial Diets

The natural diet of eel leptocephali has long been a mystery. Several hypotheses have been proposed: leptocephali protrude their large anteriorly projecting teeth against prey organisms to suck out their body fluid, or leptocephali absorb dissolved organic matter (DOM) directly from the body surface, as food-like materials cannot be found in the digestive tract (Pfeiler 1986). However, particulate organic matter (POM), such as fecal pellets of copepods and larvacean houses, was found in the gut of eel leptocephali (Otake et al. 1993; Mochioka and Iwamizu 1996), revealing that they can feed orally like other fish larvae. Biochemical and molecular analyses have revealed that traces of a wide variety of taxa, including cnidarians, radiolarians, and plants, are present in the gut of leptocephali (Riemann et al. 2010; Chow et al. 2019). Furthermore, the trophic level of eel leptocephali, as revealed by the nitrogen stable isotope analysis of amino acids, also suggests that marine snow-like aggregates, including secretions, carcasses, and feces of various zooplankton, are the primary food for leptocephali (Miller et al. 2012).

Since Tanaka et al. (2001) identified eggs of spiny dogfish *Squalus* sp. as a functional diet for allowing eel leptocephali to grow, some modifications and new ingredients have been reported (Table 15.1). Tanaka et al. (2003) fed *A. japonica* leptocephali a slurry-type diet (Fig. 15.3b) made from dogfish egg powder supplemented with soybean peptide, krill extract, vitamins, and minerals (Table 15.1), and made them to grow into glass eels. However, the spiny dogfish in the Atlantic are listed as vulnerable species on the IUCN Red List of Threatened Species in 2019. Considering the large-scale commercial production of glass eels in the future, low-cost and sustainable feed ingredients are required. Thus, chicken egg yolk as a substitute for dogfish eggs has also been tested for rearing leptocephali (Table 15.1), allowing them to metamorphose into glass eels; however, their growth performance was relatively lower than that of dogfish eggs (Okamura et al. 2013). Another diet containing fish protein hydrolysate has been proposed (Table 15.1), which also causes leptocephali to grow into glass eels (Masuda et al. 2016).

Table 15.1 Examples of proposed diets for eel larvae and their constituents

Tanaka et al. (2001)		Okamura et al. (2013)		Masuda et al. (2016)	
Ingredients	%	Ingredients	%	Ingredients	%
Spiny-dogfish egg powder	21.3	Chicken egg yolk powder	46.0	Fish protein hydrolysate	37.3
Krill extract	72.1	Krill tail muscle extract	46.0	Krill and anchovy larvae (1:1)	3.5
Soybean oligo peptide	5.3	Chicken albumen oligo peptide	7.7		
Vitamin mixture[a]	0.9	Vitamin mixture[b]	0.3		
Mineral mixture[c]	0.4			Soybean lecithin	1.2
				Vitamin mixture[c]	0.4
				Seaweed powder	0.4
				Water	56.0

[a]Vitamins A, B1, B2, B6, B12, C, D3, E, K3, nicotinamide, pantothenic acid, choline chloride, folic acid, biotin, inositol, p-aminobenzoic acid
[b]Vitamins A, B1, B2, B6, C, D3, E, K, pantothenic acid, niacin, folic acid, inositol
[c]Na, K, Mg, P, Ca, Cl, Fe, Zn, Mn, Cu, Co, Al

Feeding methods have also improved. Recently, a liquid-type diet with lower viscosity (20–1000 mPa·s) has been proposed for leptocephali because of its poor ingestion capabilities, especially in the early developmental stages (Yamada et al. 2019). Using leptocephali's negative phototactic behavior, a light above the tank directs them to the bottom, where they encounter the liquid-type diet. Leptocephali then enter the diet liquid and drink it directly. Despite these efforts, the daily growth rates of captive *A. japonica* leptocephali (0.1–0.3 mm/day) are still much lower than those of wild larvae (0.3–0.5 mm/day) (Okamura et al. 2014).

Therefore, various approaches for improving the artificial diet of eel leptocephali are still underway. Analyses of the gut contents of leptocephali and marine snow-like materials collected from the western North Pacific showed that it contained many sugars possibly derived from transparent exopolymer particles (TEPs) (Tomoda et al. 2018). In the ocean, many TEPs are produced from acidic polysaccharides secreted by phytoplankton, and these particles aggregate as substrates to form marine snow. Leptocephali can utilize these sugars. However, feeding trials of actual POM collected around the spawning area of *A. japonica* (Chow et al. 2017) or phytoplankton and its secreted polysaccharides, identifying possible elements of marine snow (Tomoda et al. 2015) have not been successful to date. According to a transcriptome analysis of wild-caught leptocephali, the transcript levels of glucose and fructose transporters increased in the leptocephalus stage (Hsu et al. 2015). In captive-bred eel larvae, the expression level of a glucose transporter (sodium/glucose co-transporter member 1) is also high, and amylase expression is maintained at a specific level in the leptocephalus stage (Murashita et al. 2013; Shin et al. 2022). These results suggest that carbohydrates may be important for the dietary nutrition of eel larvae. However, the lower expression of carbohydrate digestive enzymes in wild-caught leptocephali may indicate that they absorb monosaccharides or low-molecular-weight sugars rather than polysaccharides (Hsu et al. 2015). Furthermore, the addition of glucose or maltose to artificial diets promotes leptocephali growth. Diets supplemented with *N*-acetylglucosamine or its dimers and trimers also improve the survival and growth of *A. japonica* leptocephali (Okamura et al. 2020). Reducing dietary lipids improved survival and growth of leptocephali (Furuita et al. 2014), which is also consistent with the results in which transcript levels of lipid digestion enzymes were lower than those of protein digestion enzymes in leptocephali (Hsu et al. 2015).

15.5 Inducing Metamorphosis

Shortening the duration of the leptocephalus stage is the priority for establishing the mass production of artificial glass eels because it is difficult to maintain, costly, and time-consuming. The factors involved in initiating metamorphosis in leptocephali have remained unclear, even in nature, for many years. Wild *A. japonica* leptocephali metamorphose at 110–170 days after hatching, as determined by the analysis of the otolith daily rings (Tabeta et al. 1987; Kawakami et al. 1998) but captive larvae usually metamorphose after 200 days. Although insufficient nutrition may be the primary reason for the lower growth rate in captivity, delayed metamorphosis has been due to a lack of appropriate environmental cues for its onset. However, Okamura et al. (2012) showed that the intentional starvation of sufficiently grown leptocephali triggered the onset of metamorphosis. When the larvae of *A. japonica* > 55 mm TL were not fed for about 1 week, they subsequently all began metamorphosis within 2 weeks. This method shortens the rearing period and enables us to manipulate the timing of metamorphosis in reared leptocephali, offering advantages for commercial-scale production of glass eels.

15.6 Deformity

Controlling deformities in captive leptocephali and glass eels remains problematic. Over 50% of the reared *A. japonica* leptocephali have some type of notochord deformity (Okamura et al. 2011). The most frequent deformity is kyphosis, which is the dorsal (Λ-shaped) curvature of the notochord column. Typical kyphotic larvae have crescent-shaped bodies, with curvature frequently stretching from the head to the tail along the notochord. Severely deformed leptocephali cannot swim normally, and after metamorphosis, these deformed glass eels die because of their inability to ingest food. However, the causative factors remain unknown to date. Some feeding experiments have suggested that increasing feeding frequency and high water temperature increase the occurrence rate of this notochord deformity (Okamura et al. 2018). However, further research is necessary to control this unfavorable phenomenon and improve the yield of artificial glass eels.

15.7 Conclusions

One solution to improve sustainability of eel-related businesses is to partially replace capture-based aquaculture of eels relying on natural glass eel stocks, which have been declining rapidly in recent years, with hatchery-based aquaculture using captive-bred seedlings. Approximately 100 million glass eels per year are currently available for aquaculture in Japan; however, the current production capacity of artificial seedlings at the laboratory level is insufficient to meet this demand. Many recent studies have made significant improvements in artificial diets, rearing tanks, and feeding methods, as mentioned here. Nevertheless, maintaining fragile eel larvae is time-consuming and difficult. Therefore, the priority is further improvement of artificial diets for leptocephali to significantly shorten the rearing period. Second, it is critical to discover a consistent supply of low-cost, sustainable feed components for industrial prodction of glass eel. Finally, the mechanism underlying the incidence of deformities in captive-reared leptocephali must be elucidated, and control technology to improve glass eel yield must be developed.

References

Ahn H, Yamada Y, Okamura A, Horie N, Mikawa N, Tanaka S, Tsukamoto K (2012) Effect of water temperature on embryonic development and hatching time of the Japanese eel *Anguilla japonica*. Aquaculture 330–333:100–105. https://doi.org/10.1016/j.aquaculture.2011.12.020

Butts IAE, Sørensen SR, Politis SN, Tomkiewicz J (2016) First-feeding by European eel larvae: a step towards closing the life cycle in captivity. Aquaculture 464:451–458. https://doi.org/10.1016/j.aquaculture.2016.07.028

Chang YC, Chen YS, Lai JY, Liu FG (2004) Studies on the salinity tolerance of larvae of Japanese eel (*Anguilla japonica*). J Taiwan Fish Res 12:25–31. https://en.tfrin.gov.tw/News_Content.aspx?n=349&s=226194. Accessed 7 June 2023

Chow S, Kurogi H, Watanabe S, Matsunari H, Sudo R, Nomura K, Tanaka H, Furuita H, Nishimoto A, Higuchi M, Jinbo T, Tomoda T (2017) Onboard rearing attempts for the Japanese eel leptocephali using POM-enriched water collected in the Western North Pacific. Aquat Living Resour 30:38. https://doi.org/10.1051/alr/2017037

Chow S, Inaba N, Nagai S, Kurogi H, Nakamura Y, Yanagimoto T, Tanaka H, Hasegawa D, Asakura T, Kikuchi J, Tomoda T, Kodama T (2019) Molecular diet analysis of Anguilliformes leptocephalus larvae collected in the western North Pacific. PLoS One 14:e0225610. https://doi.org/10.1371/journal.pone.0225610

Furuita H, Murashita K, Matsunari H, Yamamoto T, Nagao J, Nomura K, Tanaka H (2014) Decreasing dietary lipids improves larval survival and growth of Japanese eel *Anguilla japonica*. Fish Sci 80:581–587. https://doi.org/10.1007/s12562-014-0713-2

Greve W (1975) The "Meteor Planktonküvette": a device for the maintenance of macrozooplankton aboard ships. Aquaculture 6:77–82. https://doi.org/10.1016/0044-8486(75)90090-3

Hsu HY, Chen SH, Cha YR, Tsukamoto K, Lin CY, Han YS (2015) De novo assembly of the whole transcriptome of the wild embryo, preleptocephalus, leptocephalus, and glass eel of *Anguilla japonica* and deciphering the digestive and absorptive capacities during early development. PLoS One 10:e0139105. https://doi.org/10.1371/journal.pone.0139105

Kawakami Y, Mochioka N, Nakazono A (1998) Immigration period and age of *Anguilla japonica* glass-eels entering rivers in northern Kyushu, Japan during 1994. Fish Sci 64:235–239. https://doi.org/10.2331/fishsci.64.235

Kuroki M, Okamura A, Takeuchi A, Tsukamoto K (2016) Effect of water current on the body size and occurrence of deformities in reared Japanese eel leptocephali and glass eels. Fish Sci 82: 941–951. https://doi.org/10.1007/s12562-016-1015-7

Kuroki M, Okamura A, Yamada Y, Hayasaka S, Tsukamoto K (2019) Evaluation of optimum temperature for the early larval growth of Japanese eel in captivity. Fish Sci 85:801–809. https://doi.org/10.1007/s12562-019-01317-z

Lee KM, Yamada Y, Okamura A, Tsukamoto K, Kaneko T (2013) Hyposmoregulatory ability and ion- and water-regulatory mechanisms during the leptocephalus stages of Japanese eel *Anguilla japonica*. Fish Sci 79:77–86. https://doi.org/10.1007/s12562-012-0576-3

Lokman PM, Young G (2000) Induced spawning and early ontogeny of New Zealand freshwater eels (*Anguilla dieffenbachii* and *A. australis*). NZ J Mar Freshw Res 34:135–145. https://doi.org/10.1080/00288330.2000.9516921

Masuda Y, Yatabe T, Matsunari H, Furuita H, Kamoshida M, Shima Y, Kuwada H (2016) Rearing of larvae of Japanese eel *Anguilla japonica* to metamorphosis into glass eel by feeding with fish protein hydrolysate-based diets. Nippon Suisan Gakkaishi 82:131–133. https://doi.org/10.2331/suisan.15-00060; in Japanese with English abstract

Matsuda H, Takenouchi T (2007) Development of technology for larval *Panulirus japonicus* culture in Japan: a review. Bull Fish Res Agen 20:77–84. https://www.fra.affrc.go.jp/bulletin/bull/bull20/12.pdf. Accessed 7 June 2023

Miller MJ, Chikaraishi Y, Ogawa NO, Yamada Y, Tsukamoto K, Ohkouchi N (2012) A low trophic position of Japanese eel larvae indicates feeding on marine snow. Biol Lett 9:20120826. https://doi.org/10.1098/rsbl.2012.0826

Mochioka N, Iwamizu M (1996) Diet of anguilloid larvae: leptocephali feed selectively on larvacean houses and fecal pellets. Mar Biol 125:447–452. https://doi.org/10.1007/BF00353257

Murashita K, Furuita H, Matsunari H, Yamamoto T, Awaji M, Nomura K, Nagao J, Tanaka H (2013) Partial characterization and ontogenetic development of pancreatic digestive enzymes in Japanese eel *Anguilla japonica* larvae. Fish Physiol Biochem 39:895–905. https://doi.org/10.1007/s10695-012-9749-3

Okamoto T, Kurokawa T, Gen K, Murashita K, Nomura K, Kim SK, Matsubara H, Ohta H, Tanaka H (2009) Influence of salinity on morphological deformities in cultured larvae of Japanese eel, *Anguilla japonica*, at completion of yolk resorption. Aquaculture 293:113–118. https://doi.org/10.1016/j.aquaculture.2009.04.005

Okamura A, Yamada Y, Horie N, Utoh T, Mikawa N, Tanaka S, Tsukamoto K (2007) Effects of water temperature on early development of Japanese eel *Anguilla japonica*. Fish Sci 73:1241–1248. https://doi.org/10.1111/j.1444-2906.2007.01461.x

Okamura A, Yamada Y, Horita T, Horie N, Mikawa N, Utoh T, Tanaka S, Tsukamoto K (2009a) Rearing eel leptocephali (*Anguilla japonica* Temminck & Schlegel) in a planktonkreisel. Aquac Res 40:509–512. https://doi.org/10.1111/j.1365-2109.2008.02127.x

Okamura A, Yamada Y, Mikawa N, Horie N, Utoh T, Kaneko T, Tanaka S, Tsukamoto K (2009b) Growth and survival of eel leptocephali (*Anguilla japonica*) in low-salinity water. Aquaculture 296:367–372. https://doi.org/10.1016/j.aquaculture.2009.08.039

Okamura A, Yamada Y, Mikawa N, Horie N, Tanaka S, Tsukamoto K (2011) Notochord deformities in reared Japanese eel *Anguilla japonica* larvae. Aquaculture 317:37–41. https://doi.org/10.1016/j.aquaculture.2011.04.024

Okamura A, Yamada Y, Mikawa N, Horie N, Tsukamoto K (2012) Effect of starvation, body size, and temperature on the onset of metamorphosis in Japanese eel (*Anguilla japonica*). Can J Zool 90:1378–1385. https://doi.org/10.1139/cjz-2012-0146

Okamura A, Yamada Y, Horie N, Mikawa N, Tanaka S, Kobayashi H, Tsukamoto K (2013) Hen egg yolk and skinned krill as possible foods for rearing leptocephalus larvae of *Anguilla japonica* Temminck & Schlegel. Aquac Res 44:1531–1538. https://doi.org/10.1111/j.1365-2109.2012.03160.x

Okamura A, Horie N, Mikawa N, Yamada Y, Tsukamoto K (2014) Recent advances in artificial production of glass eels for conservation of anguillid eel populations. Ecol Freshw Fish 23:95–110. https://doi.org/10.1111/eff.12086

Okamura A, Horie N, Mikawa N, Yamada Y, Tsukamoto K (2018) Influence of temperature and feeding regimes on growth and notochord deformity in reared *Anguilla japonica* leptocephali. Fish Sci 84:505–512. https://doi.org/10.1007/s12562-018-1188-3

Okamura A, Yamada Y, Mikawa N, Horie N, Tsukamoto K (2020) Dietary supplementation with chitin hydrolysates for *Anguilla japonica* leptocephali. Fish Sci 86:685–692. https://doi.org/10.1007/s12562-020-01440-2

Okamura A, Shimamura A, Mikawa N, Yamada Y, Horie N, Tsukamoto K (2021) Taste evaluation of grilled eel produced by hatchery-based aquaculture. Nippon Suisan Gakkaishi 87:393–399. https://doi.org/10.2331/suisan.20-00070; in Japanese with English abstract

Oliveira K, Hable WE (2010) Artificial maturation, fertilization, and early development of the American eel (*Anguilla rostrata*). Can J Zool 88:1121–1128. https://doi.org/10.1139/Z10-081

Otake T, Nogami K, Maruyama K (1993) Dissolved and particulate organic matter as possible food sources for eel leptocephali. Mar Ecol Prog Ser 92:27–34. https://doi.org/10.3354/meps092027

Otake T, Inagaki T, Hasumoto H, Mochioka N, Tsukamoto K (1998) Diel vertical distribution of *Anguilla japonica* leptocephali. Ichthyol Res 45:208–211. https://doi.org/10.1007/BF02678565

Pfeiler E (1986) Towards an explanation of the developmental strategy in leptocephalous larvae of marine teleost fishes. Environ Biol Fish 15:3–13. https://doi.org/10.1007/BF00005385

Politis SN, Mazurais D, Servili A, Zambonino-Infante JL, Miest JJ, Sørensen SR, Tomkiewicz J, Butts IAE (2017) Temperature effects on gene expression and morphological development of European eel, *Anguilla anguilla* larvae. PLoS One 12:e0182726. https://doi.org/10.1371/journal.pone.0182726

Politis SN, Mazurais D, Servili A, Zambonino-Infante J-L, Miest JJ, Tomkiewicz J, Butts IAE (2018) Salinity reduction benefits European eel larvae: insights at the morphological and molecular level. PLoS One 13:e0198294. https://doi.org/10.1371/journal.pone.0198294

Riemann L, Alfredsson H, Hansen MM, Als TD, Nielsen TG, Munk P, Aarestrup K, Maes GE, Sparholt H, Petersen MI, Bachler M, Castonguay M (2010) Qualitative assessment of the diet of European eel larvae in the Sargasso Sea resolved by DNA barcoding. Biol Lett 6:819–822. https://doi.org/10.1098/rsbl.2010.0411

Shin MG, Ryu Y, Choi HY, Kim SK (2022) Ontogenetic digestive physiology and expression of nutrient transporters in *Anguilla japonica* larvae. Aquac Reports 25:101218. https://doi.org/10.1016/j.aqrep.2022.101218

Tabeta O, Tanaka K, Yamada J, Tzeng WNW (1987) Aspects of the early life history of the Japanese eel *Anguilla japonica* determined from otolith microstructure. Nippon Suisan Gakkaishi 53:1727–1734. https://doi.org/10.2331/suisan.53.1727

Tanaka H (2015) Progression in artificial seedling production of Japanese eel *Anguilla japonica*. Fish Sci 81:11–19. https://doi.org/10.1007/s12562-014-0821-z

Tanaka H, Kagawa H, Ohta H (2001) Production of leptocephali of Japanese eel (*Anguilla japonica*) in captivity. Aquaculture 201:51–60. https://doi.org/10.1016/S0044-8486(01)00553-1

Tanaka H, Kagawa H, Ohta H, Unuma T, Nomura K (2003) The first production of glass eel in captivity: fish reproductive physiology facilitates great progress in aquaculture. Fish Physiol Biochem 28:493–497. https://doi.org/10.1023/B:FISH.0000030638.56031.ed

Tanaka H, Nomura K, Yamamoto T, Oku H (2006) Development of artificial diets and rearing systems for eel larvae: the first success of production of glass eel in captivity. Bull Fish Res Agency suppl 5:63–69. http://www.fra.affrc.go.jp/bulletin/bull/bull-b5/15.pdf. Accessed 7 June 2023; in Japanese with English abstract

Tomoda T, Kurogi H, Okauchi M, Kamoshida M, Imaizumi H, Jinbo T, Nomura K, Furuita H, Tanaka H (2015) Hatchery-reared Japanese eel *Anguilla japonica* larvae ingest various organic matter formed as part of marine snow. Nippon Suisan Gakkaishi 81:715–721. https://doi.org/10.2331/suisan.81.715; in Japanese with English abstract

Tomoda T, Chow S, Kurogi H, Okazaki M, Ambe D, Furuita H, Matsunari H, Nagai S, Yokouchi K, Swayama S, Nomura K, Tanaka H, Sudou R, Hasegawa D, Inaba N (2018) Observations of gut contents of anguilliform leptocephali collected in the western North Pacific. Nippon Suisan Gakkaishi 84:32–44. https://doi.org/10.2331/suisan.17-00025; in Japanese with English abstract

Tsukamoto K (1992) Discovery of the spawning area for Japanese eel. Nature 356:789–791. https://doi.org/10.1038/356789a0

Tsukamoto K, Chow S, Otake T, Kurogi H, Mochioka N, Miller MJ, Aoyama J, Kimura S, Watanabe S, Yoshinaga T, Shinoda A, Kuroki M, Oya M, Watanabe T, Hata K, Ijiri S, Kazeto Y, Nomura K, Tanaka H (2011) Oceanic spawning ecology of freshwater eels in the western North Pacific. Nat Commun 2:179. https://doi.org/10.1038/ncomms1174

Yamada Y, Okamura A, Mikawa N, Utoh T, Horie N, Tanaka S, Miller M, Tsukamoto K (2009) Ontogenetic changes in phototactic behavior during metamorphosis of artificially reared Japanese eel *Anguilla japonica* larvae. Mar Ecol Prog Ser 379:241–251. https://doi.org/10.3354/meps07912

Yamada Y, Okamura A, Mikawa N, Horie N, Tsukamoto K (2019) A new liquid-type diet for leptocephali in mass production of artificial glass eels. Fish Sci 85:545–551. https://doi.org/10.1007/s12562-019-01295-2

Yamamoto K, Yamauchi K (1974) Sexual maturation of Japanese eel and production of eel larvae in the aquarium. Nature 251:220–222. https://doi.org/10.1038/251220a0

Chapter 16
Breeding

Kazuharu Nomura

Contents

Domestication is the process of adapting a group of animals to both captive conditions and human interactions over many generations, with the goal of modifying selected traits to produce more productive and efficient individuals (Teletchea 2021). It is a powerful tool for addressing the economic, social, and environmental challenges in the food production sector. Domestication of fish species requires controlling their entire life cycle in captivity without introducing wild individuals, which is an extensive and ongoing process. This process began years to decades ago for most aquaculture species. Thus, <100 of the 250 aquaculture fish species have been domesticated (Teletchea 2021). Among these fish species, several economically important traits, particularly growth, have improved during domestication (Gjedrem and Rye 2018).

Many other aquaculture fish species still rely on wild populations, such as the Japanese eel *Anguilla japonica*, which is completely dependent on wild-caught glass eels. However, the successful completion of the eel life cycle in captivity has paved the way for its domestication, allowing genetic studies using artificially propagated families and the implementation of practical breeding programs. These programs

K. Nomura (✉)
Nansei Field Station, Fisheries Technology Institute, Japan Fisheries Research and Education Agency, Mie, Japan
e-mail: nomura_kazuharu57@fra.go.jp

K. Tsukamoto et al. (eds.), *Eel Science*, Fisheries Science Series,
https://doi.org/10.1007/978-981-99-5692-0_16

must consider both market and non-market values, while maintaining sufficient genetic diversity to ensure fish productivity and resilience to various environmental changes (Teletchea 2021). Ensuring a sustainable future for eel aquaculture requires ongoing improvements through long-term breeding programs. This chapter discusses the current research on breeding eels and their prospects and challenges.

16.1 Breeding Program

Many economically important traits are quantitative traits that can be improved using selective breeding programs. In aquaculture, these programs often start by focusing on improving a single major trait before gradually expanding to improve multiple traits simultaneously (Gjedrem and Baranski 2009).

The breeder's equation can be used to predict the response or phenotypic change in the next generation as a result of selecting a particular trait (Falconer 1996). This equation is expressed as follows:

$$R = h^2 S$$

where R is the response to selection, h^2 is the narrow-sense heritability, and S is the selection differential, or the difference between the population average and the average of the parental population. Because h^2 is less than 1, R is always smaller than S, meaning that the trait value of the offspring tends to regress towards the population mean and be closer to the mean than that of the selected parent. The average offspring generation is genetically changed by R, known as genetic gain, which is permanent and accumulates in the next generation. As more generations progress and genetic gains accumulate, significant changes in traits can be achieved. Traits with high heritability are easier to improve, whereas those with low heritability are more complex and take longer to improve. Traits with extremely low heritability are unsuitable for breeding programs. While maximizing long-term genetic gain is important, it is also important to be careful about increasing S too much, as this can reduce the genetic variance associated with a trait and decrease the genetic gain in the next generation. Additionally, increasing the inbreeding rate can negatively affect offspring performance, a phenomenon known as inbreeding depression.

To maximize long-term genetic gain and avoid inbreeding depression, most selective breeding programs in aquaculture aim to carefully control the inbreeding rate (Gjedrem and Baranski 2009). A breeding program involves recording phenotypes for selected traits, estimation of breeding values, selection of potential parents, and mating programs for the chosen parents (Oldenbroek and van der Waaij 2015). Key issues in a selective breeding program for new species of fish, such as eels, include:

1. Artificial reproduction techniques to produce offspring from any desired mating combination of parents.
2. Knowledge of the genetic characteristics of the target trait and tools for genetic analysis.
3. A base population with sufficient genetic variability.

The following subsections (16.3–16.6) provide an overview of current research on these critical issues in eels.

16.2 Sperm Cryopreservation

Sperm cryopreservation is a valuable tool for optimizing reproduction, improving breeding programs, and enabling gamete availability throughout the year. It also allows the transport of gametes to remote areas and supports artificial fertilization procedures (Judycka et al. 2019). With cryopreserved sperm, the need to synchronize male maturation with the timing of female ovulation is eliminated and valuable genetic resources can be stored and used semi-permanently, thereby enabling flexible mating programs.

Tanaka et al. (2002) were the first to successfully cryopreserve eel sperm using Japanese eels, and since then, many advances have been made in cryopreservation protocols for Japanese and European eels *A. anguilla* (Herranz-Jusdado et al. 2019a). The latest protocol for *A. japonica* uses methanol as a cryoprotectant and has been adapted to cryopreserve larger volumes of sperm (up to 2.5–5 mL, using large-volume straws) (Fig. 16.1; Nomura et al. 2018b). The protocol for *A. anguilla* also uses methanol as a cryoprotectant and has been adapted for the scale-up of cryopreserved sperm volume using 2 and 5 mL cryotubes (Herranz-Jusdado et al. 2019b). In the protocol for *A. japonica*, long-term storage in liquid nitrogen did not affect the post-thaw sperm motility rate, even after 2 years of storage. Fertilizing eggs with cryopreserved sperm resulted in hatching and survival rates that were

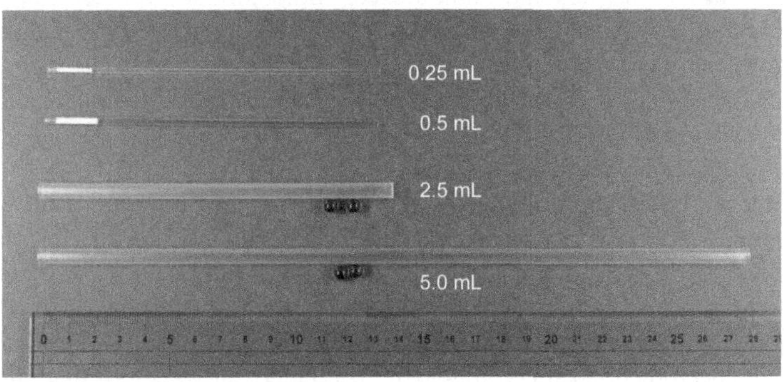

Fig. 16.1 Large-volume straws (2.5 mL and 5.0 mL for the lower two straws, respectively) for sperm cryopreservation in *Anguilla japonica*

comparable to those of fresh sperm at 7 days post-hatching, and the larvae were confirmed to grow into normal glass eels. This demonstrates that cryopreservation of *A. japonica* sperm using large-volume straws is a practical technique for large-scale fertilization programs and can improve the probability of progeny from selected males in eel breeding programs.

16.3 Genomics and Tools

Genomic resources and tools can significantly improve the efficiency of selective breeding programs and broodstock management in many aspects. While mass and family selection based on phenotypic value can be performed without genomic resources, genomic prediction and genome-wide association studies (GWASs) require the collection of genotype data using a large number of DNA markers evenly distributed throughout the whole genome. Single nucleotide polymorphisms (SNPs) are often used as DNA markers in this context. SNPs are the most abundant type of polymorphism in the genome (occurring nearly once every 1000 base pairs) and occur when a single nucleotide in the genome differs at a particular site. Identifying SNPs in the genome and determining their chromosomal locations requires highly accurate draft genome assemblies and linkage maps. At present, draft genome assemblies of freshwater eels are available in public databases for *A. anguilla* (GenBank assembly accession: GCA_013347855.1), *A. rostrata* (GenBank assembly accession: GCA_018555375.2), and *A. japonica* (GenBank assembly accession: GCA_003597225.1), and a linkage map has been reported for *A. japonica* (Nomura et al. 2018a).

Genotyping technologies for genome-wide detection of a large number of SNPs can be broadly divided into SNP arrays and genotyping by sequencing (GBS) using next-generation sequencing (NGS). SNP arrays are a genotyping method that involves hybridizing fragmented DNA sequences from a sample with a plate on which thousands to hundreds of thousands of known SNP allele-specific oligonucleotide probes are aligned and immobilized. Although this approach can provide highly accurate genotype data on fixed SNPs each time, it is expensive. Commercial SNP arrays are only available for a few major aquaculture species and have not yet been developed for eel species. In contrast, GBS is a genotyping method that involves creating DNA libraries enriched with parts of the genome, sequencing them using NGS, and searching for and detecting SNP loci. It can be used without SNP sequence information and is less expensive than the SNP arrays. However, it has the disadvantages of inconsistently available SNPs at the same loci and having a large amount of missing data. There are many variations in GBS, and the use of double-digest restriction-site-associated DNA (ddRAD) (Kai et al. 2014) and genotyping by random amplicon sequencing-direct (GRAS-Di) (Nomura et al. 2022) have been reported for *A. japonica*. The latter study showed that GRAS-Di is valuable for the genetic analysis of *A. japonica* and can be an inexpensive genotyping platform compared with other GBS and SNP arrays.

16.4 Base Population and Broodstock Management

One of the essential steps in starting a new breeding program is the creation of a base population with broad genetic diversity (Gjedrem and Baranski 2009). The genetic diversity of this population is typically expressed in terms of the effective number of mating animals (i.e., the effective population size, N_e). When the breeding population has unequal numbers of males and females, N_e can be approximated using the following equation (Falconer 1996):

$$N_e = \frac{4N_mN_f}{(N_m + N_f)}$$

where N_m and N_f are the number of males and females, respectively.

In a new breeding program, $N_e > 100$ is recommended for the base population, with a minimum of >50 (Gjedrem and Baranski 2009). Therefore, the broodstock used to form the base population should include at least 100 males and 100 females from the wild populations. If possible, using a larger number of parental fish would be beneficial for the breeding program, as it would help avoid a rapid increase in inbreeding rates and maximize the potential for long-term genetic gain. Many previous cases and simulation studies have highlighted the importance of increasing the N_e value of a base population. Some breeding programs and experiments in aquaculture species may have failed in the past because of the low genetic variation in the base population (Gjedrem and Baranski 2009).

Maintaining such a large broodstock can be challenging in many fish breeding programs because of financial and equipment constraints. However, managing the broodstock of freshwater eels is relatively easy, as they can be reared in freshwater or seawater in inexpensive facilities, and neither males nor females mature naturally in captivity (Nomura et al. 2022). Additionally, eels have the advantage of being able to easily control the sex ratio of the broodstock. Sex differentiation in eels occurs after the glass eel stage, and is largely determined by environmental factors. There is no known genetic sex determination mechanism in eels. If estradiol-17β (E2) is administered orally during this period, nearly 100% of eels become female, while more than 90% become male if E2 is not administered (Inaba et al. 2021). In eel-breeding programs, using a combination of an optimal sex allocation strategy for selected parental fish and an artificial mating program using cryopreserved sperm can be a more cost-effective approach for broodstock management. This may involve a smaller number of elite males and a larger number of females, similar to livestock animal breeding programs. However, further research and simulation studies are needed to determine the optimal approach for broodstock management.

In new breeding programs, it is recommended to incorporate as much genetic variation as possible into the founder individuals used to form the base population. To achieve this, it is recommended to collect individuals from as many subpopulations as possible if wild populations are available and use domesticated populations if they are already available (Gjedrem and Baranski 2009). In the case of eels, the wild populations of *A. japonica* (Yu et al. 2020), *A. anguilla* (Als et al. 2011), and *A. rostrata* (Côté et al. 2013) are thought to have a single panmictic population and do not appear to have genetically differentiated regional populations. Additionally, the only eel species that has completed its life cycle in captivity is *A. japonica*, which has been limited to a few countries and institutions. As a result, domesticated artificial populations are few, and just beginning. In the future, once several breeding programs and domesticated populations of eels have been created, they need to be integrated to form a new base population that can be developed into a large-scale breeding program.

16.5 Genetic Analysis of Metamorphic Traits

The estimation of genetic parameters is essential for making decisions regarding the design and implementation of selective breeding programs. Genetic parameters include various indicators. In particular, narrow-sense heritability and genetic correlation between multiple traits may be important in aquaculture breeding programs.

Heritability is a crucial parameter in quantitative genetics that can be broadly and narrowly defined. Broad-sense heritability is defined as the ratio of the total genetic variance (including additive and non-additive genetic variance) to the phenotypic variance, and narrow-sense heritability is defined as the ratio of additive genetic variance to phenotypic variance. Because only additive genetic variance is transmitted to the offspring, breeders are more interested in narrow-sense heritability. Genetic correlation is the correlation between genotypic effects of multiple traits. It is essential to know the magnitude of genetic correlations, because they reveal how the selection of 1 trait may affect correlated traits. It is important to note that estimates of these parameters may vary based on the species, population, phenotypic measures, and genetic models used (Gjedrem and Baranski 2009). Understanding the genetic architecture of a trait, including the number and frequencies of genetic variants and their interactions with each other and with the environment (Timpson et al. 2018), can also be helpful in determining the appropriate genetic model and estimating breeding values. Currently, there are 2 strategies to understand the genetic architecture of complex quantitative traits: quantitative trait loci (QTL) mapping and GWAS. It is advisable to have a good understanding of both genetic parameters and genetic architecture, particularly when implementing a breeding program for a species or trait.

Research on the development of mass production techniques for artificial glass eels of *A. japonica* has been actively conducted. However, the major challenge is the long larval period and low yield of healthy glass eels (Okamura et al. 2014; Tanaka 2015). A range of environmental factors, including diet quality and rearing conditions (water temperature, salinity, and larval density) can affect the length of the larval period in captivity. In addition to these environmental factors, genetic factors have been found to influence the timing of metamorphosis in captivity and body size at that time (Nomura et al. 2018a; Nomura et al. 2022). The narrow-sense heritability for traits related to the timing of metamorphosis from leptocephali into glass eels (age at the onset and completion of metamorphosis) was estimated to be 0.41 and 0.36, respectively, in 746 individuals from 86 full-sib families (Nomura et al. 2018a). The heritability for traits related to body size at metamorphosis (total length, pre-anal length, body depth at metamorphosis onset and completion, and body weight of glass eels) ranged from 0.16 0.33. The genetic correlations between traits related to body size (total length, pre-anal length, and body height) and the timing of metamorphosis (age) were 0.52, 0.63, and 0.19 at onset and 0.55, 0.37, and 0.69, respectively, at completion. These estimates were consistent with those obtained using both the pedigree-based best linear unbiased prediction (ABLUP) and genomic best linear unbiased prediction (GBLUP) models, with the dataset including more families and individuals in subsequent studies (Nomura et al. 2022). These results suggest that traits related to timing and body size at metamorphosis have additive genetic variance in the population and can be improved through selective breeding, with modest genetic correlations between them. In an F_1 full-sib family, QTL mapping for these traits identified 1 significant (genome-wide $P < 0.05$) and 5 suggestive (chromosome-wide $P < 0.05$) QTL (Nomura et al. 2018a). These results suggest that these metamorphic traits have a polygenic or oligogenic genetic structure comprising many QTL with minor effects and that some of these QTL are involved in both timing and body size. Therefore, it may be appropriate to use genomic selection with high-density SNP markers to capture the effects of desirable minor genes associated with these traits distributed across multiple ancestors and effectively enrich them through selective breeding.

16.6 Genomic Selection

Genomic prediction is a method for estimating the breeding value of a test population composed of individuals without recorded phenotypic data. This is done by linking genotype data from a large number of genome-wide DNA markers that cover all chromosomes with phenotype data from a reference population that includes genotyped and phenotyped individuals (Meuwissen et al. 2001). Genomic selection is a breeding program that is based on genomic predictions. This approach has become important in plant and livestock breeding because it allows for improved prediction accuracy, controlled inbreeding, and, in some cases, a shorter generation interval than traditional pedigree-based approaches. The use of genomic selection in aquaculture breeding programs has been rapidly expanding and has been shown to increase the accuracy of breeding values for several traits in various species compared with conventional selection based on pedigree information (Houston et al. 2020).

The accuracy of genomic prediction is affected by various factors, including the heritability and genetic architecture of the target trait, marker density, training population size, both family number and family size, genetic model, and relatedness between the training and validation datasets (Hayes et al. 2009). Although the heritability and genetic architecture of the target traits are complex properties that can be artificially manipulated, some factors, including marker density, training population size, and relatedness between the training and validation datasets, can be quickly and systematically controlled. The impact of marker density and training population size on the actual prediction accuracy can be tested *in silico* for each dataset, and the results can help inform the decision-making process when designing appropriate breeding programs.

Nomura et al. (2022) conducted a study to investigate the potential value of genomic selection for 10 traits in *A. japonica* related to the timing of metamorphosis and body size. The study used 1689 glass eels from partial factorial crosses (43 sires × 32 dams) and the GRAS-Di genotyping method. Results from a cross-validation analysis showed that genomic predictions of breeding value using 17,017 SNPs outperformed pedigree-based predictions for almost all traits, with a prediction accuracy improvement of 3–19%. These results suggest that using genomics to construct more accurate relationship matrices that take into account Mendelian sampling effects (i.e., genetic variation between full-sibs due to the random distribution of alleles from the same parents) can improve the prediction accuracy of breeding value. The impact of marker density on prediction accuracy was also examined, and it was found that marker densities of over 8000 SNPs were sufficient to achieve a selection accuracy similar to that obtained using higher SNP densities for all 10 traits. However, when the number of samples in the training set was limited to reduce genotyping and phenotypic recording costs, selecting only the top individuals resulted in significantly lower prediction accuracy compared to randomly selecting the same number of samples. These findings may be useful in reducing the number of SNPs and genotyped samples in the training set when applying genomic selection to metamorphic traits in *A. japonica*. Although the genotyping cost has been decreasing rapidly, it is still too high for routine use in genomic selection. Some aquaculture species have attempted to genotype target populations using low-coverage sequencing or low-density SNP arrays, and then obtain genome-wide genotypes through imputation (Meuwissen 1997).

16.7 Optimum Contribution Selection

Optimum contribution selection (OCS, Meuwissen 1997) is a breeding strategy that maximizes genetic gain while controlling inbreeding. OCS involves selecting parents in a way that limits the genetic relationships between them to control the inbreeding rate of offspring. This approach is sustainable for genetic improvement in aquaculture breeding. When genomic information is used to make OCS decisions, it is known as genomic OCS (GOCS, Gebregiwergis et al. 2020) or genomic mating (GM, Akdemir and Sánchez 2016).

Genomic selection, which involves using estimated breeding values and random mating to select candidates, may increase inbreeding and quickly deplete the genetic

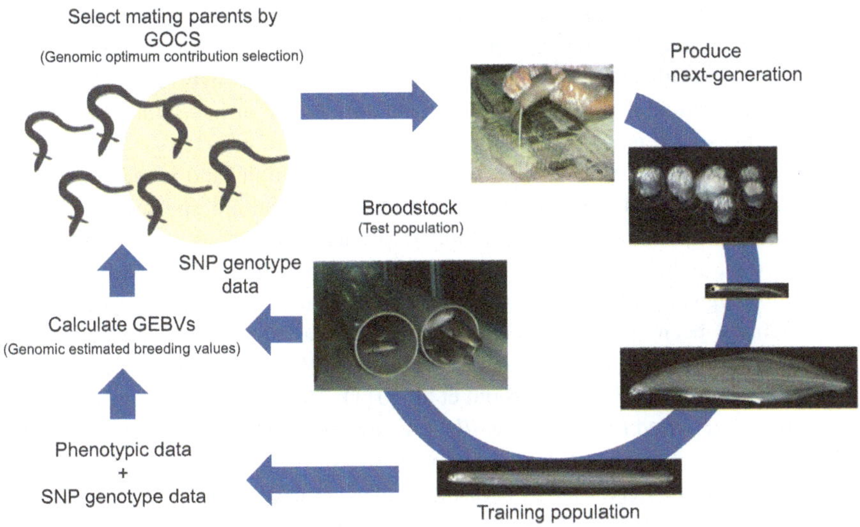

Fig. 16.2 Conceptual diagram of the breeding scheme using GOCS for *Anguilla japonica*

diversity of selected traits. In contrast, the GOCS allows for a flexible balance between genetic gain and diversity.

A conceptual diagram of the breeding scheme using the GOCS for *A. japonica* is shown in Fig. 16.2. To implement the GOCS for *A. japonica*, a training population was first created from the base population, and phenotypic and SNP genotype data were collected. The more families, individuals per family, and markers included, the higher the prediction accuracy of the genomic estimated breeding value (GEBV). The training population included data for single or multiple traits.

Next, broodstock candidates genetically related to the training population were used as the test population, and SNP genotype data were collected. The GEBVs were calculated for each individual using a prediction model created from the training population data. If multiple target traits are considered, the GEBVs are linearly combined with a selection index.

The GOCS is then used to determine the combination of parental fish selection and mating plans that balance the expected genetic gain and inbreeding in the next generation. The next generation was produced based on the determined mating plan.

16.8 Other Breeding Schemes

In addition to selective breeding, which involves exploiting genetic variation to produce genetic gain in each generation, other approaches to genetic improvement in aquaculture species include crossbreeding, chromosome manipulation techniques, and more recently, genome editing technology.

Crossbreeding involves mating different strains, inbred lines, or even different species (Gjedrem and Baranski 2009). The mating of 2 different species is called interspecific hybridization, and their offspring are known as hybrids. Hybrids may have a combination of traits from the 2 parental species, or have traits superior to both parents, which is a phenomenon called hybrid vigor or heterosis. Heterosis is expressed through the non-additive genetic variance of traits and is maximal in the F_1 generation. Hybrids are produced in various aquaculture species to increase growth and survival, transfer or combine desirable traits between 2 species into a single group, reduce unwanted maturation through sterile fish production, and provide biological protection to genetically improved stocks. Interspecific hybrids of eels have been artificially created as viable larvae using combinations of *A. anguilla* × *A. japonica* (Okamura et al. 2004; Müller et al. 2012) and *A. anguilla* × *A. australis* (Burgerhout et al. 2011). The existence of natural hybrids has also been reported for the *A. anguilla* × *A. rostrata* combination (Frankowski and Bastrop 2010). As various other combinations of interspecific hybrid eels are possible, it may be necessary to investigate the characteristics of hybrid eel aquaculture in the future. Chromosomal manipulation techniques are often used in combination with interspecific hybrids. In some cases, viability is restored by triploidization of lethal interspecific hybrids. It has been reported that triploidization is possible in *A. japonica* through heat-shock treatment after fertilization (Nomura et al. 2004).

Genome editing technologies, particularly those utilizing CRISPR/Cas9, have experienced rapid development in recent years and show great potential to accelerate the genetic improvement of aquaculture species (Yang et al. 2022). These technologies have been used to precisely edit genes to identify gene functions and generate desired traits in over 20 aquaculture species. However, genome editing in aquaculture breeding faces several technical challenges and regulatory issues. One challenge is the lack of genomic and genetic information for many aquaculture species, as well as the limited sequence information for the target genes of genome editing. Another challenge is the limited application of genome editing for complex traits. While the main focus of genome editing has been on improving traits through the loss of function of a single gene, improving complex traits and integrating with conventional selective breeding may be possible in the future. For example, introducing new mutations involved in complex traits, removing harmful alleles from the population, and combining genome editing with surrogate techniques to accelerate generations have been proposed (Houston et al. 2020); however, these attempts are still in their infancy and face technical challenges. A third challenge is the need to foster consumer acceptance and develop a global regulatory framework for using animals produced by genome-editing technologies as food products. Despite these challenges, it is clear that research on integrating genome-editing technologies with conventional selective breeding to accelerate genetic improvement in aquaculture species will continue to progress rapidly. Research using genome-editing technology will also be conducted on eels in the future.

16.9 Conclusion and Future Remarks

The results of eel breeding studies are encouraging, although the application of these techniques is still in the early stages. Previous studies have shown that selective breeding can accelerate the onset of metamorphosis and shorten the larval period, as indicated by heritability estimates. However, technology for rearing leptocephali of *A. japonica* is still being developed, and certain environmental factors, such as food and rearing methods, continue to be refined. Therefore, while promoting the use of genomic selection to shorten the larval period, it is important to gather phenotypic data on metamorphic traits from the post-selection generation and continually evaluate the genetic parameters of these traits, such as realized genetic gain, genetic trends, heritability, genetic correlation, and G × E interactions. Further research is needed to understand the genetic parameters and genetic architecture associated with growth rate, disease resistance, and quality-related traits in aquaculture conditions, and to expand breeding programs to include these and other traits as targets in addition to metamorphic traits. Additionally, combining traditional selective breeding with genome-editing technologies may accelerate the improvement of eel genetics. To sustainably utilize eel resources, domestication levels must be increased to improve populations, encouraging better adaptation to aquaculture conditions to reduce their reliance on wild populations.

References

Akdemir D, Sánchez JI (2016) Efficient breeding by genomic mating. Front Genet 7:210. https://doi.org/10.3389/fgene.2016.00210

Als TD, Hansen MM, Maes GE, Castonguay M, Riemann L, Aarestrup K, Munk P, Sparholt H, Hanel R, Bernatchez L (2011) All roads lead to home: panmixia of European eel in the Sargasso Sea. Mol Ecol 20:1333–1346. https://doi.org/10.1111/j.1365-294X.2011.05011.x

Burgerhout E, Brittijn SA, Kurwie T, Decker P, Dirks RP, Palstra AP, Spaink HP, Van den Thillart GEEJM (2011) First artificial hybrid of the eel species *Anguilla australis* and *Anguilla Anguilla*. BMC Dev Biol 11:16. https://doi.org/10.1186/1471-213X-11-16

Côté CL, Gagnaire PA, Bourret V, Verrault G, Castonguay M, Bernatchez L (2013) Population genetics of the American eel (*Anguilla rostrata*). *F*ST = 0 and North Atlantic oscillation effects on demographic fluctuations of a panmictic species. Mol Ecol 22:1763–1776. https://doi.org/10.1111/mec.12142

Falconer DS (1996) Introduction to quantitative genetics. Pearson Education India

Frankowski J, Bastrop R (2010) Identification of *Anguilla anguilla* (L.) and *Anguilla rostrata* (Le Sueur) and their hybrids based on a diagnostic single nucleotide polymorphism in nuclear 18S rDNA. Mol Ecol Resour 10:173–176. https://doi.org/10.1111/j.1755-0998.2009.02698.x

Gebregiwergis GT, Sørensen AC, Henryon M, Meuwissen T (2020) Controlling coancestry and thereby future inbreeding by optimum-contribution selection using alternative genomic-relationship matrices. Front Genet 11:345. https://doi.org/10.3389/fgene.2020.00345

Gjedrem T, Baranski M (2009) Selective breeding in aquaculture: an introduction. Springer, Netherlands

Gjedrem T, Rye M (2018) Selection response in fish and shellfish: a review. Rev Aquac 10:168–179. https://doi.org/10.1111/raq.12154

Hayes BJ, Bowman PJ, Chamberlain AJ, Goddard ME (2009) Invited review: genomic selection in dairy cattle: progress and challenges. J Dairy Sci 92:433–443. https://doi.org/10.3168/jds.2008-1646

Herranz-Jusdado JG, Gallego V, Morini M, Rozenfeld C, Pérez L, Müller T, Horváth Á, Ohta H, Asturiano JF (2019a) Eel sperm cryopreservation: an overview. Theriogenology 133:210–215. https://doi.org/10.1016/j.theriogenology.2019.03.033

Herranz-Jusdado JG, Gallego V, Rozenfeld C, Morini M, Pérez L, Asturiano JF (2019b) European eel sperm storage: optimization of short-term protocols and cryopreservation of large volumes. Aquaculture 506:42–52. https://doi.org/10.1016/j.aquaculture.2019.03.019

Houston RD, Bean TP, Macqueen DJ, Gundappa MK, Jin YH, Jenkins TL, Selly SLC, Martin SAM, Stevens JR, Santos EM, Davie A, Robledo D (2020) Harnessing genomics to fast-track genetic improvement in aquaculture. Nat Rev Genet 21:389–409. https://doi.org/10.1038/s41576-020-0227-y

Inaba H, Hara S, Horiuchi M, Ijiri S, Kitano T (2021) Gonadal expression profiles of sex-specific genes during early sexual differentiation in Japanese eel *Anguilla japonica*. Fish Sci 87:203–209. https://doi.org/10.1007/s12562-020-01491-5

Judycka S, Nynca J, Ciereszko A (2019) Opportunities and challenges related to the implementation of sperm cryopreservation into breeding of salmonid fishes. Theriogenology 132:12–21. https://doi.org/10.1016/j.theriogenology.2019.03.022

Kai W, Nomura K, Fujiwara A, Nakamura Y, Yasuike M, Ojima N, Masaoka T, Ozaki A, Kazeto Y, Gen K, Nagao J, Tanaka H, Kobayashi T, Ototake M (2014) A ddRAD-based genetic map and its integration with the genome assembly of Japanese eel (*Anguilla japonica*) provides insights into genome evolution after the teleost-specific genome duplication. BMC Genomics 15:233. https://doi.org/10.1186/1471-2164-15-233

Meuwissen TH (1997) Maximizing the response of selection with a predefined rate of inbreeding. J Anim Sci 75:934–940. https://doi.org/10.2527/1997.754934x

Meuwissen TH, Hayes BJ, Goddard ME (2001) Prediction of total genetic value using genome-wide dense marker maps. Genetics 157:1819–1829. https://doi.org/10.1093/genetics/157.4.1819

Müller T, Horváth Á, Takahashi E, Kolics B, Bakos K, Decsi K, Kovács B, Taller J, Urbányi B, Bercsényi M, Horváth L, Adachi S, Arai K, Yamaha E (2012) Artificial hybridization of Japanese and European eel (*Anguilla japonica* × *A. anguilla*) by using cryopreserved sperm from freshwater reared males. Aquaculture 350:130–133. https://doi.org/10.1016/j.aquaculture.2012.04.007

Nomura K, Nakajima J, Ohta H, Kagawa H, Tanaka H, Unuma T, Yamauchi K, Arai K (2004) Induction of triploidy by heat shock in the Japanese eel *Anguilla japonica*. Fish Sci 70:247–255. https://doi.org/10.1111/j.1444-2906.2003.00798.x

Nomura K, Fujiwara A, Iwasaki Y, Nishiki I, Matsuura A, Ozaki A, Sudo R, Tanaka H (2018a) Genetic parameters and quantitative trait loci analysis associated with body size and timing at metamorphosis into glass eels in captive-bred Japanese eels (*Anguilla japonica*). PLoS One 13: e0201784. https://doi.org/10.1371/journal.pone.0201784

Nomura K, Koh ICC, Iio R, Okuda D, Kazeto Y, Tanaka H, Ohta H (2018b) Sperm cryopreservation protocols for the large-scale fertilization of Japanese eel using a combination of large-volume straws and low sperm dilution ratio. Aquaculture 496:203–210. https://doi.org/10.1016/j.aquaculture.2018.07.007

Nomura K, Ishikawa T, Sudo R, Fujiwara A (2022) Genomic prediction of 10 metamorphic traits of captive-bred Japanese eels (*Anguilla japonica*) using the GRAS-Di genotyping method. Aquaculture 548:737671. https://doi.org/10.1016/j.aquaculture.2021.737671

Okamura A, Zhang H, Utoh T, Akazawa A, Yamada Y, Horie N, Mikawa N, Tanaka S, Oka HP (2004) Artificial hybrid between *Anguilla anguilla* and *A. japonica*. J Fish Biol 64:1450–1454. https://doi.org/10.1111/j.0022-1112.2004.00409.x

Okamura A, Horie N, Mikawa N, Yamada Y, Tsukamoto K (2014) Recent advances in artificial production of glass eels for conservation of anguillid eel populations. Ecol Freshw Fish 23:95–110. https://doi.org/10.1111/eff.12086

Oldenbroek K, van der Waaij L (2015) Textbook animal breeding and genetics for BSc students. Centre for Genetic Resources and Animal Breeding and Genomics Centre, Wageningen University and Research, The Netherlands. Groen Kennisnet: https://wiki-groenkennisnet.atlassian.net/wiki/spaces/TAB/overview. Accessed 28 May 2022

Tanaka H (2015) Progression in artificial seedling production of Japanese eel *Anguilla japonica*. Fish Sci 81:11–19. https://doi.org/10.1007/s12562-014-0821-z

Tanaka S, Zhang H, Horie N, Yamada Y, Okamura A, Utoh T, Mikawa N, Oka HP, Kurokura H (2002) Long-term cryopreservation of sperm of Japanese eel. J Fish Biol 60:139–146. https://doi.org/10.1111/j.1095-8649.2002.tb02393.x

Teletchea F (2021) Fish domestication in aquaculture: 10 unanswered questions. Anim Front 11: 87–91. https://doi.org/10.1093/af/vfab012

Timpson NJ, Greenwood CMT, Soranzo N, Lawson DJ, Richards JB (2018) Genetic architecture: the shape of the genetic contribution to human traits and disease. Nat Rev Genet 19:110–124. https://doi.org/10.1038/nrg.2017.101

Yang Z, Yu Y, Tay YX, Yue GH (2022) Genome editing and its applications in genetic improvement in aquaculture. Rev Aquac 14:178–191. https://doi.org/10.1111/raq.12591

Yu L, Liu Y, Liu J (2020) Gene-associated microsatellite markers confirm panmixia and indicate a different pattern of spatially varying selection in the endangered Japanese eel *Anguilla japonica*. J Ocean Limnol 38:1572–1583. https://doi.org/10.1007/s00343-020-0048-z

Chapter 17
Diseases

Tomoyoshi Yoshinaga

Contents

Eels are susceptible to various pathogens that can cause infectious diseases both on farms and in the wild. There are several characteristics of infection and disease control in eels. First, changes in eel culture methods, especially the transitions of aquaculture facilities from outdoor tanks to heated greenhouses and of eel feeds from live or low-processed feed to compound feed, have significantly affected disease outbreaks. Second, the international trade of eels causes the spread of pathogens, resulting in host expansion from natural host species to new host species. In addition, pathogens affect both wild and farmed eels. This chapter introduces significant diseases in eels, with a focus on the Japanese and European eels.

17.1 Brachionephritis with Unknown Etiology

In 1969, European eel *Anguilla anguilla* elvers began to be imported and delivered to eel farms nationwide in Japan to compensate for the shortage of elvers for eel production. Since November of that year, mass mortalities of Japanese eels *A. japonica* due to a disease with unknown etiology were reported from outdoor ponds, which were accompanied with pathological abnormalities such as thickening and adhesion of epithelial cells in the gill lamellae, hyaline droplet degeneration of

T. Yoshinaga (✉)
Graduate School of Agricultural and Life Sciences, The University of Tokyo, Tokyo, Japan
e-mail: yoshinaga-tomoyoshi@g.ecc.u-tokyo.ac.jp

K. Tsukamoto et al. (eds.), *Eel Science*, Fisheries Science Series,
https://doi.org/10.1007/978-981-99-5692-0_17

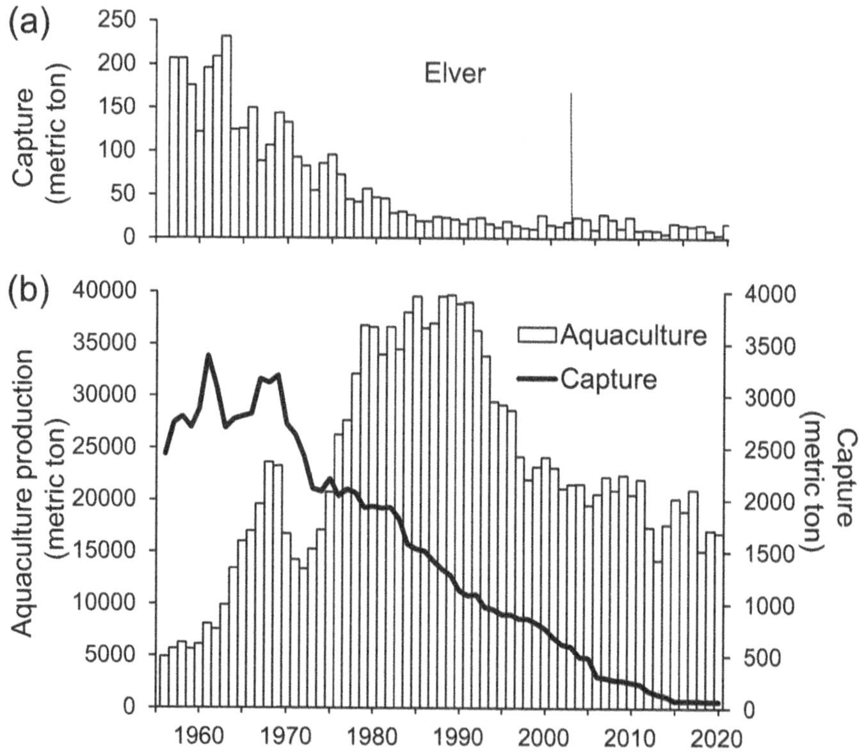

Fig. 17.1 Elver capture (**a**) and aquaculture and capture eel production (**b**) in Japan. The data collection method for elver capture has changed since 2003 (dotted line). Based on statistical data issued from Ministry of Agriculture, Forestry and Fisheries, Japan

the renal tubules, and low concentration of chloride ions in the peripheral blood (Egusa 1970; Nishio et al. 1971). Based on these pathological changes, the disease was named branchionephritis (Egusa 1970).

Branchionephritis frequently occured in eels cultured in outdoor ponds during the winter. Because it could be prevented and cured with increased water temperature, heated greenhouse eel ponds rapidly spread for eel culture in Japan and occupied most of the eel culture ponds. While the disease caused a dramatic decrease in eel culture production in Japan in the early 1970s, it was eradicated owing to the replacement of outdoor ponds with greenhouse ponds, and eel culture production recovered by the late 1970s (Fig. 17.1).

Although intensive research has previously been conducted to determine the etiology of this disease, particularly potential pathogens, the cause has not been elucidated. No research is currently being conducted because of the eradication of the disease. It is speculated that the disease may be related to the change in feed from pupae generated by silk production and boiled fish to fishmeal-based compound feeds, as the change occurred around the same period as the disease outbreak.

Another explanation is that branchionephritis is a pathological change caused by cold-water stress, which is based on results of a previous study wherein nephroses and low serum chloride levels were observed in eels with cold-water stress (Kobayashi et al. 2000).

17.2 Viral Diseases

17.2.1 Eel Virus European (EVE)

To elucidate the etiology of branchionephritis, researchers in Japan attempted to isolate causative viruses. In 1973, an aquabirna virus named eel virus European (EVE) was isolated from European eels imported into Japan (Sano et al. 1981). Later, the virus was isolated from Japanese eels in Japan and subsequently from other East Asian countries, including Taiwan and Korea. Since the first isolation of the virus from European eels in Europe in the late 1970s, it has been repeatedly isolated from European eels and elvers in Europe, including the UK, Germany, Denmark, the Netherlands, and Greece (van Ginneken et al. 2004). It has also been isolated from American eels *A. rostrata* in the USA. It is suggested that the disease originated from the European eel and was introduced into East Asian countries with the import of European eel elvers.

EVE has been isolated from both apparently healthy and diseased eels (van Ginneken et al. 2004). Diseased Japanese eels clinically exhibit abnormal trunk shapes and congestion on the body surface. Pathologically, enlargement of the kidneys, necrosis or petechial hemorrhages in the liver, hematoma formation, and pillar cell necrosis in the gill lamellae were observed.

17.2.2 Eel Virus European X (EVEX)/Eel Virus American (EVA)

In 1974, a rhabdovirus was isolated from a young American eel imported from Cuba into Japan, and named Eel Virus American (EVA) (Sano 1976). Another rhabdovirus, named Eel Virus European X (EVEX), was isolated from European eels imported from France to Japan in 1977, and named Eel Virus European X (EVEX) (Sano et al. 1977). Although both viruses were once regarded as single-virus species, molecular comparisons revealed that they belonged to different genetic groups. EVEX and EVA are likely to originate in Europe and America, respectively.

EVEX has been isolated from farmed and wild European eels in European countries, including France, the UK, Denmark, the Netherlands, and Italy (van Ginneken 2004). In Japan, EVEX/EVA was isolated from Japanese eels during postharvest stocking in cages (Kobayashi and Miyazaki 1996), when the 2 viruses were not yet distinguished.

When EVA was isolated from an American eel in Japan, the affected eel clinically showed congestion in the pectoral and anal fins and in the abdominal skin, and histopathologically showed hemorrhages and degeneration in the skeletal muscles, hyperemia in the branchial vessels, and hemorrhages in the Bowman's space and tubules in the kidney (Sano 1976). EVEX has been isolated from diseased and apparently healthy farmed and wild European eels. In a swimming tunnel experiment, silver European eels infected with EVEX developed petechial hemorrhages, bloody body fluid, and anemia after swimming 1000–1500 km. However, uninfected eels successfully swam 5500 km, which is the estimated distance from Europe to the Sargasso Sea (the probable spawning area of European eels) (van Ginneken et al. 2005). This suggests that EVEX obstructs the spawning migration of European eels.

17.2.3 Anguillid Herpesvirus 1

A herpes virus was isolated from Japanese and European eels in Japan in 1985, named *Herpesvirus anguillae*, or anguillid herpes virus 1 (AngHV-1) (Sano et al. 1990). Since the first isolation of AngHV-1 in Japan, herpes viruses have been isolated from Japanese eels with cutaneous lesions or gill diseases, and are occasionally accompanied by mass mortalities in Japan and Taiwan. In Europe, the virus has been repeatedly isolated from wild and farmed European eels. The virus has also been isolated from the American eel.

Clinically, viral infection is accompanied by apathy, loss of appetite, and various stress-induced symptoms on the skin and fins, including erythema, hemorrhage, and ulcers. Mortalities ranged from 0–30%.

van Beurden et al. (2012) analyzed diagnostic data on viral infection in wild and farmed European eels in the Netherlands from 1990 to 2011. AngHV-1, EVE, and EVEX were the most commonly observed. Among them, AngHV-1 was the most frequently observed in wild eels.

17.2.4 Viral Endothelial Cell Necrosis

In the mid-1980s, a disease characterized by intense congestion in the central venus sinuses of gill filaments appeared in Japanese eels cultured in heated greenhouse eel farms and subsequently prevailed in greenhouse eel farms nationwide (Egusa et al. 1989). Based on the findings of virions in various tissues, including endothelial cells, by TEM observation and the successful results in challenge experiments, the disease was confirmed to be a viral disease and was named viral endothelial cell necrosis (VECNE) (Inoue et al. 1994). The causative virus was first isolated by Ono et al. (2007), almost 25 years after the first occurrence of the disease, using the endothelial cell line of Japanese eels established in their study. Later, the virus was identified as a novel DNA virus, based on molecular analysis, and named Japanese eel endothelial cell-infecting virus (JEECV) (Mizutani et al. 2011).

This disease is the only viral disease that causes significant mortality in Japanese eels cultured in heated greenhouse eel farms. It is likely to be adapted to high temperatures, as experiments have shown higher mortality rates in eels reared at 28 °C and 31 °C than those reared at 20 °C and 24 °C (Tanaka et al. 2008). JEECV has been detected not only in farmed eels but also in wild elvers and yellow eels caught in rivers in Japan and in 2/5 apparently healthy mature eels caught in the West Mariana Ridge (Okazaki et al. 2015, 2016; Okazaki-Terashima et al. 2016). These findings suggest that JEECV has prevailed in wild eels without apparent symptoms and is introduced by infected elvers into greenhouse eel farms, where eels are reared at high temperatures around 28 °C and develop VECNE. Recently, however, waterborne transmission from older eels to newly introduced eels has been suggested as another major transmission route of the virus to eel farms (Umeda K, personal communication).

17.3 Bacterial Diseases

17.3.1 Pseudomonas angulliseptica

A disease characterized by petechial hemorrhage on the skin of Japanese eels occurred on eel farms in Tokushima and Shizuoka Prefectures in Japan in the spring of 1971. Disease outbreaks spread to outdoor ponds, and was referred to as the red spot disease of eels. Its causative bacterium was subsequently isolated and identified as a new species, *Pseudomonas angulliseptica,* by Wakabayashi and Egusa (1972). In addition to petechial hemorrhage, the disease causes various symptoms, such as blood congestion in the liver, decoloring and atrophy of the spleen, atrophy of the kidney, reddening in the intestine and stomach, and petechial hemorrhage in the peritoneum, which leads to mortality in heavily affected eels. The bacterium needs salt (0.1–4%) for growth and does not grow in media without it. The bacterium grows between 5–30 °C with an optimal growth temperature of 15–20 °C. Therefore, outbreaks of the disease occur more frequently in eels in brackish water containing low salt concentrations than those in freshwater. Currently, the disease seldom occurs in heated greenhouse facilities that have been extensively used for eel culture in Japan since the late 1970s, where eels are reared at high temperatures of 28–30 °C.

Since the discovery of *P. angulliseptica* in Japanese and European eels in Japan and Asia, the bacterium has been recorded from a variety of marine and freshwater fish in fish farms and in the wild in Europe and Asia, including European eels, salmonid fish, cod, and gilthead seabream. The strains of the bacterium can be separated into 2 genotypes: 1 type is mostly isolated from eels, including Japanese and European eels, and the other type is from other fish species, indicating that the bacteria isolated from eels have the same origin (López-Romalde et al. 2003).

Pathogenicity of the bacterium is low in the European eel (Muroga et al. 1975; Haenen and Davidse 2001). It is believed that the bacterium causing red spot disease in Japanese eel was originally distributed in Europe, introduced to Japan accompanying the seedlings of European eel imported into Japan in 1969 and the 1970s, and expanded the host range to the Japanese eel, which has higher susceptibility to the bacterium than its original natural host, the European eel.

17.3.2 Edwardsiella tarda/piscicida/anguillarum

Edwardsiella tarda is the causative agent of edwardsiellosis in various fish species, including freshwater and marine fish, and in terrestrial animals, including humans, and is distributed worldwide in Europe, Asia, and North America. However, most bacterial strains isolated from fish and previously identified as *E. tarda* were reclassified and identified as 1 of 2 novel species, *E. piscicida* and *E. anguillarum*, based on molecular methods, although *E. tarda* has also been isolated from fish (Buján et al. 2018). Both *E. piscicida* and *E. anguillarum* have been isolated from Japanese, European, American, and mottled eels, causing mass mortality.

Eels affected by edwardsiellosis exhibit erratic swimming, reddening, and petechiae on the fins, belly, and tissue around the anus, as well as enlargement and extrusion of the anus. Ulcers and abscesses are formed in the kidneys and liver. Edwardsiellosis develops at high water temperatures and occurs only in the summer in the outdoor ponds. Once greenhouse eel ponds became widespread and common in Japan in the 1970s, eswardsiellosis occurred regardless of season. However, outbreaks of edwardsiellosis have become rare after tubificids given to elvers for food acclimatization were replaced with artificial feed, for unknown reasons.

In Europe, outbreaks of edwardsiellosis are rare on eel farms. In contrast, wild European eels showed prevalence of edwardsiellosis at a 12–33% in warm seasons in a lake in the Mediterranean (Alcaide et al. 2006). Thus, edwardsiellosis may be considered as a possible cause of worldwide eel population reductions.

17.4 Parasitic Diseases

17.4.1 Anguillicola crassus *(syn.* Anguillicoloides crassus*)*

The Japanese eel is the final natural host of the nematode *A. crassus*. Adults of this parasite live in the lumen of the swim bladder of the eels. The parasite causes little damage to Japanese eels, in which the prevalence, infection intensity, and pathogenicity are low; therefore, research on this parasite has been limited in Japanese eels. However, when elvers of the European eel were introduced to Japan to compensate for the shortage of elvers of Japanese eel for eel culture in the 1970s, the introduced European eels showed high mortality due to the high prevalence and intensity of parasitic infection (Egusa 1979). The parasite accidentally entered Europe with the import of Japanese eels in the early 1980s, rapidly expanded its distribution through natural migration and artificial transportation, and subsequently affected European eels on eel farms and in the wild (Kirk 2003; Kennedy 2007). There have been reports of mortality rates reaching 65% due to the parasite in aquaculture farms in the Netherlands and Denmark. Mass mortalities of infected European eels have also been reported in Lake Balaton in Hungary and the Vranov Reservoir in the Czech Republic, where the densities of European eel populations were high.

Adult nematodes lay eggs in the lumen of the swim bladder of eels. Nematodes molt four times before becoming adults, passing through the first, second, third, and fourth stages of larval development. In *A. crassus*, the first molt is completed in the egg and the larva hatches from the egg as the second-stage larva, which is surrounded by the cuticle of the first-stage larva. Eggs or hatched larvae are released into the environment via the pneumatic duct and gastrointestinal tract of the eel. The larvae hatched from the released eggs are ingested by copepods and migrate into the copepod lumen, where they become third-stage larvae after the second molting. When infected copepods are ingested by eels, the third-stage larvae quickly leave the copepods, penetrate the digestive tract, and migrate to the swim bladder, where they grow to the fourth-stage larvae and subsequently to adults after molting twice.

In a challenge experiment in which the third-stage larvae were orally administered to Japanese and European eels, only 14% of the larvae successfully invaded and ~60% of larvae recovered from Japanese eels were found dead because of the host defense, whereas in European eels, 33% invaded and no larvae were found dead (Knopf and Mahnke 2004). This study indicated that the new host species, the European eel, is less adapted to nematode infection than the natural host, the Japanese eel. The successful invasion and survival of the nematode could be the reason for the high pathogenicity and infection levels in European eels.

Various pathological changes occur in infected European eels, including hemorrhage and inflammation in the swim bladder, fibrosis, and thickening of the swim bladder wall. In heavily infected European eels, the volume of the swim bladder lumen is reduced owing to live worms, and adults that die after reproduction fill the lumen (Kirk 2003). In addition, physiological disturbances such as impaired gas gland function, reduced swimming speed, and low stress tolerance have been reported in heavily infected European eels. Eels are also more susceptible to secondary bacterial infections. Mortalities of infected European eels in aquaculture farms and lakes are thought to be attributable to the combination of infection with environmental stresses such as high temperatures, low dissolved oxygen, and secondary bacterial infection.

Furthermore, the life cycle of this parasite has changed since its introduction in Europe. In Europe, freshwater fish, mollusks, amphibians, and aquatic insects have been reported as paratenic or extended intermediate hosts. When the paratenic host animal ingest copepods harboring the third-stage larvae, the third-stage larvae do not progress further. European eels are susceptible to multiple parasitism by feeding on a paratenic host that has ingested the third stage. Small eels are thought to be infected by feeding on the intermediate host, copepods, whereas large eels are infected by feeding on paratenic hosts, especially fish. In Japan, however, no paratenic host has been found thus far, although its existence cannot be ruled out. This change in the infection cycle is thought to be one of the reasons for the higher number and rate of parasitism in the European eels.

Since infection with this nematode causes significant damage to the swim bladder, which plays an important role in swimming, it is assumed, although not directly proven, that the infection inhibits the migration of silver eels to the Sargasso Sea, their probable spawning grounds. Together with severe physiological damage and high mortality, this nematode has been suspected to be strongly involved in the decrease in European eel abundance since the 1980s.

17.4.2 **Pseudodactylogyrus** *spp.*

The monogenean parasites *Pseudodactylogyrus anguillae* and *Pseudodactylogyrus bini* have a similar morphology and life cycle (Fig. 17.2). They are small (total length: 1–2 mm) and are attached to the gill filaments of eels with an anchor-shaped structure called humli. They feed on cells, debris, and mucus. Onchomiracidia hatch from eggs (in water) and swim to reach the host gill filaments through the mouth.

The Japanese eel is considered a natural host of parasites. When the elvers of the European eel were introduced in the 1970s and later, European eels showed much higher susceptibility to and were significantly more damaged by the parasites than Japanese eels (Egusa 1979).

The two parasite species have been reported from European countries since the 1980s. Both monogeneans are thought to have been introduced to Europe and the Americas from Asia with the transportation of live eels from Asia, and spread throughout the continent by transportation of elvers (Buchmann et al. 1987). In an experiment in which European and Japanese eels were cohabitated and challenged by two monogeneans (Fang et al. 2008), European eels showed higher susceptibility than Japanese eels to the parasites. The results also suggest that the Japanese eel was the original host of these parasites. These parasites have caused damage and economic losses in eel farms in Europe, especially in intensive pond cultures, whereas significant damage has not been reported in wild eels.

17.5 Characteristics of Eel Diseases

17.5.1 Changes in Farming Methods and Diseases

In the 1970s, eel farms shifted from being outdoor pond farms to being heated greenhouse farms. In heated greenhouses, eels are reared at high temperatures >28 ° C all year, where they can be raised more rapidly without experiencing low winter temperatures. Heated greenhouses were effective in controlling branchionephritis,

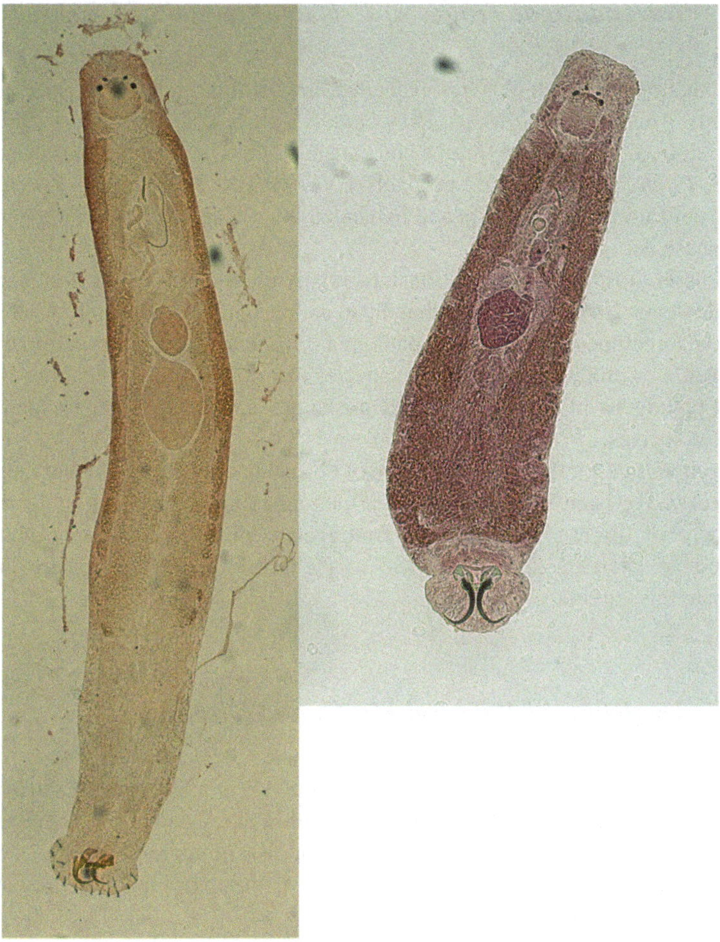

Fig. 17.2 *Pseudodactylogyrus bini* (left) and *P. anguillae* (right). Provided by Laboratory of Fish Diseases, the University of Tokyo

which led to the rapid spread of utilizing greenhouse ponds. Apart from branchionephritis, high-temperature farming dramatically affects the composition of eel diseases in farms. Infections with *Pseudomonas anguilliseptica* and atypical *Aeromonas salmonicida* have almost disappeared from eel farms, while Japanese eel endothelial cells-infecting virus (JEECV) has become a major problem.

Changes in diet for food acclimatization also affected the occurrence of edwardsiellosis. After the diet for food acclimatization changed from tubificids to artificial diets, the disease has seldom occurred; however, the relationship between edwardseillosis and tubificids has not been fully clarified.

17.5.2 International Trades of Eels and Spread of Diseases

The international trade of eels has expanded the distribution and host species of eel pathogens. *Anguillicola carassus, Pseudodactylogyrus anguillae,* and *P. bini*, were introduced from East Asia to Europe and expanded their hosts to the European eel. In contrast, *P. anguilliseptica* and probably EVE, EVEX, and AngHV-1 were introduced from Europe to East Asia and expanded their hosts from the European eel to the Japanese eel.

In general, host-pathogen relationships are symbiotic in their evolutionary contexts. However, new pathogens that host animals have not experienced during evolution sometimes become very pathogenic to new host species, although they cannot settle or propagate in new host species that are unsusceptible. However, the risks of pathogens in new host species are unpredictable before the expansion to other host species.

To compensate for the recent shortage of elver Japanese eel aquaculture, elvers of exotic eels have been imported from various areas into Japan, without prior risk assessment of disease invasion and contingency planning in case of invasion. Considerable attention should be paid to the potential risk of disease invasion accompanying imports.

17.5.3 Depletion of Wild Eel Stocks and Diseases

The stocks of the 3 major eel species, *A. anguilla*, *A. japonica*, and *A. rostrata*, were dramatically depleted. Potential factors involved in stock depletion include climate change, artificial environmental changes (habitat loss, migration barriers, and chemical pollution), overfishing, and combinations of these factors. Diseases are also involved in the eel stock depletion.

In the European eel, much attention has been paid to the adverse effects of infectious diseases on eel stock (Haenen et al. 2012), especially on their long spawning migration (FAO 2006), and various studies have been conducted to estimate the impact of eel pathogens. *A. crassus*, which infects the swim bladder and significantly affects the swimming ability of eels, can be a major obstacle to spawning. Furthermore, EVEX can negatively impact migration because infected eels are severely affected during swimming in swim tunnels. EVE and AngHV-1 are also suspected to be involved in stock depletion.

In contrast, in Japan, studies on eel diseases have mainly focused on eel farms, and few studies have been performed in wild eels, except for the JEECV. Considering that the decline of eel capture and aquaculture began in the 1970s, when new pathogens potentially affecting eel stocks were found and isolated in Japan, namely EVE, EVEX, AngHV-1, and *P. anguiliseptica*, the possibility that invasive diseases significantly contributed to the stock depletion of the Japanese eel should not be ignored. Integrated studies should be conducted to elucidate the impact of diseases on Japanese eel stocks.

References

Alcaide E, Herraiz S, Esteve C (2006) Occurrence of *Edwardsiella tarda* in wild European eels *Anguilla anguilla* from Mediterranean Spain. Dis Aquat Organ 73:77–81. https://doi.org/10.3354/dao073077

Buchmann K, Mellergaard S, Køie M (1987) *Pseudodactylogyrus* infections in eel: a review. Dis Aquat Organ 3:51–57. https://doi.org/10.3354/dao003051

Buján N, Toranzo AE, Magariños B (2018) *Edwardsiella piscicida*: a significant bacterial pathogen of cultured fish. Dis Aquat Organ 131:59–71. https://doi.org/10.3354/dao03281

Egusa S (1970) Brabchionephritis prevailed among eel populations in farm-ponds in the winter of 1969-70. Fish Pathol 5:51–66. https://doi.org/10.3147/jsfp.5.51

Egusa S (1979) Notes on the culture of the European eel (*Anguilla anguilla* L.) in Japanese eel-farming ponds. Rapp P-v Reun Cons Int Explor Mer 174:51–58

Egusa S, Tanaka M, Ogami H, Oka H (1989) Histopathological observations on an intense congestion of the gills in cultured Japanese eel, *Anguilla japonica*. Fish Pathol 24:51–56. https://doi.org/10.3147/jsfp.24.51; in Japanese with English abstract

Fang J, Shirakashi S, Ogawa K (2008) Comparative susceptibility of Japanese and European eels to infections with *Pseudodactylogyrus* spp. (Monogenea). Fish Pathol 43:144–151. https://doi.org/10.3147/jsfp.43.144

FAO (2006) Report of the 2006 session of the Joint EIFAC/ICES Working Group on Eels. EIFA Occasional Paper. No. 38, ICES, 352 pp

Haenen OLM, Davidse A (2001) First isolation and pathogenicity studies with *pseudomonas anguilliseptica* from diseased European eel *Anguilla anguilla* (L.) in The Netherlands. Aquaculture 196:27–36. https://doi.org/10.1016/S0044-8486(00)00566-4

Haenen OLM, Mladineo I, Konecny R, Yoshimizu M, Groman D, Munoz P, Saraia A, Bergmann SM, van Beurden SJ (2012) Diseases of eels in an international perspective: workshop on eel diseases at the 15th international conference on diseases of fish and shellfish, Split, Croatia, 2011. Bull Eur Assoc Fish Pathol 32:109–115. https://eafp.org/download/2012-volume32/issue_3/109-Haenen.pdf

Inoue K, Miwa S, Aoshima H, Oka H, Sorimachi M (1994) A histopathological study on the etiology of intense congestion of the gills of Japanese eel, *Anguilla japonica*. Fish Pathol 29:35–41. https://doi.org/10.3147/jsfp.29.35

Kennedy CR (2007) The pathogenic helminth parasites of eels. J Fish Dis 30:319–334. https://doi.org/10.1111/j.1365-2761.2007.00821.x

Kirk RS (2003) The impact of *Anguillicola crassus* on European eels. Fish Manag Ecol 10:385–394. https://doi.org/10.1111/j.1365-2400.2003.00355.x

Knopf K, Mahnke M (2004) Differences in susceptibility of the European eel (*Anguilla anguilla*) and the Japanese eel (*Anguilla japonica*) to the swim-bladder nematode *Anguillicola crassus*. Parasitology 129:491–496. https://doi.org/10.1017/S0031182004005864

Kobayashi T, Miyazaki T (1996) Rhabdoviral dermatitis in Japanese eel, *Anguilla japonica*. Fish Pathol 31:183–190. https://doi.org/10.3147/jsfp.31.183

Kobayashi T, Goto K, Miyazaki T (2000) Pathological changes caused by cold-water stress in Japanese eel *Anguilla japonica*. Dis Aquat Org 40:41–50. https://doi.org/10.3354/dao040041

López-Romalde S, Magariños B, Núñez S, Toranzo AE, Romalde JL (2003) Phenotypic and genetic characterization of *Pseudomonas anguilliseptica* strains isolated from fish. J Aquat Anim Health 15:39–47. https://doi.org/10.1577/1548-8667(2003)015%3C0039:PAGCOP%3E2.0.CO;2

Mizutani T, Sayama Y, Nakanishi A, Ochiai H, Sakai K, Wakabayashi K, Tanaka N, Miura E, Oba M, Kurane I, Saijo M, Morikawa S, Ono S (2011) Novel DNA virus isolated from samples showing endothelial cell necrosis in the Japanese eel, *Anguilla japonica*. Virology 412:179–187. https://doi.org/10.1016/j.virol.2010.12.057

Muroga K, Jo Y, Sawada T (1975) Studies on red spot disease of pond-cultured eels—II Pathogencity of the causative bacterium, *Pseudomonas anguilliseptica*. Fish Pathol 9:107–114. https://doi.org/10.3147/jsfp.9.107; in Japanese with English abstract

Nishio K, Hioki M, Takeno N, Shiraishi Y, Takano H, Shiraishi S, Kawamura E, Toshida S (1971) A report of investigations on branchionephritis of cultured eels—I. Fish Pathol 6:47–56. https://doi.org/10.3147/jsfp.6.47; in Japanese

Okazaki S, Manabe H, Omats T, Tsuchiaka S, Yamamoto T, Chow S, Mizutani T (2015) Detection of Japanese eel endothelial cells-infecting virus (JEECV) in the Japanese eel *Anguilla japonica* (Temminck & Schlegel), living in natural habitats. J Fish Dis 38:849–852. https://doi.org/10.1111/jfd.12294

Okazaki S, Yasumoto S, Koyama S, Tsuchiaka S, Naoi Y, Omatsu T, Ono S, Mizutani T (2016) Detection of Japanese eel endothelial cells-infecting virus in *Anguilla japonica* elvers. J Vet Med Sci 78:705–707. https://doi.org/10.1292/jvms.15-0515

Okazaki-Terashima S, Kurogi H, Chow S, Yamamoto T, Matsuya N, Ijiri S, Mochioka N, Tsuchiaka S, Naoi Y, Sano K, Omatsu T, Ono S, Kuwada H, Mizutani T (2016) Detection of Japanese eel endothelial cells-infecting virus (JEECV) in mature Japanese eel *Anguilla japonica* caught from their spawning area. Fish Pathol 51:64–66. https://doi.org/10.3147/jsfp.51.64

Ono SI, Wakabayashi K, Nagai A (2007) Isolation of the virus causing viral endothelial cell necrosis of eel from cultured Japanese eel *Anguilla japonica*. Fish Pathol 42:191–200. https://doi.org/10.3147/jsfp.42.191; in Japanese with English abstract

Sano T (1976) Viral Diseases of Cultured Fishes in Japan. Fish Pathol 10:221–226

Sano T, Nishimura T, Okamoto N, Fukuda H (1977) Studies on viral diseases of Japanese fishes. VII. A rhabdovirus isolated from European eel, *Anguilla anguilla*. Nippon Suisan Gakkaishi 43: 491–495. https://doi.org/10.2331/suisan.43.491

Sano T, Okamoto N, Nishimura T (1981) A new viral epizootic of *Anguilla japonica* Temminck and Schlegel. J Fish Dis 4:127–139. https://doi.org/10.1111/j.1365-2761.1981.tb01117.x

Sano M, Fukuda H, Sano T (1990) Isolation and characterization of a new herpesvirus from eel. In: Perkins FO, Cheng TC (eds) Pathology in marine sciences. Academic, pp 15–31

Tanaka M, Satoh T, Ma WJ, Ono SI (2008) Effectiveness of increasing temperature of rearing water and non-feeding against viral endothelial cell nectosis of eel. Fish Pathol 43:79–82. (in Japanese with English abstract)

van Beurden SJ, Engelsma MY, Roozenburg I, Voorbergen-Laarman MA, van Tulden PW, Kerkhoff S, Haenen OL (2012) Viral diseases of wild and farmed European eel *Anguilla anguilla* with particular reference to The Netherlands. Dis Aquat Org 101:69–86. https://doi.org/10.3354/dao02501

van Ginneken VJT, Haenen OLM, Coldenhoff K, Willemze R, Antonissen E, van Tulden PW, Dijkstr S, Wagenaar F, van den Thillart G (2004) Presence of virus infections in eel species from various geographic regions. Bull Eur Assoc Fish Pathol 24:268–272. https://edepot.wur.nl/198496

van Ginneken V, Ballieux B, Willemze R, Coldenhoff K, Lentjes E, Antonissen E, Haenen O, van den Thillar G (2005) Hematology patterns of migrating European eels and the role of EVEX virus. Comp Biochem Physiol C Toxicol Pharmacol 140:97–102. https://doi.org/10.1016/j.cca.2005.01.011

Wakabayashi H, Egusa S (1972) Characteristics of a *Pseudomonas* spp. from an epizootic of pond-cultured eels (*Anguilla japonica*). Nippon Suisan Gakkaishi 38:577–587. https://doi.org/10.2331/suisan.38.577

Part V
Resources and Conservation

Chapter 18
Fisheries

Kazuki Yokouchi and Mari Kuroki

Contents

The relationship between freshwater eels and humans has a long history, dating back thousands of years. Eels have flexible ecological traits specific to the situation in which each individual is placed at each life history event, and they inherently have a relatively high adaptive capacity. Each continental developmental stage of this remarkable life cycle is the target of various forms of fisheries. In estuaries of Asia, glass eels are collected as seedlings for aquaculture in areas influenced by the Kuroshio Current and its extension in the Pacific. In Europe, along the Atlantic coast, there was once a tradition of frying glass eels in olive oil for consumption; however, nowadays, substitutes are often used. Yellow- and silver-stage eels are caught in waters close to human living areas such as streams, rivers, ponds, lakes, estuaries, and inner bays, mainly for direct consumption. The diversity of capture methods used for yellow and silver eels is substantial compared to other fish species. Depending on the developmental stage and habitat, eel fisheries exhibit a variety of ingenuity. Several different fishing gears and methods have been devised for each developmental stage. However, some traditional eel fishing methods have become obsolete because of declining stocks and a decrease in the number of fishers. This chapter introduces the main fishing gear and methods, mainly in Japan, devised for eels with their life history stages over the years.

K. Yokouchi (✉)
Fisheries Resources Institute, Japan Fisheries Research and Education Agency, Kanagawa, Japan
e-mail: yokouchi_kazuki56@fra.go.jp

M. Kuroki
Graduate School of Agricultural and Life Sciences, The University of Tokyo, Tokyo, Japan

K. Tsukamoto et al. (eds.), *Eel Science*, Fisheries Science Series,
https://doi.org/10.1007/978-981-99-5692-0_18

18.1 Glass Eel Fisheries

Glass eels migrate ashore to estuaries and coastal areas from early winter to late spring in the temperate zone. The timing of their arrival varies depending on their location from the spawning grounds in the ocean. In Europe, *Anguilla anguilla* glass eels arrive in estuaries along the Atlantic Coast in winter (southwestern areas) and spring (eastern Mediterranean, western, and northwestern areas) (Dekker 2003). In East Asia, the fishing season for the *A. japonica* glass eel is early winter in the south and spring in the north (Kuroki and Tsukamoto 2012; Mochioka 2019). In subtropical Taiwan, facing the Philippine Sea located in the southern part of its distribution, glass eel fishing starts in October; in southern Japan, it starts in December; and in northeast Japan, the northern limit of the fishing area begins in February. In the Seto Inland Sea, these areas do not directly face the Kuroshio; thus, it takes time for the glass eel to reach the coast, resulting in a later fishing season than that in the estuarine areas along the Pacific coast. Recently, glass eels have also been reported in southern Hokkaido, which is the northern limit of its distribution (Morita and Kuroki 2021). In addition, the peak of the fishing season varies greatly from year to year, even at the same location (Aoyama et al. 2012; Yamamoto 2019).

In the glass eel fishery, transparent glass eels measuring ~50–70 mm in size gather at night in coastal estuarine waters during flood tide and are collected from the surf or riverbanks using hand nets or small set nets (Fig. 18.1) (see Chap. 6). When scooping glass eels from riverbanks, fishers use the small scoop net with lights to enable visualization while collecting. Large scoop nets have been used to catch eels in the sandy beach area near the mouth of the river (Mochioka 2019). In Europe, glass eels have been caught using handheld or ship-based nets, and a wide range of dipnet types have previously been applied, such as trawls, stow nets, and fyke nets (Dekker 2003). From these glass eel fisheries, catch per unit effort (CPUE) data

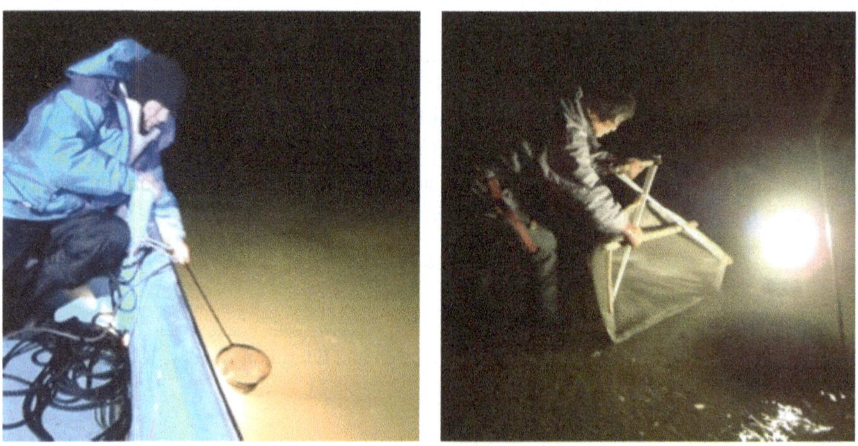

Fig. 18.1 Glass eel fishing using hand nets at river mouths in winter in Kochi prefecture (left) and Shizuoka prefecture (right), Japan. Photograph courtesy Mari Kuroki

could be obtained and used as important indices for stock assessment (see Chap. 19). In Asia, most of the collected glass eels are used as seedlings for aquaculture or stocking, depending on the country. In Japan, a special harvest permit is required for the harvest of glass eels. The use of boats or manpower to pull fishing gear and the use of secondary guiding fishing gear (commonly known as *kaki-ami* in Japanese) are prohibited in Japan (Mochioka 2019). In Taiwan, however, in addition to hand-scooping and set nets, small boats are used to catch fish by towing fine-mesh nets.

There are 2 types of eel cultivation cycles in Japan: single-year cultivation and year-round cultivation (Shiraishi and Crook 2015; Yokouchi 2019). In the single-year culture cycle, glass eels are placed in culture ponds from December to January and cultured for ~6 months until the traditional day of *Doyou-no-Ushi* in summer (late July to early August) of the same year. This Japanese custom of eating highly nutritious eels at the hottest times of the year has continued since the Edo era. Therefore, ~30–40% of the annual consumption in Japan occurs during this season. In the year-round cultivation method, glass eels are placed in ponds from February to April and cultivated for shipment from September to July. As a result of the large consumption of cultured eels in Japan on the day of *Doyou-no-Ushi* in the summer, demand and prices are generally higher for glass eels caught earlier.

The glass eel fishery is affected not only by fluctuations in glass eel recruitment and stock status but also by the historical development of aquaculture techniques and special socioeconomic circumstances. Japanese eel aquaculture began in 1879, when all seedlings were natural juveniles caught by traditional fisheries (Tanaka 2014 and references therein). With the development of the eel industry in the Tokai region of central Japan, around 1960, a large number of elver eels were introduced to the aquaculture industry from the large Tone River system near Tokyo. During the fishing season from mid-March to late October, seedlings collected from the lower reaches of the Tone River contained elvers and yellow eels, of which 60% were supplied to the Tokai region as seedlings for eel farming (Matsui 1972). As a result, the statistics of glass eels caught around 1960 include more elver eels and small yellow eels (young-of-the-year) than recent catches (see Chap. 19). The rate of decrease in the glass eel catch is likely to be less than that shown in the total seed catch in the sea and inland waters in Japan. Furthermore, estimating the annual demand for glass eels by converting the aquaculture production of commercial-sized eels to a certain survival and growth rate indicated that Japanese statistics may have overreported the catch during that period (Kishida and Kanto 2013) (see Chap. 19).

In the early 1990s, East Asia experienced a severe shortage of *A. japonica* glass eel seedlings for aquaculture, causing a rapid increase in imports of *A. anguilla* glass eels from Spain, France, and the Netherlands to replace the *A. japonica* glass eels. In the mid-1990s, a huge number of glass eels recruited annually in Europe were caught by fisheries, most of which were sent to aquaculture ponds in Asia (see Chap. 20). Thus, the glass eel fishery and aquaculture industries are socioeconomically linked on a global scale. The situation for glass eel fisheries has changed since the export of European eel became regulated by the Washington Convention (CITES: Convention on International Trade of Endangered Species of Wild Fauna) and conservation measures are being implemented in EU countries (see Chap. 23). As a result, part of the recent demand has shifted to other anguillid eel species.

18.2 Yellow Eel Fisheries

The yellow eel stage is the longest in the life history stages of eels, typically lasting ~5–15 years until the onset of maturity for *A. japonica* (see Chap. 3). Yellow eels inhabit environments familiar to humans, such as small rivers, ponds, lakes, estuaries, tidal flats, and inner bays (Yokouchi et al. 2012), with anthropogenic impacts (Righton et al. 2021; Yokouchi et al. 2022). Fishing methods targeting yellow eels are diverse and have a long history of use. Yellow eels are highly sedentary and feed on a variety of organisms, depending on their body size and habitat (e.g., Nishimoto et al. 2023) (see Chap. 8). In addition, yellow eels have a habit of hiding in structures such as stone piles, stone crevices, and vegetation roots. Therefore, as a fishing method, eels are captured using these habits to attract them by hiding places and bait. In estuarine areas and tidal flats, eels build burrows to spend the daytime in (Aoyama et al. 2005), and thus, there are also methods of capturing eels without attracting them, such as eel scraping.

The accurate monitoring of yellow eel catches requires certain considerations. In Japan, because the main method of stocking eels is to release small yellow eels (originating from eel farms) into rivers and lakes, farmed and stocked individuals are common in inland waters (Kaifu et al. 2014, 2018). As a result, it is difficult to collect accurate data on yellow and silver eel fisheries in Japan's inland waters; however, relatively accurate information on the dynamics of wild Japanese eel stocks can be obtained from coastal and estuarine areas where wild eels are dominant (Kaifu and Yokouchi 2019).

Fishing methods that use bait to attract eels include hook-and-line fishing, longline, hole fishing, and eel pots. There is also a traditionally specialized way of fishing eels known as bobbing or *juzu-tsuri* in Japanese or *la pêche à la vermée* in French (Kuroki and Tsukamoto 2012; Kuroki et al. 2014a; Feunteun and Robinet 2014). Among these, the longline method is still used locally. Longlining is a common fishing method that is used in many areas. The bait used varies depending on the fishing area and season, and may include pond snails, earthworms, shrimp, mud shrimp, hornet larva, and cultured worms. Hole fishing is performed with a hook tied to the end of a bamboo rod approximately 1-m long with a bait attached. The hook is inserted into a crevice where eels are likely to hide. After the bait is fully chewed and swallowed by the eel, the hook is gradually pulled out. In Japan, the bait should be earthworms, ayu, or loach (Mochioka 2019). Fishing methods using bamboo traps exist in many areas of the world. Bamboo strips woven into a tube or basket are used as traditional fishing gear; this method also utilizes the elongated shape of eels to hide in narrow places (Righton and Roberts 2014; McCarthy 2014; Tzeng 2014; Jellyman 2014) (Figs. 18.2 and 18.3). They can be submerged in the

Fig. 18.2 Photo of an eel pot woven with wicker at the Eel Festival in Ely, England. Photograph courtesy Mari Kuroki

water alone, connected together in a long line, or guided to the gear by a weir built into the riverbed.

Even if quite specific, traditional fishing methods can have similar plans between distant locations. In Japan, *Juzu-tsuri* (beads fishing in Japanese) is a fishing method that does not use hooks. It was popular in parts of the Chugoku region of western Japan in the 1950s; however, this method is rarely seen today. A 5-m long string of bait was made by threading dozens of polychaetes (bristle worms) and earthworms vertically with a sewing needle, winding it into a 3–5 cm loop and tying it with a weight in the center of the loop to form a spherical mass of bait. *Juzu-tsuri* is done in brackish water at night during flood tide, and when a fish strikes, it is taken into the boat. Although large fish are caught, most of the catch consists of juveniles weighing less than 20 g (Mochioka 2019). A similar fishing method was found in some countries, known as bobbing or *la pêche à la vermée* in French (Feunteun and Robinet 2014). In France, "*la pêche à la vermée*" was commonly used to hold an

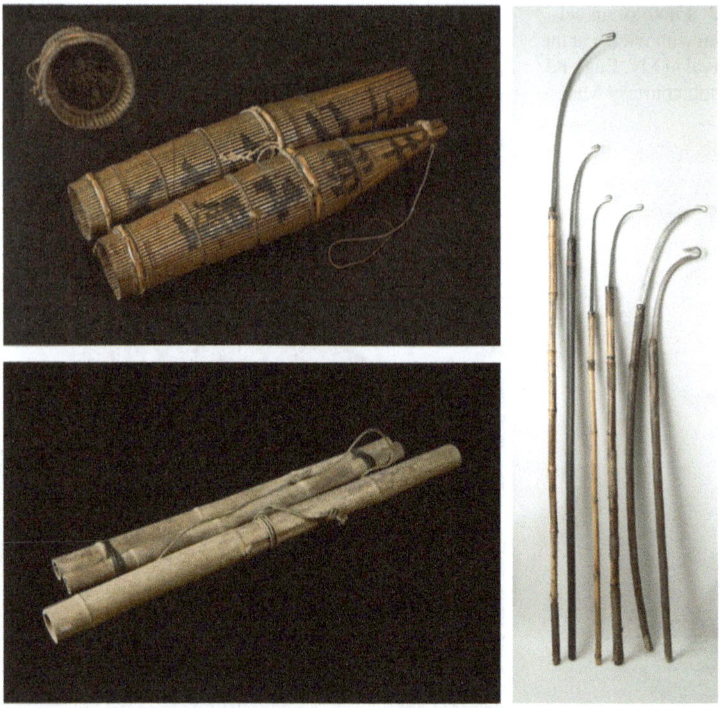

Fig. 18.3 Traditional eel fishing gears, Japan. Bamboo tubes with return trap (top left), bamboo tubes without return (bottom left), and eel sickles in the possession of Shunsuke Sanda (right). Kuroki and Tsukamoto (2012) reproduced with permission

umbrella upside down under the bait, enabling the eel dropped off from the bait to be captured (Feunteun and Robinet 2014).

Methods of luring eels by providing artificial shelters without bait allow them to come and go at night, including several techniques such as baitless eel tubes, bush dip, *shibazuke* in Japanese, and a traditional Japanese method using stones known as *ishikura*. Fishing methods still in use today include baitless eel tubes and Japan's *ishikura* fisheries. The baitless eel tubes are traditionally made of bamboo; however, PVC tubes are currently used often (5–15 cm in diameter, 2–3 tubes bundled together) with knots cut out and ropes attached to both ends and submerged in water (Fig. 18.3). The tube can be gradually raised to the surface horizontally during the day to collect eels hiding in the tubes by tilting the tube inside the hand net. In the bush dip method, bundles of bamboo, chinquapin, oak, or other twigs are immersed in water for at least one night, then slowly pulled out of the water; small eels that have entered it are collected using a large scoop net at the water surface.

Ishikura fishing is a traditional fishing method with unique procedures for catching eels (Fig. 18.4) that still remains in the Kyushu and Shikoku regions in Japan. The *ishikura* fishing is done in the tidal zone near the river mouth, where stones of the size of a human head are piled up in the water to create a shelter for eels.

Fig. 18.4 Traditional *Ishikura* (rock pile) fishing for yellow eels (top) and migrating silver eel fishing with scoop-nets (bottom) from *Mie-ken Suisan Zukai*, published in 1883. Source, Mie Prefectural Museum

At low tide, the stone shelter is surrounded by a net, and eels that hide in the gaps between stones at high tide are caught by removing the rock pile and/or driving them into a fish-catching section of the net with a return. Some fishers do not use a net for

Fig. 18.5 Fyke net fishing in Lough Ennell, Ireland. Photograph courtesy Kazuki Yokouchi

the *ishikura* fishing, and instead use box glasses or other equipment to visualize the inside of the stone piles, pick up stones, and catch eels with eel shears or other tools.

Fishing methods that do not use attractants include fyke nets, scraping, poking, hand fishing, *kaibori* (drain-up), small-bottom trawling, push nets, and small-set nets. With the exception of some net fishing methods, including small set nets and fyke nets, scraping, poking, and hand fishing or *kaibori* methods are rarely used commercially today. Yellow eels are often caught using small set nets in brackish lakes and are sometimes caught as bycatch using small trawl nets in semi-closed bays in Japan. Fyke nets and draft nets are a type of fishing net commonly used in Europe (i.e., Ireland) for collecting yellow eels from estuaries and lakes (Fig. 18.5) (McCarthy 2014; Rindom et al. 2014). The Fyke net is a long cylindrical bag net with several inner returns that facilitates entry and makes escape difficult. It is also equipped with a wing or leader to guide the eels towards the entrance of the bag. Draft nets 80–100 m in length with a cod-end are deployed from boats in open water (Aprahamian et al. 2021).

Scraping and poking use special spears, forks, rakes, or sickles to catch eels. These fishing methods are used worldwide. Eel poking is a fishing method in which eels are poked using a fish spear with box glasses on a riverboat (Kuroki et al. 2014a; Righton and Roberts 2014; McCarthy 2014; Rindom et al. 2014; Jellyman 2014). For eel scraping, warped iron fishing gear with several claws on the tip is used to collect eels hidden in the mud (Fig. 18.3). At river mouths and tidal flats, eels hiding in the mud bottom are caught by inserting an eel scraper with a long handle into the bottom sediment from a boat. In the past, this method was actively used in small irrigation canals and reservoirs after water was drained in Japan. Hand fishing, or noodling, is a method of catching eels that hide in burrows on tidal flats or natural stone piles by inserting their hands into them. Eel grips and shears are often used. Unfortunately, scraping and hand fishing are no longer commercially viable.

Other Japanese traditional fishing methods in which all or part of a small swamp, pond, or river is partitioned using mud, stones, or wood, the water is drained out to collect the stranded fish (Kuroki and Tsukamoto 2012; Mochioka 2019). The ponds and swamps are then sun-dried, which is a traditional management method used during the off-season to maintain the reservoirs for agriculture in Japan. Similar fishing methods have also been used in Europe (Feunteun and Robinet 2014); however, these methods are not commonly employed.

18.3 Silver Eel Fisheries

It is known that the size at silvering and age at maturity vary widely among eels in the temperate zone, and that a certain minimum size must be achieved for eel silvering maturity (see Chap. 3). Since silver eels that have begun spawning migration rarely catch food, fishing methods that use bait are not used. They are caught mainly using fish weirs, waiting nets, and small set nets, or caught as bycatch using small trawl nets. Silver eels migrating towards the sea along the shoreline are collected using small set nets from late autumn to winter. However, recently in the inland waters of Japan, silver eel fisheries have stopped fishing to conserve eel resources.

Female eels tend to attain larger body sizes when growth conditions are sufficient to spend extra time in their growth habitats. When growth rates decline, the onset of maturation and spawning migration is triggered (Yokouchi et al. 2018). Eels caught by small set nets in coastal areas are often large females, while male silver eels are small and may be able to slip through the set nets (Mochioka 2019). Small set nets can be found in semi-closed or inner bays for eel fishing. Both male and female silver are typically successfully collected using small set nets (Yokouchi et al. 2009). Although silver eels tend to burrow in holes less often, they can also be captured by the aforementioned-eel scrapers and *ishikura*. In Europe, silver eels are often caught in set nets lowered into the water at eel weirs in several river systems, a long-established fishing method (McCarthy 2014; Aprahamian et al. 2021). In Indonesia, fish weirs catch silver-phase tropical eels of *A. marmorata* and *A. celebesensis* heading downstream to each spawning area (Hagihara et al. 2018).

Silver eels are collected from late autumn to early winter when the river water levels increase from autumn rains. To catch silver eels swimming down the river, stones are piled up to gather river water, with nets set at their ends, or large eel pots woven from bamboo are installed (McCarthy 2014; Jellyman 2014). Other silver eel fishing gear includes traditional scoop nets (Fig. 18.4) and beam nets that lined up two boats facing the current. The beam-net method was previously used in Japan when rainfall and runoff occurred. However, these methods are rarely used today.

Silver eel biology is essential for understanding eel life history and contributing to conservation and management efforts. To conserve eel resources effectively, it is crucial to monitor maturing silver eels in estuarine and marine areas, where it is possible to target all migratory types assumed to contribute to reproduction. In addition, accurate information on wild Japanese eels can be obtained from coastal and estuarine areas, where wild eels are dominant. Therefore, it is essential to accumulate knowledge on the habitat and ecology of yellow eels in brackish/marine habitats and migrating silver eels in the waters around the coast (Righton et al. 2021) (see Chap. 23). Furthermore, the use of several existing fishing methods to capture silver eels is important not only from a scientific point of view but also from a social perspective. For example, in Lake Hamana, a private-sector-led conservation initiative targeting silver eels has been implemented ahead of others to purchase and release wild silver eels, which involves stakeholders such as fishers, distributors, and restaurants in Japan (Iida et al. 2017).

As noted in this chapter, there is a great diversity of traditional eel fishing methods worldwide. Surprisingly, some traditional methods have high similarity among distant locations, even if the methods are quite specific. This case is an interesting case study of the cultural convergent development between eels and humans (Kuroki et al. 2014b; Righton and Roberts 2014). Various fishing methods are introduced in this section, each of which targets a different stage, size, or habitat. To accurately understand eels, research-based surveys alone, such as those conducted by recent electrofishing, bio-logging, and environmental DNA methods, are costly and limited in terms of technical requirements. As such, it is considered most effective to utilize the characteristics of various existing fishing methods and conduct monitoring of the current status of eel resources by combining these fishing methods and research-based surveys.

There is concern that without stable relationships between eels and humans through an interest in eels or fishing of eels, the environments they inhabit may decline. Without people's awareness and intervention, the decline of eel resources and the degradation of the surrounding ecosystem will unintentionally accelerate. Therefore, the various existing fishing methods are of cultural value and great value to natural science in maintaining a stable relationship between eels and humans. However, many of the traditional eel fisheries mentioned above have already disappeared because of declining stocks and a decrease in the number of fishers. Thus, learning and utilizing culturally distinctive, ecologically efficient traditional fishing methods in various activities, such as monitoring, research, education, and recreation, will become increasingly crucial for conserving ecosystem diversity and connectivity for eels and other species.

References

Aoyama J, Shinoda A, Sasai S, Miller MJ, Tsukamoto K (2005) First observations of the burrows of *Anguilla japonica*. J Fish Biol 67:1534–1543. https://doi.org/10.1111/j.1095-8649.2005.00860.x

Aoyama J, Shinoda A, Yoshinaga T, Tsukamoto K (2012) Late arrival of *Anguilla japonica* glass eels at the Sagami River estuary in two recent consecutive year classes: ecology and socio-economic impacts. Fish Sci 78:1195–1204. https://doi.org/10.1007/s12562-012-0544-y

Aprahamian MW, Evans DW, Briand C, Walker AM, McElarney Y, Allen M (2021) The changing times of Europe's largest remaining commercially harvested population of eel *Anguilla anguilla* L. J Fish Biol 99:1201–1221. https://doi.org/10.1111/jfb.14820

Dekker W (2003) Status of the European eel stock and fisheries. In: Aida K, Tsukamoto K, Yamauchi K (eds) Eel biology. Springer, Tokyo, pp 237–254

Feunteun E, Robinet T (2014) Freshwater eels and people in France. In: Tsukamoto K, Kuroki M (eds) Eels and humans. Springer, Tokyo, pp 75–89

Hagihara S, Aoyama J, Limbong D, Tsukamoto K (2018) Interspecific difference in downstream migratory season between two tropical eels, *Anguilla celebesensis* and *Anguilla marmorata*. J Fish Biol 93:729–732. https://doi.org/10.1111/jfb.13750

Iida M, Tanaka T, Nishimoto A, Yokouchi K (2017) Biological characteristics and conservation of Japanese eel *Anguilla japonica* inhabiting seawater habitats - based on survey in the Miyakoda River system. KAIYOU Monthly 49:556–559; in Japanese

Jellyman DJ (2014) Freshwater eels and people in New Zealand: a love/hate relationship. In: Tsukamoto K, Kuroki M (eds) Eels and humans. Springer, Tokyo, pp 143–154

Kaifu K, Yokouchi K (2019) Increasing or decreasing? - current status of the Japanese eel stock. Fish Res 220:105348. https://doi.org/10.1016/j.fishres.2019.105348

Kaifu K, Maeda H, Yokouchi K, Sudo R, Miller MJ, Aoyama J, Yoshida T, Tsukamoto K, Washitani I (2014) Do Japanese eels recruit into the Japan Sea coast?: a case study in the Hayase River system, Fukui, Japan. Environ Biol Fish 97:921–928. https://doi.org/10.1007/s10641-013-0193-8

Kaifu K, Yokouchi K, Higuchi T, Itakura H, Shirai K (2018) Depletion of naturally recruited wild Japanese eels in Okayama, Japan, revealed by otolith stable isotope ratios and abundance indices. Fish Sci 84:757–763. https://doi.org/10.1007/s12562-018-1225-2

Kishida T, Kanto I (2013) Reconsideration on the catch of glass eel in Japan. Bull Japan Soc Fish Oceanogr 77:164–166; in Japanese with English abstract

Kuroki M, Tsukamoto K (2012) Eels on the move—mysterious creatures over millions of years. Tokai University Press, Hadano

Kuroki M, van Oijen MJP, Tsukamoto K (2014a) Eels and the Japanese: an inseparable, long-standing relationship. In: Tsukamoto K, Kuroki M (eds) Eels and humans. Springer, Tokyo, pp 91–108

Kuroki M, Righton D, Walker AM (2014b) The importance of Anguillids: a cultural and historical perspective introducing papers from the World Fisheries Congress. Ecol Freshw Fish 23:2–6. https://doi.org/10.1111/eff.12089

Matsui I (1972) Mangaku. An eel science. Kouseisha-Kouseikaku, Tokyo; in Japanese

McCarthy KT (2014) Eels and people in Ireland: from mythology to international eel stock conservation. In: Tsukamoto K, Kuroki M (eds) Eels and humans. Springer, Tokyo, pp 13–40

Mochioka N (2019) Fishing gear and fishing methods. In: Tsukamoto K (ed) Science of eels. Asakura Publishing, Tokyo, pp 120–125. (in Japanese)

Morita K, Kuroki M (2021) Japanese eel at the northern edge: glass eel migration into a river on Hokkaido, Japan. Ichthyol Res 68:217–221. https://doi.org/10.1007/s10228-020-00771-5

Nishimoto A, Iida M, Yokouchi K, Fukuda N, Yamamoto T (2023) Eels as natural samplers highlight spatial heterogeneity in energy flow in an estuary. Estuar Coast Shelf Sci 281:108215. https://doi.org/10.1016/j.ecss.2023.108215

Righton D, Roberts M (2014) Eels and people in the United Kingdom. In: Tsukamoto K, Kuroki M (eds) Eels and humans. Springer, Tokyo, pp 1–12

Righton D, Piper A, Aarestrup K, Amilhat E, Belpaire C, Casselman J, Castonguay M, Diaz E, Doerner H, Faliex B, Feunteun E, Fukuda N, Hanel R, Hanzen C, Jellyman D, Kaifu K, McCarthy K, Miller MJ, Pratt T, Sasal P, Schabetsberger R, Shiraishi H, Simon G, Sjoberg N, Steele K, Tsukamoto K, Walker A, Westerberg H, Yokouchi K, Gollock M (2021) Important questions to progress science and sustainable management of Anguillid eels. Fish Fish 22:762–788. https://doi.org/10.1111/faf.12549

Rindom S, Tomkiewicz J, Munk P, Aarestrup K, Als TD, Pedersen MI, Graver C, Anderberg A (2014) Eels in culture, fisheries and science in Denmark. In: Tsukamoto K, Kuroki M (eds) Eels and humans. Springer, Tokyo, pp 41–60

Shiraishi H, Crook V (2015) Eel market dynamics: *Anguilla* production, trade and consumption in East Asia. TRAFFIC, Tokyo

Tanaka E (2014) Stock assessment of Japanese eels using Japanese abundance indices. Fish Sci 80: 1129–1144. https://doi.org/10.1007/s12562-014-0807-x

Tzeng W-N (2014) Freshwater eels and humans in Taiwan. In: Tsukamoto K, Kuroki M (eds) Eels and Humans. Springer, Tokyo, pp 129–142

Yamamoto T (2019) 4.4. Stock dynamics of the glass eel. In: Tsukamoto K (ed) Unagi no kagaku. Asakura Publishing, Tokyo, pp 132–137; in Japanese

Yokouchi K (2019) 2.6. Growth. In: Tsukamoto K (ed) Unagi no Kagaku. Asakura Publishing, Tokyo, pp 37–41; in Japanese

Yokouchi K, Sudo R, Kaifu K, Aoyama J, Tsukamoto K (2009) Biological characteristics of silver-phase Japanese eels, *Anguilla japonica*, collected from Hamana Lake, Japan. Coast Mar Sci 33: 54–63. https://doi.org/10.15083/00040704

Yokouchi K, Fukuda N, Miller MJ, Aoyama J, Daverat F, Tsukamoto K (2012) Influences of early habitat use on the migratory plasticity and demography of Japanese eels in central Japan. Estuar Coast Shelf Sci 107:132–140. https://doi.org/10.1016/j.ecss.2012.05.009

Yokouchi K, Daverat F, Miller MJ, Fukuda N, Sudo R, Tsukamoto K, Elie P, Poole WR (2018) Growth potential can affect timing of maturity in a long-lived semelparous fish. Biol Lett 14: 20180269. https://doi.org/10.1098/rsbl.2018.0269

Yokouchi K, Itakura H, Wakiya R, Yoshinaga T, Mochioka N, Kimura S, Kaifu K (2022) Cumulative effects of low-height barriers on distributions of catadromous Japanese eels in Japan. Anim Conserv 25:137–149. https://doi.org/10.1111/acv.12725

Chapter 19
Resources

Naohito Okazoe, Leanne Faulks, and Hiroshi Hakoyama

Contents

Temperate anguillid eels are enigmatic species that have captured the curiosity of scientists for hundreds of years, and have also historically been exploited as a fisheries resource. In particular, fishing of American, European, and Japanese eels has been consistent, and severe declines in population size have been observed since the 1970s. Scientists and resource managers have been studying eel biology and fisheries to understand the status of eel resources and the causes of eel population decline. However, many knowledge gaps remain owing to the limited scientific data and knowledge on specific aspects of eel biology, ecology, and the interacting effects of changes in marine and freshwater environments and human activities. In this chapter, we provide an overview of the current state of Japanese eel resources and their management through the regulation of fishing, aquaculture, and trade. The first half summarizes the latest assessment of eel resources and describes the challenges for data collection and knowledge generation owing to the complex life cycle of the Japanese eel and numerous uncertainties related to its ecology. The key challenge is to improve the accuracy and timeliness of the temporal and spatial data obtained from catch statistics and field research. The second half outlines the updated domestic regulations for Japanese eel catch/fishery and aquaculture and the ongoing international efforts to conserve Japanese eel resources. Resource managers have implemented a variety of policies and tools based on the best scientific knowledge

N. Okazoe
Fisheries Agency of Japan, Tokyo, Japan

L. Faulks (✉) · H. Hakoyama
Institute of Freshwater Biology, Nagano University, Nagano, Japan

K. Tsukamoto et al. (eds.), *Eel Science*, Fisheries Science Series,
https://doi.org/10.1007/978-981-99-5692-0_19

available while generating a collective will among countries/regions to adopt a coordinated approach to conserving Japanese eel resources.

19.1 Current State of Resources

Japanese eel has long been a popular commercially harvested species in eastern Asia, and the history of eel fisheries in Japan extends back to the Edo era (1603–1868). Official data of eel catches in Japan have been available since 1894, and they indicate that the domestic yellow eel catch was stable at 3000–4000 tons in the early nineteenth century (Hakoyama et al. 2016). However, this decreased during World War II, and although it recovered temporarily to the 3000-ton range in the 1960s, the catch has been declining since 1970 (Fig. 19.1a). Although the domestic catch exceeded 600 tons in the early 2000s, it fell to 500 tons after 2005, then to less than 100 tons after 2015, and in 2020 the catch plummeted to 65 tons. While a decrease in inland fishermen during this period may have contributed to this trend, it cannot be estimated because of the lack of necessary data. However, a survey that assessed fishing effort (catch per unit effort, CPUE) revealed that during 2003–2016, the CPUE of yellow eels in Okayama Prefecture, western Japan, decreased by 1/3 in both longline and small set-net fisheries (Kaifu et al. 2018). It is important to note that, although farmed eels account for most of the domestic supply in Japan, wild yellow eels (developmental stage) are still caught by longlines or traps set in freshwater and brackish waters (Mochioka 2019).

Although domestic eel farming using wild-caught elver eels (juvenile stage) began in 1879, the eel farming industry only became viable in the 1920s as the methods for raising glass eels advanced. Consequently, the number of inland fishermen (the number of fishery management entities in lakes that mainly target yellow/silver eels) decreased remarkably (Hakoyama et al. 2016) and aquaculture production in the 1930s exceeded the wild catch of yellow eels (Tanaka 2019). Aquaculture production peaked at 39,704 tons in 1988 and has been stable at ~15,000–20,000 tons since 2000.

Aquaculture production relies on the catch of wild glass eels, which, like the yellow eel catch, has declined. Glass eels are harvested in 4 jurisdictions: Japan, China, Korea, and Chinese Taipei. In Japan, the catch period generally extends from December to April. Harvesters catch eels by scooping or by using set nets in coastal estuaries and river mouths. Glass eel catches are managed by prefectural governments under a permit system. Prefectural governors typically restrict the catch period, fishing gear, and fishing areas. In the 1960s, domestic total seed catch was often between 150 and 200 tons; however, catch declined in the 1970s, and since the 1980s, the annual catch has been <25 tons (Fig. 19.1b). In these early catch statistics, seeds caught in the sea had "glass eel" labels as their item names or footnotes that denoted them as glass eels. The eels caught in inland waters had no clear description or labels, but the quantities collected from rivers and lakes were recorded. According to the Minister's Secretariat Statistics Department, most seeds

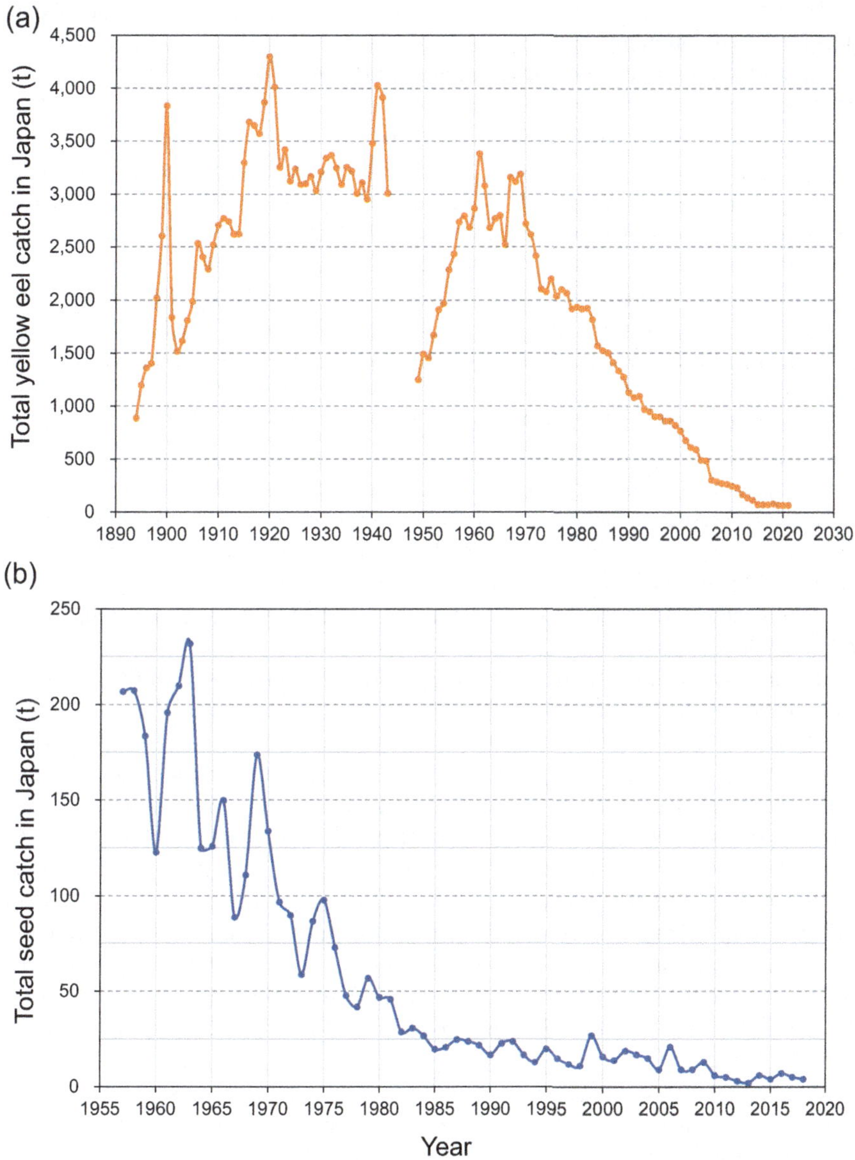

Fig. 19.1 (**a**) The catch of Japanese yellow eels in Japan based on fisheries statistics of the Government of Japan. (**b**) Total seed catch of Japanese eel (including glass eels and elver eels) in the sea and inland waters of Japan. Data from fisheries statistics of the Government of Japan (Hakoyama et al. 2016; The annual report of catch statistics on fishery and aquaculture in 2018)

were assumed to be glass eels; however, some elvers might have been included in the data (personal communication).

Around 1960, the developing eel aquaculture industries of Shizuoka, Aichi, and Mie prefectures, central Japan, introduced large numbers of elvers from the Tone River system in Ibaraki and Chiba, north of Tokyo. During the fishing season, from mid-March to late October, the seeds collected from the lower Tone River included elver eels that were 5–20 g and 15–25-cm long. Of these, 60% were supplied to Shizuoka, Aichi, and Mie prefectures as seeds for eel farming (Matsui 1972). Since 1978, the catch season for seeds in 9 major aquaculture prefectures has been limited to glass eels during winter (Eel Culture Research Council 1980). This suggests that eels caught around 1960 included more elver eels than recent catches. If this is the case, the rate of decrease in glass eel catch is likely to be smaller than that shown in Fig. 19.1b.

Recent glass eel catch data show a decrease to <10 tons in 2010–2013, a slight increase to 15 tons in 2014–2017, a reversion to <10 tons during the 2018–2019 fishing season, and then 3.7 tons in the 2019 fishing season (Fig. 19.2a). Although the catch increased considerably in the 2020 fishing season to 17.1 tons, the catch for the 2021 fishing season decreased to 11.1 tons. These fluctuations in catch are small compared to the considerable decline evident since the 1960s (even if a proportion of the 1960s catch contained elvers), and indicate that the Japanese eel now remains at a relatively low population size.

The trend for decline in the catch of Japanese eel in Japan is mirrored in the data from other jurisdictions across the species distribution. The global catch of Japanese eel decreased from 3619 tons in 1969 to 121 tons in 2019 (Fig. 19.2b). Regarding recent glass eel catches (from 2009 to 2021), China had the highest annual catch, followed by Japan (Fig. 19.2a); these 2 countries account for most of the total catch. During the last several decades, the total number of glass eels caught in the 4 jurisdictions has fluctuated from year to year, ranging from 20 to 90 tons. In the 2019 fishing season, the catches were 3.7 tons in Japan, 14.5 tons in China, 0.6 tons in Korea, and 2.75 tons in Chinese Taipei, but the catches in the 2020 fishing season increased significantly to 17.1 tons in Japan, 50 tons in China, 4.5 tons in Korea, and 5.2 tons in Chinese Taipei. In the 2022 catch season (November 1, 2021 to April 30, 2022), the catches were 8.3 tons in Japan, 2.2 tons in Korea, and 1.6 tons in Chinese Taipei.

Although catch datasets may not provide the full picture, they are the most comprehensive data available for the Japanese eel and clearly reflect an overall long-term decline in population size, which remains low. The Japanese eel is listed as an endangered species by both the Japanese Ministry of the Environment and IUCN (categorized as endangered IB in 2013 and 2014, respectively). Although it is difficult to identify the causes of the population decrease in this species, changes in the marine environment, overfishing, and the deterioration of freshwater and estuarine habitats are regarded as important factors. The population assessment of this species is challenging, mainly because of its complex life cycle and numerous ecological uncertainties; only one population assessment has been made to date (Tanaka 2014). However, the decline in the Japanese eel population continues to be

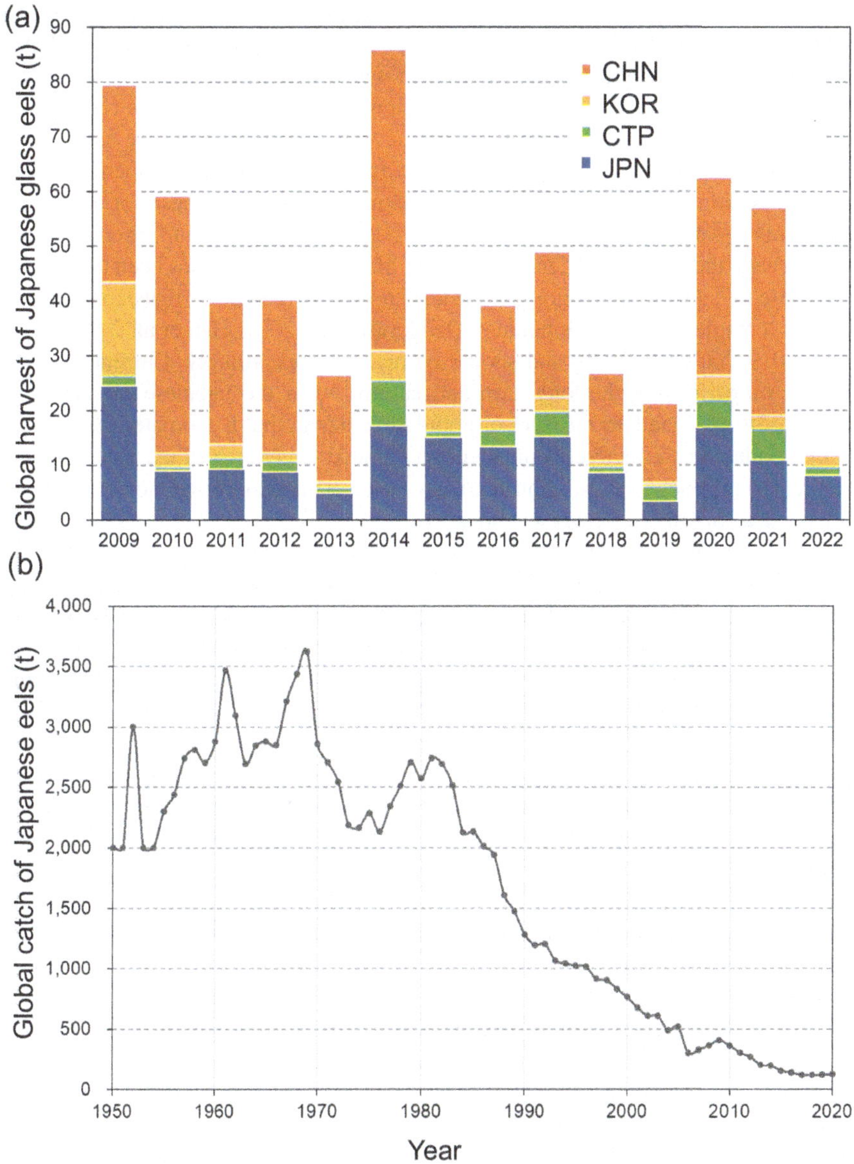

Fig. 19.2 (**a**) Global harvest of Japanese glass eels. These data are based on the Joint Press Release of the Informal Consultation on International Cooperation for Conservation and Management of Japanese Eel Stock and Other Relevant Eel Species (Fisheries Agency of Japan, https://www.jfa.maff.go.jp/j/press/sigen/attach/pdf/170711-2.pdf and https://www.jfa.maff.go.jp/j/press/sigen/attach/pdf/210727-7.pdf, accessed on August 31 2023). CHN China, KOR Republic of Korea, CTP Chinese Taipei, JPN Japan. (**b**) Global catch of wild Japanese eel (including all stages). These data are based on FAO 2021 statistics

of concern to Japanese and international communities, especially those from the eel industry and conservation managers. Thus, in 2019, the Fisheries Agency of Japan launched a multidisciplinary research project with the goal of developing a comprehensive assessment of Japanese eel populations. This project includes research on population genomics and effective population size (N_e) of the Japanese eel.

Elucidating the population genetic structure of organisms is indispensable, not only for defining the management units of natural resources, but also for accurately estimating N_e, which can be used to evaluate population viability and perenniality. In recent years, the population genetic structure of the Japanese eel has been the subject of intense debate, but the majority of studies now clearly indicate that Japanese eels exist as a single panmictic population (Ishikawa et al. 2001; Han et al. 2010; Gong et al. 2019; Yu et al. 2020) and should be managed accordingly. Current research focuses on estimating the long-term and current N_e of the Japanese eel, which is different from a standard measure of population size in that it estimates the number of individuals that effectively contribute to the next generation (thus, it is generally smaller than the actual population size), and when estimated from genomic data, it is independent of fisheries catch statistics.

Assessing historical changes in N_e can provide an evolutionary perspective on current population dynamics. Whole genome level analyses using pairwise and multiple sequential Markovian coalescent (PSMC and MSMC) methods (Mather et al. 2020) have indicated that 1–4 million years ago (Ma) N_e decreased, then from ~1 Ma up until ~22,000–30,000 years ago, the Japanese eel population steadily increased in N_e, peaking at ~80,000 individuals (Faulks et al. 2022; Fig. 19.3). During the Last Glacial Maximum (LGM;19,000–33,000 years ago), N_e decreased to ~60,000 individuals. Owing to the restricted power of the PSMC and MSMC methods to detect changes within the last 20,000 years, changes in the population size of the Japanese eel following the LGM are still unknown. However, ongoing studies using single nucleotide polymorphism data and linkage disequilibrium analyses (Waples and Do 2010) indicate that the current N_e is ~20, 000 (Sekino M, personal communication). Overall, these results indicate that the Japanese eel has experienced several population bottlenecks and that levels of genetic diversity are relatively low. This background indicates that the Japanese eel may be sensitive to further declines in population size, which may affect the ability of the species to adapt to future environmental changes.

A lack of progress has been made in developing mathematical methods to predict the population dynamics of the Japanese eel and contribute to management strategies because it is difficult to fully understand the species' biology and identify the causes of population decline. Future research must ascertain population trends and work towards the sustainable harvest of Japanese eels. To achieve this, it is necessary to improve the accuracy and timeliness of temporal and spatial data, expand our knowledge regarding fluctuations in the N_e of the species, and use the best available data for the development of a mathematical model for population management. Additionally, enhancing scientific cooperation and communication with other jurisdictions within the distributional range of the Japanese eel is important.

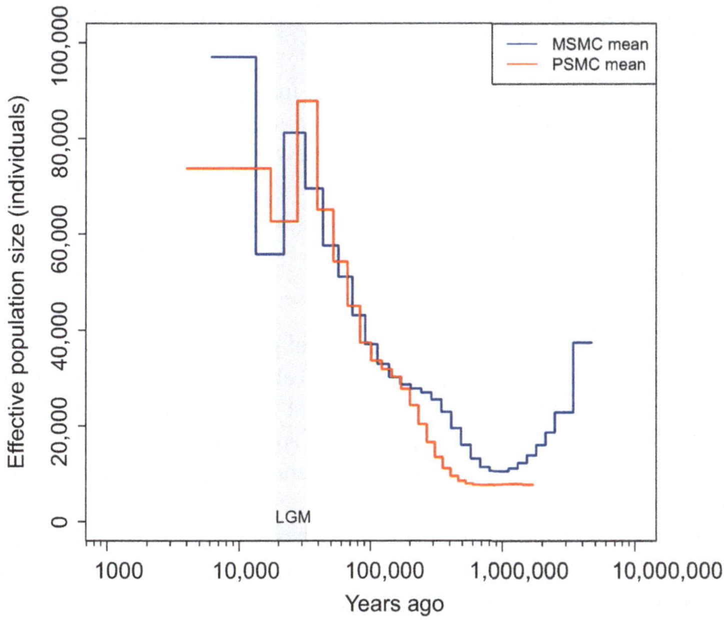

Fig. 19.3 Plot of the changes in effective population size N_e over the last ten million years for Japanese eel. N_e was estimated from the whole genome re-sequencing data of 11 individuals by using the pairwise and multiple sequentially Markovian coalescent (PSMC and MSMC) methods. Plot depicts the mean values of N_e for all 11 analyzed individuals; PSMC = red, MSMC = blue. LGM = Last Glacial Maximum 19,000–33,000 years ago

19.2 Resource Management

19.2.1 Precautionary Approach

As mentioned above, while there are concerns about the decrease in the Japanese eel population, scientific knowledge is still insufficient with regard to the mechanism of population decline, which would, if sufficiently understood, contribute to developing mathematical methods to predict population dynamics and contribute to management strategies. Researchers have posited various potential factors for the population decline, including environmental change and pollution in the marine environment, overfishing, habitat destruction due to river construction and segregation, disease by parasites, and an increase in predators; however, the importance of each factor and their interactions remains unclear (Knights 2003; Friedland et al. 2007; Bonhommeau et al. 2008; Chang et al. 2018; Chang et al. 2019; Drouineau et al. 2018; Righton et al. 2021). Therefore, it is necessary to adopt a precautionary approach (UNCED 1992; FAO 1996). Japan is doing this by implementing comprehensive measures, including population management and habitat restoration, for the sustainable use of Japanese eel resources.

19.2.2 Domestic Conservation and Management Measures

Japan has taken the following conservation and management measures: (1) regulating the catch of glass eels and adult eels; (2) granting aquaculture permits, including regulating inputs of eel seeds (glass eel and elver) into aquaculture ponds; and (3) regulating international trade.

19.2.2.1 Eel Catch

The catch of glass eels is regulated by prefectural governments according to prefectural fisheries adjustment rules. Although the catch of glass eels is prohibited in principle, prefectural governors can issue special catch permits for particular catch seasons because glass eels are necessary as aquaculture seeds. By setting the conditions of these permits, such as the catch period, gear, and location, prefectural governors can regulate the glass eel catch.

Prefectural governments also regulate catches of adult eels. They implement conservation and management measures, such as gear restriction, upper limits of catches, and catch suspension, based on prefectural fisheries adjustment rules, and licenses are granted for the class 5 common fishery right to catch eels in accordance with the Fishery Act. Each prefecture may have its own unique circumstances. For example, prefectures where eel aquaculture is active have introduced measures to protect and conserve spawning stocks, such as mandatory or voluntary suspension of eel fishing from October to March, when eels descend rivers to undergo spawning migration in the ocean. Recently, almost all prefectures where wild adult eels are distributed have prohibited the catch of silver eels that are descending to spawn.

To support stakeholders and ensure the implementation of eel resource management at the prefectural level, the national government provides technical advice to prefectural governments regarding the management of Japanese glass eel catch and adult eel fishing (Fisheries Agency of Japan, https://www.jfa.maff.go.jp/j/saibai/unagi.html, accessed on June 11, 2022). For example, the national government provides technical advice regarding the implementation of special glass eel catch permits issued by prefectural governments, including how to implement an appropriate system to report catch quantity and period, as well as how to provide guidance and law enforcement to effectively control the catch of glass eels when the upper limit is reached. For adult eel fishing, the national government provides advice regarding the restriction of fishing eels that descend the river to spawn in the ocean. In the technical advice, the national government also highlights the condition of granting a license for the class 5 common fishery right to catch adult eels; that is, those who are licensed are required to engage in activities to assist in the reproduction of eels (Article 168, Fishery Act; Fisheries Agency of Japan), mainly through the release of Japanese eel seeds into the wild.

19.2.2.2 Eel Aquaculture

The national government regulates eel aquaculture in Japan. Eel aquaculture was specified as “designated aquaculture” under the Act on the Promotion of Inland Fisheries in June 2015, which requires those seeking to engage in eel aquaculture to obtain permission from the Ministry of Agriculture, Forestry, and Fisheries (MAFF). Under this Act, the total input of eel seeds for the entire country (upper limit: 21.7 tons) is allocated across aquaculture farms. The quantity of eel seeds allowed in each aquaculture farm is specified by the aquaculture permit. Farmers who receive an allocation are required to report the amount of eel seeds they input and their eel production to the national government every month. This process enables Japan to quantitatively manage domestic eel aquaculture. Thus, the international agreement on conservation and management measures for the upper limit of eel seeds to be input into aquaculture ponds is effectively implemented.

19.2.2.3 International Trade

To ensure the effectiveness of domestic management measures to protect Japanese eel resources, the Japanese national government specifies conditions for issuing approval for the export of eel seeds up to 13 g in weight: (1) the jurisdiction where eel seeds are exported must take conservation and management measures based on the outcomes of the Informal Consultation (see Sect. 19.3), which must be appropriately implemented; (2) the eel seeds must be caught in accordance with domestic laws and regulations; and (3) for the export of glass eels that have never been farmed in aquaculture ponds, their origin and trade must be traceable, and the overall input of glass eels into aquaculture ponds in Japan must exceed 50% of the Japanese upper limit (Fisheries Agency of Japan, https://www.jfa.maff.go.jp/j/saibai/unagi/export_unagi.html, accessed on June 11, 2022).

19.2.3 International Arrangements for Conservation and Management of Japanese Eels

Japan, China, Korea, and Chinese Taipei (hereafter referred to as “Members”) take collaborative actions at the Informal Consultation on International Cooperation for Conservation and Management of Japanese Eel Stock and Other Relevant Eel Species (Informal Consultation) under the framework of the APEC Ocean and Fishery Working Group (OFWG) to advance the regional management of Japanese eel resources, and ensure that the conservation and management efforts of each member have positive effect on the entire resource in the region. In 2014 at the seventh meeting, members issued the joint statement on international cooperation for conservation and management of Japanese eel stock and other relevant eel species

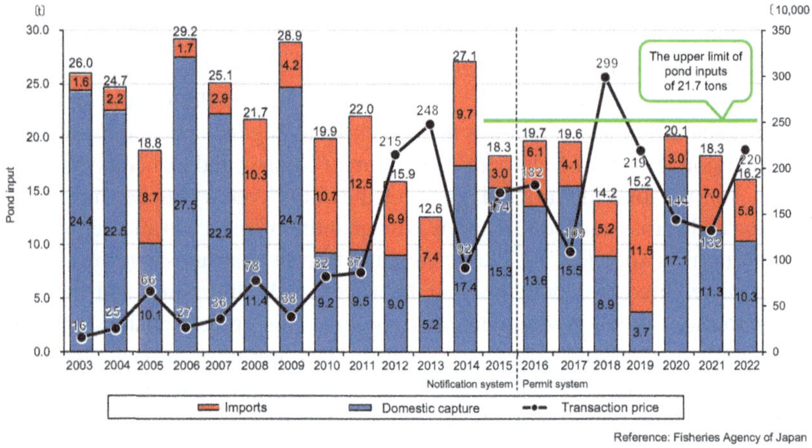

Fig. 19.4 Inputs of Japanese eel seeds into aquaculture ponds in Japan and transaction prices in each catch season from 2003–2022. The annual pond input represents the total volume from November in the previous year to May. The data for 2003–2013 and for 2014–2022 are originated from the industry research and the research by the Fisheries Agency, respectively. Transaction prices are originated from industry data. Imports are calculated from Trade Statistics of Japan

(Joint Statement 2014, http://www.jfa.maff.go.jp/j/saibai/pdf/140917jointstatement.pdf, accessed on June 11, 2022) as a compilation of their commitments. In the Joint Statement, it was articulated that for Japanese eel stock, the initial input of eel seeds for the 2014–2015 catch season would be no more than 80% of that of the 2013–2014 input season (from November 1, 2013 to October 31, 2014), and for other relevant eel species, each Member would take every possible measure to maintain the initial input levels of eel seeds from the previous 3 years. The upper limits of the input of eel seeds into aquaculture ponds were maintained as a result of the annual review of conservation and management measures and eel statistics by Members at the Informal Consultation (Fig. 19.4).

This Joint Statement has also promoted collaboration among stakeholders in the private sector. Based on the Joint Statement, the "Alliance for Sustainable Eel Aquaculture" (ASEA) was founded as an international non-governmental group of eel management organizations for each Member to discuss eel resource management. The ASEA plays an important role as a platform for discussing the regional management of eel resources in the private sector.

19.2.4 Recent Situation Surrounding Management of Japanese Eel Resources

19.2.4.1 Domestic Efforts

Glass eels cannot be caught without special catch permits issued in accordance with prefectural fisheries adjustment rules. In 2020, under the new Fishery Act, eels ≤13 cm in total length were designated as "specified aquatic animals and plants" (Article 41, Regulation for Enforcement of the Fishery Act; Fisheries Agency of Japan), which refers to aquatic animals and plants that are likely to be gathered or caught for the purpose of acquiring unlawful economic benefits. Any person is prohibited from gathering or catching specified aquatic animals and plants without specific permits from the relevant authorities. The designation for eels of ≤13 cm in total length will come into effect in December 2023. As a result of this improvement in the Fishery Act, catching eels ≤13 cm in total length will be regulated as a fishery permitted by the governor, and there will be a strict penalty for noncompliance (imprisonment for ≤3 years or a fine of ≤30 million yen) (Article 189, Fishery Act; Fisheries Agency of Japan). Furthermore, under the Act on Ensuring the Proper Domestic Distribution and Importation of Specified Aquatic Animals and Plants, eels ≤13 cm in total length are recognized as aquatic animals and plants that are in particular need of conservation and management because of their significant risk of illegal and excessive catching or gathering in Japan (excluding catching or gathering by foreign fishing vessels). After December 2025, business operators who catch/gather or distribute such eels will be required to (1) notify the administrative authorities in advance, (2) inform other business operators of catch numbers and other information when transferring the eels, and (3) prepare and keep transaction records for the transfers (Article 1, Regulation for Enforcement of the Act on Ensuring the Proper Domestic Distribution and Importation of Specified Aquatic Animals and Plants; Fisheries Agency of Japan).

In addition to fisheries management, continuous efforts have been made towards the creation and conservation of favorable riverine environments for the Japanese eel (see Chap. 22). The concept of "nature-oriented river works" has been adopted in river management to conserve and create intrinsic river habitats. One example is *ishikura*, an artificial stone-filled cage (see Chap. 18) that is placed instream to provide structural habitat and refuge for Japanese eels. Approximately 440 cages have been placed in rivers and lakes across Japan, and their design and placement are under continuous evaluation and improvement (Mochioka N, personal communication). In addition, a study from Lake Shinji, Shimane, suggested that the use of neonicotinoid pesticides since 1993 has caused declines in Japanese eel and Japanese smelt *Hypomesus nipponensis* populations by altering food web structure and dynamics (Yamamuro et al. 2019). Thus, the management of pesticides and other chemicals is also an important issue in inland habitats of Japanese eel.

19.2.4.2 International Efforts

Following the 14th annual Informal Consultation (Fisheries Agency of Japan, ttps://www.jfa.maff.go.jp/j/press/sigen/210727.html, accessed on June 11, 2022), it was agreed that a scientific meeting on the Japanese eel would be held to promote regional communication and research collaboration. Therefore, as a regional initiative, the first scientific meeting on Japanese eel and other relevant eels, under the framework of the Informal Consultation, was held online in April 2022 (Fisheries Agency of Japan, https://www.jfa.maff.go.jp/j/press/sigen/220415.html, accessed on June 11, 2022). The meeting was attended by representatives from China, Japan, the Republic of Korea, and Chinese Taipei, as well as invited experts on European eels. Members shared and exchanged scientific knowledge of eels, particularly the Japanese eel, and discussed ways to enhance scientific activities and collaboration. They agreed on a "Roadmap for Scientific Activities and Collaborative Research on Japanese Eel" which focuses on: (1) developing close relationships among scientists in the Northeast Asia region and collecting and organizing long-term time-series data on Japanese eel to understand and forecast the stock trend of Japanese eel in that region, and (2) exchanging information on tracking techniques to track migration paths of Japanese eels, and the other relevant eels, from rivers to spawning grounds in Northeast Asia and other regions, and subsequently analyzing/evaluating tracking data. Members also exchanged views on establishing standard working formats for the statistics of the glass eel, elver, and adult eel (catch, input into aquaculture ponds, and aquaculture and trade). Input from each Member was considered to ensure efficient collection and collaborative use of the statistics at the Informal Consultation.

Discussions at the global level are ongoing. The Convention on International Trade in Endangered Species of Wild Fauna and Flora (CITES) was designed to protect endangered wild species by controlling their capture and exploitation for international trade. European eels were listed in the CITES Appendix II at the 14th Conference of Parties (CoP). Currently, the EU does not issue export permits for the European eel, and European countries base conservation and management measures on management advice from the Advisory Committee of the International Council for the Exploration of the Sea (ICES). In August 2019, the CITES CoP18 decided to ensure sustainable trade of eels not listed in the CITES appendices, including the Japanese eel. In the decision, range states are encouraged to: (1) cooperate with other countries and regions that share resources to set common management objectives and improve understanding of biological information; (2) introduce a monitoring system for resource status; (3) improve traceability in international trade; and (4) report on the above efforts and measures to the CITES Secretariat (CITES Decision 18.198., 2019, https://cites.org/eng/dec/index.php/42080, accessed on June 11, 2022). Since 2012, regional efforts through the Informal Consultation, which are compiled in the Joint Statement in 2014, have been in line with this decision, and it is expected that Members will continue to enhance efforts through this framework.

19.2.5 Management of European Eel in Europe

The European eel *Anguilla anguilla* has a complex life cycle: the spawning grounds are in the Sargasso Sea in the Caribbean and from there the glass eels reach European shores by drifting on the Gulf Stream current. Eels spend 5–20 years in fresh and/or brackish waters and then return to the marine environment to spawn (European Commission, https://oceans-and-fisheries.ec.europa.eu/ocean/marine-biodiversity/eel_en, accessed on October 10, 2022). A drastic decline in the recruitment of European eels after the 1980s raised concerns over the status of eel stock, highlighting the need to improve scientific understanding of the stock. The European eel is listed as critically endangered by the IUCN; fishing is regulated, including 3-month fishing closures, and trade outside the EU is banned.

To manage European eel, the European Commission (EC) requested scientific advice from the ICES, and based on this advice, the EC established measures for the recovery of eel stocks. In 2007, the European Commission Council Regulation 1100/2007 "establishing measures for the recovery of the stock of European eel" was published, requiring EU Member states to establish and implement eel management plans (EMPs). These EMPs require EU countries to restrict fishing, undertake stocking activities, facilitate eel migration, and increase the escape of silver eels to 40% of the pristine escapement levels.

19.3 Future Perspectives

Eel resource management, including Japanese and European eels, is addressed through both domestic regulations and regional arrangements (for the Japanese eel, Fig. 19.5), while the management frameworks differ depending on regional contexts

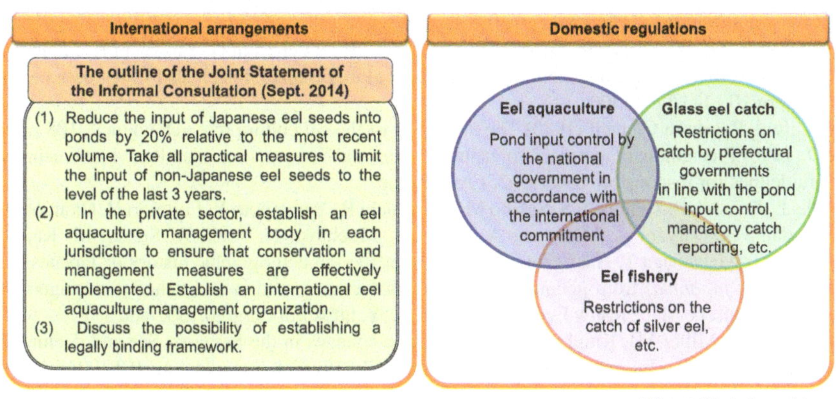

Fig. 19.5 Regional management of Japanese eel resources

and scientific knowledge of stocks. For the management of Japanese eel, the Informal Consultation under the framework of the APEC OFWG functions as a place for regionally coordinating conservation policies, including the regulation of glass eel inputs into aquaculture ponds. Under the consultation mechanism, a scientific body was established in 2022 to encourage scientists from range countries/regions to cooperate to bridge knowledge gaps and provide scientific advice for the conservation and management of the species. For the management of European eel, ICES provides scientific advice to the EC based on the knowledge synthesis by WGEEL, and the EC regulation specifies regulatory principles, such as the 40% of pristine escape levels of silver eel, for conservation of the species, taking into account the scientific recommendations.

In Japan, it is the responsibility of resource users to continue improving conservation and management measures for the Japanese eel, which is the reason why the national and prefectural governments manage eel fisheries, and the distribution and input of eel seeds into aquaculture ponds. The implementation of these measures could be more effective if scientific knowledge of Japanese eel is advanced. In addition to monitoring biomass levels and trends, future research should focus on producing additional scientific information and data from the Northeast Asia region and to advance the stock assessment of the Japanese eel.

References

Bonhommeau S, Chassot E, Planque B, Rivot E, Knap AH, Le Pape O (2008) Impact of climate on eel populations of the Northern Hemisphere. Mar Ecol Prog Ser 373:71–80. https://doi.org/10.3354/meps07696

Chang Y-LK, Miyazawa Y, Miller MJ, Tsukamoto K (2018) Potential impact of ocean circulation on the declining Japanese eel catches. Sci Rep 8:5496. https://doi.org/10.1038/s41598-018-23820-6

Chang Y-LK, Miyazawa Y, Miller MJ, Tsukamoto K (2019) Influence of ocean circulation and the Kuroshio large meander on the 2018 Japanese eel recruitment season. PLoS One 14:e0223262. https://doi.org/10.1371/journal.pone.0223262

Drouineau H, Durif C, Castonguay M, Mateo M, Rochard E, Verreault G, Yokouchi K, Lambert P (2018) Freshwater eels: a symbol of the effects of global change. Fish Fish 19:903–930. https://doi.org/10.1111/faf.12300

Eel Culture Research Council (1980) The 9th report of the eel culture research council, 108 pp

FAO (1996) Precautionary approach to capture fisheries and species introductions. FAO technical guidelines for responsible fisheries 2. FAO, Rome

Faulks L, Kaushik P, Taniguchi S, Sekino M, Nakamichi R, Yamamoto Y, Fujimori H, Okamoto C, Kodama S, Daryani A, Manwong A, Galang I, Mochioka N, Araki K, Suzuki M, Kaji Y, Ichiki T, Matsunaga T, Hakoyama H (2022) Inferring the demographic history of Japanese eel (*Anguilla japonica*) from genomic data: insights for conservation and fisheries management. Aquat Conserv Mar Freshw Ecosyst 32:1092–1098. https://doi.org/10.1002/aqc.3810

Friedland KD, Miller MJ, Knights B (2007) Oceanic changes in the Sargasso Sea and declines in recruitment of the European eel. ICES J Mar Sci 64:519–530. https://doi.org/10.1093/icesjms/fsm022

Gong X, Davenport ER, Wang D, Clark AG (2019) Lack of spatial and temporal genetic structure of Japanese eel (*Anguilla japonica*) populations. Conserv Genet 20:467–475. https://doi.org/10.1007/s10592-019-01146-8

Hakoyama H, Hiroka F, Chiaki O, Kodama S (2016) Compilation of Japanese fisheries statistics for the Japanese eel, *Anguilla japonica*, since 1894: a historical dataset for stock assessment. Ecol Res 31:153. https://doi.org/10.1007/s11284-015-1332-9

Han Y-S, Hung C-L, Liao Y-F, Tzeng W-N (2010) Population genetic structure of the Japanese eel *Anguilla japonica*: panmixia at spatial and temporal scales. Mar Ecol Prog Ser 401:221–232. https://doi.org/10.3354/meps08422

Ishikawa S, Aoyama J, Tsukamoto K, Nishida M (2001) Population structure of the Japanese eel *Anguilla japonica* as examined by mitochondrial DNA sequencing. Fish Sci 67:246–253. https://doi.org/10.1046/j.1444-2906.2001.00227.x

Kaifu K, Yokouchi K, Higuchi T, Itakura H, Shirai K (2018) Depletion of naturally recruited wild Japanese eels in Okayama, Japan, revealed by otolith stable isotope ratios and abundance indices. Fish Sci 84:757–763. https://doi.org/10.1007/s12562-018-1225-2

Knights B (2003) A review of the possible impacts of long-term oceanic and climate changes and fishing mortality on recruitment of anguillid eels of the Northern Hemisphere. Sci Total Environ 310:237–244. https://doi.org/10.1016/S0048-9697(02)00644-7

Mather N, Traves SM, Ho SYW (2020) A practical introduction to sequentially Markovian coalescent methods for estimating demographic history from genomic data. Ecol Evol 10: 579–589. https://doi.org/10.1002/ece3.5888

Matsui I (1972) Mangaku. An eel science. Kouseisha-Kouseikaku, Tokyo; in Japanese

Mochioka N (2019) Fishing gear and fishing method. In: Tsukamoto K (ed) Science of eels. Asakura Publishing, Tokyo, pp 120–125; in Japanese

Righton D, Piper A, Aarestrup K, Amilhat E, Belpaire C, Casselman J, Castonguay M, Díaz E, Dörner H, Faliex E, Feunteun E, Fukuda N, Hanel R, Hanzen C, Jellyman C, Kaifu K, McCarthy K, Miller MJ, Pratt T, Sasal P, Schabetsberger R, Shiraishi H, Simon G, Sjöberg N, Steele K, Tsukamoto K, Walker A, Westerberg H, Yokouchi K, Gollock M (2021) Important questions to progress science and sustainable management of anguillid eels. Fish Fish 22:762–788. https://doi.org/10.1111/faf.12549

Tanaka E (2014) Stock assessment of Japanese eels using Japanese abundance indices. Fish Sci 80: 1129–1144. https://doi.org/10.1007/s12562-014-0807-x

Tanaka H (2019) History and current status of eel aquaculture. In: Tsukamoto K (ed) Science of eels. Asakura Publishing, Tokyo, pp 157–161; in Japanese

UNCED (1992) Principle 15, Rio Declaration on Environment and Development

Waples RS, Do C (2010) Linkage disequilibrium estimates of contemporary ne using highly variable genetic markers: a largely untapped resource for applied conservation and evolution. Evol Appl 3:244–262. https://doi.org/10.1111/j.1752-4571.2009.00104.x

Yamamuro M, Komuro T, Kamiya H, Kato T, Hasegawa H, Kameda Y (2019) Neonicotinoids disrupt aquatic food webs and decrease fishery yields. Science 366:620–623. https://doi.org/10.1126/science.aax3442

Yu L, Liu Y, Liu J (2020) Gene-associated microsatellite markers confirm panmixia and indicate a different pattern of spatially varying selection in the endangered Japanese eel *Anguilla japonica*. J Ocean Limnol 38:1572–1583. https://doi.org/10.1007/s00343-020-0048-z

Chapter 20
Trading and Distribution

Tetsuji Ida

Contents

Eels have long been an important food source for the Japanese people. The first written mention of "*kabayaki*," a method of cooking eel by broiling it with sweet and spicy soy-based sauce, is from a book written in 1399. By the Edo period (1603–1867) in the eighteenth century, *kabayaki* utilization was similar to that of today, popularizing restaurants that specialized in it. Indeed, eel restaurants specializing in serving *kabayaki* are still popular today. Undoubtedly, the business of providing customers with a single species of fish cooked in a single way (*kabayaki*) is one of Japan's unique culinary traditions, rarely seen in other parts of the world.

20.1 The Emergence and Expansion of Eel Farming

Currently, the common eel-farming practice is to use glass eels and raise them to adult size in aquaculture ponds. Eel farming began in Japan in 1879, when a freshwater fish farmer in Tokyo built fishponds in what is now Koto Ward, Tokyo, and caught juvenile eels that had migrated up the city rivers and subsequently fed and raised them to adult size. Eel farming expanded rapidly around urban areas where demand was high, leading to further expansion of eel farming all over Japan. This was a major turning point for the eel trade and business in Japan, as many farmed eels began to be distributed instead of wild-caught eels, which were

T. Ida (✉)
Kyodo News, Tokyo, Japan

K. Tsukamoto et al. (eds.), *Eel Science*, Fisheries Science Series,
https://doi.org/10.1007/978-981-99-5692-0_20

originally caught in small quantities. It is said that the production of farmed eels surpassed that of wild adult eels in 1930. After World War II, farmed and wild eels coexisted in the market; however, the eel business and trade in Japan changed dramatically during the high growth period from 1955. It is characterized by an increase in demand, a decrease in natural resources, and an expansion of international trade.

Eel fisheries have long existed in Japan where wild adult eels are harvested and served to consumers. According to statistics from the Fisheries Agency of Japan, the catch of adult eels reached nearly 4500 tons in the 1910s. After World War II, the catch peaked around 1960 at ~3500 tons, or ~18 million eels. However, the catch of adult eels has declined rapidly since then. One of the main reasons for this is believed to be that estuary weirs and dams have been constructed in many parts of rivers in Japan, blocking the upstream migration of glass eels. For example, the Tone River, which flows along the border between Ibaraki and Chiba prefectures, and the Kasumigaura and Kitaura areas upstream of the Tone River were major sources of natural adult eels until the mid-1960s. However, after a series of floodgates and estuary weirs were constructed in the middle reaches of the river in the late 1960s, the catch sharply declined, and now wild eel fishery is nearly obsolete. Similar cases have been reported in rivers across Japan, particularly in urban areas. For about a decade after the Fukushima-Daiichi nuclear power plant accident in 2011, shipments of wild-caught eels had been restricted in some rivers of Tohoku and Kanto regions, and there has been a growing movement to refrain from fishing for adult eels to protect resources. Therefore, the catch plummeted to a record low of 66 tons in 2019 (see Chap. 19; Fig. 19.1a).

In 1960, the supply of farmed eels exceeded that of wild-caught eels in the Japanese market, but since the catch of adult eels has almost disappeared, the vast majority of eels are supplied to the Japanese market today, and almost all of the eels consumed by the Japanese people are farmed eels. It is important to note, however, that the artificial propagation technology for raising eels from eggs to adults has not yet been commercialized; thus, even if the eels are farmed, they depend entirely on harvesting glass eels for their seedlings. Since the 1960s, when the aquaculture industry began to expand, the domestic catch of glass eels, which are used as seedlings for aquaculture, began to decline, as did the catch of natural adult eels. The largest catch of glass eels occurred around 1960, nearly in the same period as the adult eels. The catch of glass eels, which used to be nearly 240 tons per year, began to decline sharply, as did the catch of natural eels. By 1971, the catch dropped to below 100 tons. The serious downward trend in catches continued in subsequent years, with a record low of only 3.7 tons caught in 2019 (see Chap. 19, Fig. 19.1b).

This rapid decline in catch, both adults and glass eels, is an important rationale for Japanese eel categorization as an endangered species by the International Union for Conservation of Nature (IUCN) and the Japanese Ministry of the Environment.

20.2 Domestic and International Trade of Glass Eels

Aquaculture using glass eels as seedlings is a mainstream industry at present. In Japan, the eel trade is based on the following steps: eel fishermen catch glass eels, collectors collect them, intermediaries distribute them to eel farmers, and farmers and retailers sell them to consumers. The collectors are responsible for gathering glass eels, which, in some cases, amount to only a few grams per day, until they reach a certain amount. Eels collected by collectors usually pass through several distribution intermediaries before being supplied to the aquaculture industry, thus creating a complex, multi-step domestic trade structure. This significantly reduces the transparency of the domestic eel trade and distribution non-transparent, which has resulted in it becoming a major breeding ground for many illegal, unreported, and unregulated (IUU) fishery-related activities.

Although Japanese eels migrate to the coasts of Korea, China, and Taiwan in addition to Japan, Japan has consistently remained the largest consumer of eels. The growth of Japan's eel market, where eels are traded at high prices, has led to an increase in imports of adult Japanese eels and glass eel from China, Taiwan, Korea, and other countries. In 1988, the domestic glass eel catch was ~25 tons; however, Japan imported ~15 tons of glass eels for farming from China, South Korea, North Korea, Taiwan, and other countries. Due in part to increased imports of glass eels, according to the Fisheries Agency, Japan's domestic eel production doubled from ~2000 to 40,000 tons in the 1970s and around 1980, respectively, and numbers continued to rise until the early 1990s. However, the catch of glass eels has significantly declined throughout East Asia since 1989, when the production of farmed eels in Japan peaked at ~40,000 tons and began to decline. The current farming production is ~1/2 of its peak.

As the decline in Japanese eel stocks became more significant, using non-native *Anguilla* species (American and European eels) for aquaculture gained popularity, thereby expanding their imports in Japan. In Europe and the US, there was a significant rise in eel exports in the mid-1980s, specifically targeting the Japanese market; it made national news headlines, according to import industry insiders. The focus of these exports was on European eels, but Japan was unable to establish cultivation technology for them. In contrast, China succeeded in establishing their own European eel cultivation technology, and a new business was born by importing large quantities of European eels, cultivating them in mass, and exporting them to Japan. Around 1985, imports of farmed eels, mainly from China, exceeded 40,000 tons, surpassing Japan's domestic production of farmed eels. Since then, imports of farmed eels from China have increased rapidly.

20.3 Changing Patterns of Eel Consumption

With the increased eel supply in the market, the pattern of eel consumption in Japan has changed dramatically. Eel consumption in Japan was mainly at specialty restaurants, where live eels were cooked for customers. However, during this period, the market was dominated by preprocessed packaged eels. Since then, the consumption of pre-processed packaged eels has exceeded that of specialty restaurants, and eels have become much more common to Japanese consumers than before. The "consumption explosion" of eels in Japan was triggered mainly by Chinese aquacultures and the Japanese traders who import them. According to government statistics, since the mid-1980s, the quantity of eels supplied to the Japanese market has increased rapidly. It nearly doubled in a short period, from ~80,000 to 158,000 tons in 1985 and 2000, respectively. Furthermore, in 2000, domestic aquaculture production (mainly that of Japanese eels) was ~20,000 tons. The supply of imported eels, mainly from China, amounted to 133,000 tons, and imports of eels from China, mainly European eels, accounted for nearly 85%. Consequently, the consumption of eels, at specialty restaurants at relatively high prices, decreased. Thin profit margins and the mass consumption of eels became mainstream in Japan as eels were packaged and sold in large quantities at relatively low prices. However, this was short-lived. Eel consumption in Japan increased further, worsening the deteriorating status of European eel stocks, and subsequently leading to a decline in imports.

In 2007, the European eel was listed in Appendix II of the Convention on International Trade in Endangered Species of Wild Fauna and Flora (CITES), which came into effect in 2009. In addition, the EU banned commercial trade in all commodities of European eels to and from the EU in December 2010 (see Chap. 23). In 2008, the International Union for Conservation of Nature (IUCN) designated the European eel as Critically Endangered (CR), which is the highest of the 3 levels of extinction risk, and this led to calls for the conservation of European eel resources. Since then, the supply of European eels imported to Japan via China began to decline. The decline in the supply of imports, combined with the worsening stock status of Japanese glass eels, caused market prices to skyrocket and the supply of eels to the Japanese market to decline rapidly. Supplies of eels, both domestically produced and imported, fell to ~33,000 tons in 2013, similar to the 1975 level. Supplies began to increase slightly, reaching ~63,000 tons by 2021; however, this is less than 40% of the peak level. Of this, 42,000 tons were imported, leading to Japanese domestic aquaculture production of ~21,000 tons (Fig. 20.1). It is undeniable that the short-term consumption explosion in Japan had a major negative impact on global eel resources, especially those of temperate eel species, the European and Japanese eels. These species are now listed as endangered or CR on the IUCN Red List (see Chap. 23). The Japanese eel was declared as an endangered species in 2013 by Japan's Ministry of the Environment, and in 2017, it was classified as a critically endangered species in Taiwan.

Although eel consumption in Japan has been declining due to the worsening resource availability and soaring prices associated with the declining supply, there

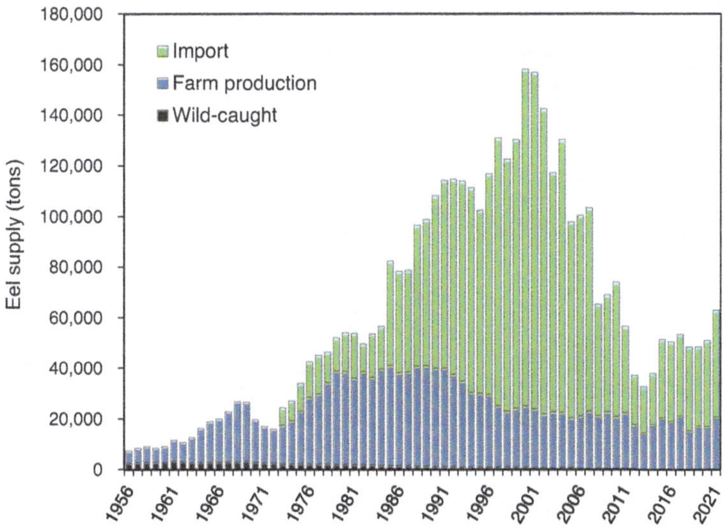

Fig. 20.1 Trends in eel supply in Japan. Data is based on Annual Report of Fishery and Aquaculture Production Statistics by the Ministry of Agriculture, Forestry and Fisheries, Japan and Fisheries and Trade Statistics by the Ministry of Finance, Japan

has been no significant change in the eel business, which is based on thin profit margins, with packaged eels sold at supermarkets and convenience stores and consumed at fast-food restaurants. While the market is increasingly dominated by large suppliers that buy eels in short supply and sell them to supermarkets and other major sellers, the number of traditional restaurants facing high prices and difficulty obtaining eels has been declining rapidly since 1990. With eel resources showing no signs of recovery, and the continued mass consumption of packaged eels (mainly on the day of *Doyou-no-Ushi* once a year), scientific verification would be required to determine the appropriate consumption level of endangered Japanese eels to sustain their resources in the long run.

20.4 Illegal Fishing of Glass Eel Fisheries and Trade

The eel trade and distribution in Japan continues to lack transparency, with illegal and unreported fishing, trade, and smuggling happening daily according to various media reports in Japan. This problem has been pointed out by many environmental groups, researchers, and media, and some progress has been made in administrative efforts. However, these basic problems remain unresolved.

One of the major problems with the eel trade in Japan is that IUU fishing is rampant in domestic glass eel fisheries. The glass eel fishery is a permit fishery, and only those who have obtained a permit are allowed to fish (see Chap. 19). However, glass eel fishing does not require large equipment, such as fishing boats; thus, it is a

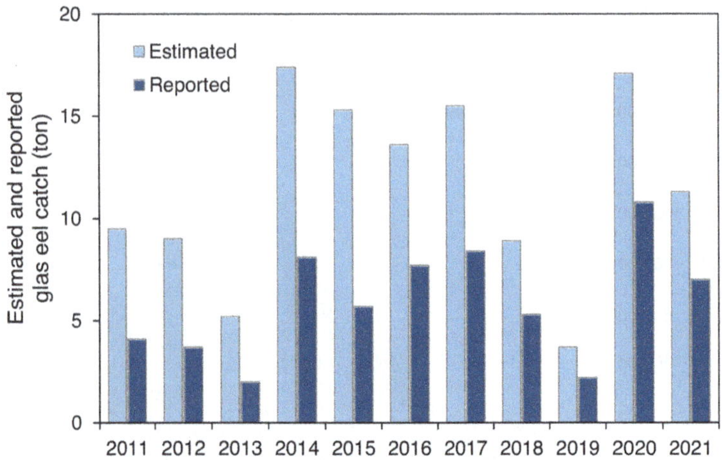

Fig. 20.2 Comparison of domestic catch of Japanese glass eel; estimated by the Fishery Agency and reported from each prefecture of Japan. The glass eel catch in each year is the total from November of the previous year to May of the year. (Modified from the data of Fisheries Agency, Japan)

primitive fishery that can be performed by anyone using only a simple fishing net. The location and season of the upwelling are specified; the fishing season is around the new moon between the fall and spring of the following year, making it easy for anyone to enter a fishery. Moreover, because fishing occurs at night, it is difficult to uncover illegal activities. As a result, poaching and unreported glass eel fishing have become rampant in many parts of Japan, creating a vicious cycle that is spurred by soaring prices due to declining resources. In areas where glass eel fishing is active, organized criminal groups are frequently found to be involved in glass eel fishing, according to various industry insiders. Government data show a large gap between the amount of glass eel catch reported to prefectural governments each year and the estimated glass eel catch based on the reported amount placed in aquaculture ponds. The Fisheries Agency estimated the catch of glass eels based on the amount of glass eels placed in aquaculture ponds each year and compared it with the catch reported to the prefectures where fishery was actually conducted. It was found that the catch estimated from the amount put in the ponds for the 2011–2021 fishing season has continued to far exceed the reported amount every year. In 2021, the estimated catch was 11.3 tons, while the reported catch was 7 tons. In 2014, the reported amount was only 8.1 tons out of an expected 17.4 tons, and in 2015, when the gap was the largest, the estimated amount was 15.3 tons, while the reported amount was only 5.7 tons. From 2011 to 2015, the estimated catch was more than twice that reported. From 2011 to 2021, there continued to be a large difference between the number of glass eels placed in aquaculture ponds and the reported catch, averaging about twice as much as the former and the latter (Fig. 20.2). These data indicate that serious unreported fishing practices continue in the Japanese domestic glass eel fishery.

Some experts and insiders also point out that even among the catches reported to prefectural governments, some were caught illegally.

According to data from the Ministry of Finance and the Ministry of Internal Affairs and Communications in Japan, the price of glass eels, which was ~150,000 yen per kg in 2000, continued to rise, reaching 2.15 million yen per kg in 2012, exceeding 2 million yen per kg for the first time. In 2018, when catch was particularly low, the price reached 3 million yen. Assuming that the weight of a glass eel is 0.2 g, this translates to 600 yen per small eel. Some prefectures require traders to trade glass eels at the official price, which is far lower than the market price, and/or to regulate sales to other municipalities. As a result, many glass eels are found on the black market where they can be sold at a higher price. It is well known that there have been cases in which the catch of glass eels has exceeded the quotas set by each municipality and has been passed on to aquaculture companies without being reported.

In principle, glass eels are considered to be seed stocks for aquaculture within the local government; therefore, if the catch exceeds the capacity of the aquaculture ponds, the price within the local government drops. In some cases, local governments set sales prices for designated destinations within municipalities. If a vendor outside the municipality is willing to buy the eel at a higher price, even if it is not reported, the vendor will sell to them rather than reporting properly. Experts point out that the existence of a patchy administration, with different regulations and systems of trade and varying trade prices among the prefectures where fishing is allowed, has led to rampant unreported fishing. In addition, organized criminal gangs and traffickers engaged in the eel trade in a criminal manner are still active in the industry, despite the police making several arrests every year. As mentioned above, administrative oversight is not readily available due to the primitive nature of the glass eel fishery and the ease with which anyone can enter the market, which is another reason why illegal fishing has not been eliminated. For example, in Okayama Prefecture, glass eels were previously harvested for the prefecture's aquaculture industry under a special permit granted by the governor. All glass eels caught were to be sold to aquaculture companies in the prefecture; however, in 2003, all aquaculture companies in the prefecture went out of business and all glass eels were banned. Since then, unauthorized glass eel fishing has continued and there have been several criminal cases. Poaching is reported almost every year in many eel-producing areas other than Okayama Prefecture. Long-term investigative reporting by the Kochi Shimbun, a local newspaper in Kochi Prefecture (a major eel-producing areas), has revealed the existence of poaching, unreported fishing, and large-scale trade in glass eels in Japan that routinely disregards regulations; this may be the tip of an iceberg.

As IUU fishing of eels is a flourishing trafficking industry in Japan, the smuggling of eels out of Japan is also a continuing problem. Although the export of glass eels caught in Japan was prohibited from 1976 to 2021, there was a continuous smuggling of glass eels to Taiwan, either through Hong Kong or directly to Taiwan. The smuggling of glass eels from Japan overseas began to increase in the mid-1990s, when the decline in East Asian glass eel stocks became an issue, and has continued to the present day. In 2020, seven men were arrested in Osaka Prefecture for illegally smuggle ~60 kg of glass eels into Hong Kong. In an interview with the Asahi Shimbun, 1 of the suspects admitted that he had tried to smuggle glass eels collected by traders from various regions in suitcases, saying: "This happens every year. A stranger gave me a suitcase containing glass eels, 50,000 yen in an envelope as a reward, and money for my stay in Hong Kong, and I flew to Hong Kong in the business class. I waited in front of a fountain at the local airport and handed the suitcase to a stranger. I was told that I would not get caught. It was a good part-time job".

20.5 Dubious International Trade

Another problem with Japan's eel trade and market distribution system is the non-transparent presence of imported eels. It has been common practice for many East Asian countries/territories to export Japanese glass eels to Japan, where huge aquaculture businesses exist. Since 2000, Taiwan has been the largest exporter of glass eels to Japan, with 4.51 tons (83% of Japan's 5.44 tons of glass eels imported in 2006) coming from Taiwan. However, in 2007, the Taiwanese government imposed a ban on the export of glass eels, highlighting the need to protect domestic glass eel resources. It was thought that the Japanese aquaculture industry, which had been suffering from a severe shortage of resources, would receive a major blow; however, the impact was minimal. Immediately after the Taiwanese embargo, Japan began importing large quantities of "Hong Kong-grown" glass eels. According to Japanese trade statistics, the volume of glass eels from Taiwan to Japan plummeted to 85 kg in 2007, whereas imports from Hong Kong increased from 0 in 2006 to ~4 tons in 2007. In 2008, this amount increased further to 11.3 tons. Although there are increases and decreases depending on the annual glass eel recruitment, imports of glass eels from Hong Kong remain the highest. However, there are no glass eel fisheries in Hong Kong. It is pointed out that many of the eels said to be from Hong Kong are illegally taken there from Taiwan and other countries, and subsequently exported to Japan and other countries; this is an open secret between many eel traders. The Hong Kong government has not participated in discussions regarding the management of eel resources in East Asia in Japan, China, Korea, and Taiwan. Furthermore, it is internationally acknowledged that continuing illegal and unreported fishing on a large scale significantly affects the maintenance of proper trade and resource management. In Taiwan, smuggled glass eels travelling to Hong Kong are frequently seized. In 2018, the CITES Secretariat also issued a report on the poaching, smuggling, and laundering fishes destined for Hong Kong. As a result of discussions between Japanese and Taiwanese authorities to address this issue, in 2021, the ban on the export of glass eels from Japan to Taiwan was lifted for the first time in 45 years; conversely, the ban on the export of glass eels from Taiwan may also be lifted in the near future. Glass eel trafficking in Hong Kong continues today. The amount of glass eels imported from Hong Kong to Japan during the fishing season from November 2021 to May 2022 amounted to ~5.5 tons, with the total amount of imported glass eels reaching 6.4 t. This accounted for 85% of the total imports. The total of imports and domestic catch during the 2021–2022 fishing season was 17.7 tons, of which 4.3 tons were unreported domestically and 5.5 tons were from Hong Kong, which is uncertain. Thus, nearly 50% of glass eels that are put into aquaculture ponds domestically are likely to be sourced from IUU fishing and/or illegal trade.

Another major problem is the continued smuggling of European eels for farming (mainly in China) from European countries where the fishing and export of glass eels is banned or strictly regulated to protect resources. Japan has been re-importing large quantities of European eel products from China, even after the CITES regulations came into effect. European law enforcement officials have pointed out that many of the European eels imported to Japan from China are believed to have originated from Morocco and other countries; however, the possibility of smuggling eels from the EU mixed with these products cannot be ruled out. Despite EU regulations banning the export of European eel fry, there has been no end to the number of smuggling busts of eels believed to be destined for aquaculture in East Asia. In July 2019, the European Police Agency (Europol) announced that more than 20% of European glass eels caught in the EU from 2018 to early 2019 were smuggled to China and other Asian countries, resulting in a black market worth ~3 billion euros (~36 billion yen). According to Europol, > 400 tons of eel fry reach Europe annually, of which ~100 tons, or Europol estimates that ~100 tons, or 300–350 million fish, were smuggled from Europe. Smuggling is often done by "couriers" posing as tourists who pack the fry collected by poachers into suitcases and take them from airports around Europe. Investigative authorities in European countries have identified 153 smugglers during the fishing season from fall 2018 to spring 2019, an increase of ~50% from the previous year. In total, 15 million fish were seized. Spanish authorities have pointed out that a network of Asians, including the Chinese living in Europe, is involved in smuggling. A European expert commented: "In monetary terms, this is the world's largest crime against wildlife. This undermines the recovery of precious species." This means it is highly likely that some of the broiled eels exported from China to Japan still contain smuggled glass eels from Europe.

20.6 Conclusion

To eliminate IUU fishing and prevent human rights violations related to fishing, governments in Europe, the US, and other countries are increasingly regulating IUU fishing, and companies are taking steps to do the same. Under such circumstances, while it is a well-known fact that such large-scale, non-transparent fishing and trading, and distribution is taking place in Japan's glass eel fishery, the government and companies have not made progress in improving traceability, and many large companies and restaurant chains continue to widely sell products in which the origin is unclear. To achieve fair trade and proper resource management, it is imperative to implement policies to increase transparency in eel fishing and trade, and to ensure the traceability of eel products in Japanese markets.

In order to prevent poaching, the Japanese government has taken measures such as designating glass eels as "specified aquatic animals and plants" in the new Fishery Act, which was revised and enforced in 2020; regulations on glass eels will begin in 2023. In addition, the Japanese government has also decided to add glass eels (excluding catching or gathering by foreign fishing vessels) to the newly enacted Act on Ensuring the Proper Domestic Distribution and Importation of Specified Aquatic Animals and Plants, which requires the issuance and maintenance of documents certifying that eels were caught legally. The regulation of glass eels begins in 2025 (see Chap. 19). While it is hoped that this will have a positive effect on the optimization of eel fishery in Japan, some experts question its effectiveness, as imported glass eels are expected to be exempted from the program. As the status of the Japanese eel stock deteriorates and the species stands on the brink of extinction, sustainability of the Japanese eel fishery, eel trade, and distribution remains a pressing issue.

Chapter 21
Circulation

Tatsuki Yoshinaga

Contents

The freshwater eel genus *Anguilla* consists of 19 species/subspecies that inhabit brackish and freshwater habitats worldwide. The distribution of the genus is mainly in the tropics, however its exact distribution range remains unclear. The freshwater eels have been often traded as living and cooked products for propagation and commercial use, which may negatively affect resource conservation and environmental protection. For example, the European eel had been frequently observed in natural waters in Japan and is suspected to have been released for stock enhancement despite the fact that this species does not naturally inhabit these waters and does not contribute to reproduction in the open ocean. In addition, in the mid-2010s, tropical species were imported to East Asian countries (mostly from the Philippines and Indonesia) as aquaculture seedlings, causing problems with species disguising due to high prices and difficulty in species identification. This chapter outlines (1) the obscurity in defining the distribution range of catadromous eels, which are born in the open ocean and grow in freshwater; (2) the occurrence of alien eel species in East Asian countries; (3) possible problems in utilizing tropical species for aquaculture; and (4) changes in the commercial use of the species in the Japanese market during the past decade.

T. Yoshinaga (✉)
School of Marine Biosciences, Kitasato University, Kanagawa, Japan
e-mail: yosinaga@kitasato-u.ac.jp

K. Tsukamoto et al. (eds.), *Eel Science*, Fisheries Science Series,
https://doi.org/10.1007/978-981-99-5692-0_21

21.1 Distribution Range of Anguillid Species: A Case of the Japanese Eel

The 19 species/subspecies of the genus *Anguilla* are widely distributed throughout the world (see Chap. 1). Of these, 13 are found in tropical countries, mainly in Southeast Asia. In the temperate zone, 3 species are distributed in the Northern Hemisphere (Europe, North America, and East Asia) and 4 in the Southern Hemisphere (Oceania). Because of the high commercial value of freshwater eels, live adults, juveniles, and processed foods are traded internationally (see Chap. 23). However, imported live eels are sometimes released into rivers for propagation, resulting in the spread of non-native species.

To distinguish between native and non-native (exotic/alien) species, it is essential to have information on their natural distribution ranges. In the case of the freshwater fish that spend their entire lives in rivers, lakes, and ponds, they may move to neighboring rivers when the water rises, but they do not travel long distances. Therefore, it is possible to determine boundaries by examining whether fish live in each locality. However, catadromous freshwater eels that spawn in the open ocean and grow in freshwater are quite complicated. Here, we first considered the natural distribution range of the Japanese eel *A. japonica*, which is one of the most well-studied species in the genus. The Japanese eel is born in the western waters of the Mariana Islands and subsequently transported by the Kuroshio Current to the coastal areas of Taiwan, Korea, China, and Japan (Fig. 21.1). The southern limit of the distribution range of Japanese eel*s* is the Philippines. In fact, 60 out of 63,237 glass eels recruited from northern Luzon Island are Japanese eels (Tabeta et al. 1976; Yoshinaga et al. 2014, unpublished data). For example, on Mindanao Island in the southern Philippines, 2 out of 4745 glass eels were Japanese eels (Shirotori et al. 2016). These facts suggest that the Philippines are involved in the distribution range of the Japanese eel. However, the Japanese eel population is extremely small compared to other species; thus, it is reasonable to assume that the Japanese eel is not substantially distributed in the Philippines, despite coming to the shore due to occasional dispersal.

Next, we considered freshwater eel species that are naturally distributed in Japanese waters. In addition to the Japanese eel, 4 species/subspecies, the Indo-Pacific eel *A. marmorata*, Pacific bicolor eel *A. bicolor pacifica*, and Luzon eel *A. luzonensis* have been recorded from the Japanese coast. *A. marmorata* is found in rivers along the Pacific coast from Kyushu to the Kii Peninsula in the middle of Japan. In contrast, *A. bicolor pacifica* and *A. luzonensis* have been confirmed only in limited southern areas, such as Yakushima, Tanegashima, Iriomote, and Okinawa Islands (Inoue et al. 2021; Kita et al. 2021). Although it is not clear whether these individuals contribute to reproduction in the open ocean, it is reasonable to assume that *A. japonica* and *A. marmorata* are the 2 main species inhabiting Japan.

In Southeast Asian countries, such as Indonesia and the Philippines, where the number of anguillid species is much larger than that in temperate zones, the distribution range of each species is not well known. For example, the New Guinea

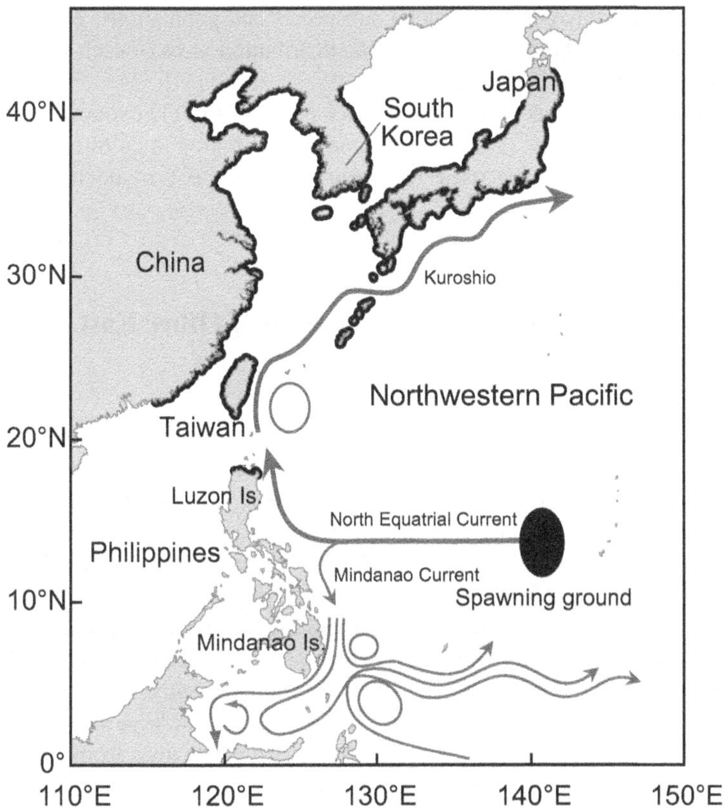

Fig. 21.1 Distribution range of the Japanese eel *Anguilla japonica* as shown by bold coastlines. The spawning area is denoted by a black oval, and the two major currents that transport the leptocephalus larvae, by thick arrows

eel *A. interioris*, which was originally thought to be from New Guinea Island, has also been found in southern Java in Indonesia facing the Indian Ocean as well as in the southern Philippines. However, the actual situation is completely unknown, and the distribution center of each tropical eel species has not been sufficiently clarified.

The existence of subspecies and genetically differentiated populations is also problematic when considering the distribution range of anguillid eels. *A. marmorata* is composed of at least 4 genetically differentiated populations (3 in the Pacific Ocean and 1 in the Indian Ocean (Minegishi et al. 2008), but is taxonomically recognized as a single species. In addition, there are 2 genetically differentiated subspecies of bicolor eels, *A. bicolor* in the Indian and Pacific Oceans, which are designated as *A. bicolor bicolor* and *A. bicolor pacifica*, respectively. Strictly speaking, *A. bicolor bicolor* and the Indian Ocean population of *A. marmorata* should be regarded as non-native species in the Pacific. However, spawning in the open ocean has limited migration barriers compared to inland waters, and thus, genetic exchange may continue among subspecies and populations. Therefore, it is difficult to define the boundary between the distribution areas of each subspecies or population of the genus *Anguilla*.

The purpose of this section is to introduce past cases of (1) exotic eel species in Japan and other countries; (2) alternative and counterfeiting in cultured eel species; and (3) changes in species used in processed foods to provide basic data for conservation.

21.2 Non-native Eel Species in Japan and Other East Asian Countries

Exotic or alien eel species (those that are not supposed to be naturally distributed in Japan) have been reported in many places within the past few decades. The most common species was the European eel *A. anguilla*. The European eel is widely distributed from northern Africa to Europe, but does not inhabit the Pacific Ocean. They were reported to be from the Uono River in Niigata Prefecture and Lake Shinji in Shimane Prefecture in 2000 (Zhang et al. 1999; Aoyama et al. 2000), but have also been found in schools of the Japanese eel migrating to spawning grounds in the East China Sea (Sasai et al. 2001). The European eel has also been observed in Mikawa Bay, Aichi Prefecture (Okamura et al. 2002), and more recently, in the upper reaches of the Tone River in 2015 (Arai et al. 2017). Because of their large size (total length: > 100 cm) it is assumed that these individuals were left over from the previous release. In addition to the European eel, the American eel *A. rostrata* has also been found in Japan, probably originally cultured in China and subsequently imported to Japan. Indeed, there is no rule restricting the release to native species; thus, the species with the lower price per weight are selected for obligatory stocking of inland fisheries. European and American eels have also been found in the natural waters of Taiwan and Korea (Han et al. 2002; Honga et al. 2017). Furthermore, Korea has utilized 6 exotic species/subspecies for aquaculture (in addition to *A. rostrata* which has been intensively cultured): *A. bicolor bicolor*, *A. bicolor pacifica*, *A. anguilla*, Mozambique eel *A. mossambica*, and *A. marmorata* (Honga et al. 2017). Thus, they can be found in natural waters if escape occurs.

Species identification is a challenge in surveys of exotic eel species. Freshwater eels are classified into 4 morphological groups: (I) variegated marking of the body surface and undivided maxillary and mandibular bands of teeth (4 species); (II) variegated with a toothless longitudinal groove in the maxillary and mandibular bands of teeth (4 species); (III) not variegated with a long dorsal fin (6 species); and (IV) not variegate with a short dorsal fin (5 species/subspecies) (Watanabe et al. 2004, 2009). *Anguilla japonica* belongs to group III, which includes 6 species: *A. anguilla*, *A. rostrata*, Borneo eel *A. borneensis*, *A. mossambica*, and New Zealand longfin eel *A. dieffenbachii*. However, the European and American eels found in Japanese waters cannot be distinguished from the Japanese eel based on morphological traits; thus, their presence is confirmed by genetic analysis. Therefore, it is possible that these exotic species were released in areas in which no surveys were conducted. To precisely identify freshwater eel species with similar morphological traits, various genetic methods based on polymerase chain reaction (PCR) have been developed (Itoi et al. 2005; Minegishi et al. 2009; Tanaka et al. 2014; Yoshitake et al. 2019). These methods are considerably expensive, but are essential for species identification of morphologically indistinguishable anguillid eels. Genetic species identification of freshwater eels collected from various parts of Japan in the mid-2010s showed that most of them were *A. japonica* except for the 4 *A. anguilla* in the upper reaches of the Tone River mentioned above. Although it is not clear how *A. anguilla* interacts with native *A. japonica*, competition for resources and habitats may explain why these species occupy similar ecological niches.

The most serious problem with exotic eel species is the occurrence of diseases caused by parasites and bacteria (see Chap. 17). In the European eel, this has been caused by the parasitic nematode, *Anguillicola crassus*. This nematode parasitizes the swim bladder and causes no noticeable symptoms in small numbers; however, in large numbers, it causes inflammation by compressing internal organs (Ogawa 2006). Originally distributed in East Asia, this nematode was thought to have been introduced to Germany with the Japanese eel exported from Taiwan around 1980 (Koops and Hartmann 1989). Since then, it has rapidly spread to most parts of Europe. This nematode also parasitizes the Japanese eel, but no significant symptoms are observed. This is due to the difference in susceptibility of European and Japanese eels to the nematode. The Japanese eel, which has coexisted with the nematode for many years, has developed partial resistance to the parasite, which inhibits its excessive growth. However, the European eel has only recently been infected with the parasite, and thus has no means of defense, resulting in severe symptoms. Dermocystidiosis, a fungal disease, has been introduced in Japan by eels imported from Europe. This is not a serious case, but shows that parasites and pathogens are inevitably introduced when eels are imported as hosts. Anthropogenic movement of eels is expected to accelerate in the future, creating concern that pathogens may be introduced. Indeed, Eel virus European (EVE), Eel virus European X (EVEX), and Anguillid herpesvirus-1 (HVA) have been found in the distribution range of *A. anguilla* and potentially threaten the stock of the species (McConville et al. 2018). The spread of pathogens by the release of non-native species into natural waters may cause catastrophic damage to the natural populations of native species. Since it is almost impossible to eliminate the pathogen once it has spread, great care must be taken not only to control artificial releases but also to prevent its dispersal from aquaculture facilities.

21.3 Alternative Species for Aquaculture Seedlings

In East Asian countries, Japanese and European eels have been utilized for aquaculture for decades, and, more recently, the American eel. However, stocks of these species have been partly decreasing because of over-exploitation. Thus, to compensate for the shortage of these temperate species, tropical species have been utilized as alternative seedlings. Among the 13 tropical anguillid subspecies, *A. bicolor* is the most notable and has been utilized on a commercial scale. However, the recruitment of glass eels in temperate and tropical regions can be quite different. In Southeast Asian countries, several species simultaneously recruit at the same place, and dominant species differ monthly; therefore, it can be technically difficult to utilize only a specific species for aquaculture.

Based on the species composition of tropical eels that recruited four localities in the Philippines and Indonesia (northern Luzon, Mindanao, Sulawesi, and Java), we considered the possibility of collecting *A. bicolor* for aquaculture (Fig. 21.2; Table 21.1). Of the 2 subspecies of *A. bicolor*, *A. bicolor pacifica* is recruited in northern Luzon, Mindanao, and Sulawesi, and *A. bicolor bicolor* on Java Island. Semi-annual to annual sampling surveys confirmed the recruitment of 4–7 species/subspecies at each locality. In northern Luzon, 5 species/subspecies have been observed (Yoshinaga et al. 2014; Aoyama et al. 2015). In November and December 2011, *A. bicolor pacifica* occupied most of the recruitment; however, in the same months of the following year, less than half was observed. In Mindanao, the Philippines, *A. bicolor pacifica* accounted for almost half of the recruitment in

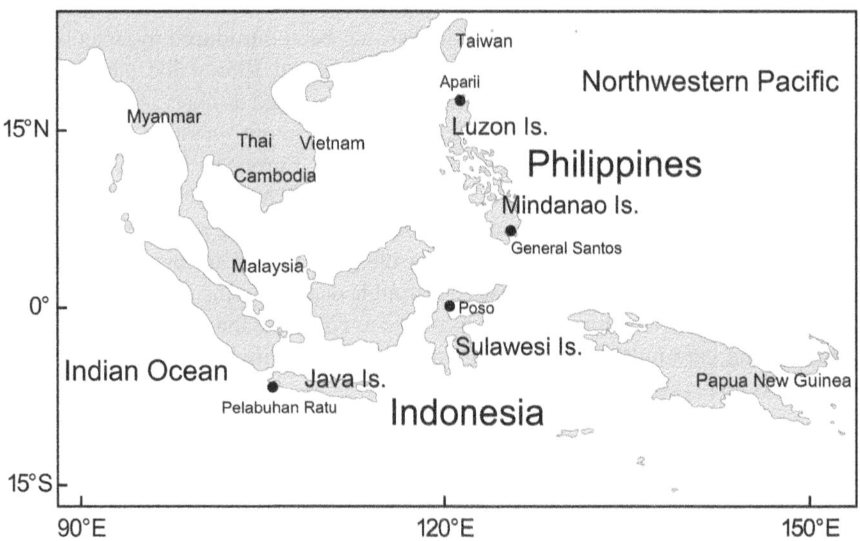

Fig. 21.2 Map showing the four localities—denoted by black dots—in the Philippines and Indonesia where the species composition of glass eels was investigated (Table 21.1)

Table 21.1 Recruitment of tropical anguillid species in East Asian Countries

Country	Locality	Species	Common name	Morphology group	Year and month																
					2009			2011		2012											
					Jan	Feb		Nov	Dec	Jan	Feb	Mar	Apr	May	Jun	Jul	Aug	Sep	Oct	Nov	Dec
Philippines	Northern Luzon	*A. luzonensis*	Luzon eel	1st				1.4	1.7				1.1	39.3	86.5	96.7	87.3	77.4	23.1	7.1	1.2
		A. celebesensis	Celebes eel	1st				0.1												0.3	
		A. marmorata	Indo-Pacific eel	2nd	73.1	43.0		3.5	2.5	68.4	93.1	97.9	98.7	59.9	11.2	2.1	2.9	9.0	35.9	51.5	54.6
		A. japonica	Japanese eel	3rd	16.8																
		A. bicolor pacifica	Pacific bicolor eel	4th	10.1	57.0		95.1	95.8	31.6	6.9	2.1	0.2	0.8	2.3	1.2	9.8	13.7	41.0	41.1	44.2
					2014									2015							
	Mindanao				Apr	May	Jun	Jul	Aug	Sep	Oct	Nov	Dec	Jan	Feb	Mar					
		A. celebesensis	Celebes eel	1st			1.4	11.9	1.1	2.0	1.2		9.1	2.2	2.8						
		A. interioris	New Guinea eel	1st		1.4	0.3			0.3	0.7	4.5	5.5	6.3	1.7						
		A. luzonensis	Luzon eel	1st		1.0		0.5	1.1												
		A. marmorata	Indo-Pacific eel	2nd	99.6	92.5	90.0	68.5	48.1	66.3	94.9	72.0	79.5	74.6	87.0	71.3					
		A. japonica	Japanese eel	3rd											0.3						
		A. borneensis	Borneo eel	3rd		0.5															
		A. bicolor pacifica	Pacific bicolor eel	4th	0.4	4.6	8.3	19.1	49.8	31.4	3.2	23.6	5.9	16.9	8.1	28.7					
Indonesia					2014		2015														
	Slawesi				Nov	Dec	Jan	Feb	Mar	Apr	May	Jun	Jul	Aug	Sep						
		A. celebesensis	Celebes eel	1st	9.5	–	–						2.0	3.4	2.0						
		A. interioris	New Guinea eel	1st	1.4	–	–														
		A. marmorata	Indo-Pacific eel	2nd	88.5	–	–	100.0	99.1	95.4	96.1	99.6	95.7	95.7	97.7						
		A. bicolor pacifica	Pacific bicolor eel	4th	0.7	–	–		0.9	4.6	3.9	0.4	2.3	0.9	0.3						

(continued)

Table 21.1 (continued)

Country	Locality	Species	Common name	Morphology group	Year and month																
					2009			2011		2012											
					Jan	Feb		Nov	Dec	Jan	Feb	Mar	Apr	May	Jun	Jul	Aug	Sep	Oct	Nov	Dec
					2015																
	Java				Feb	Mar	Apr	May	Jun	Jul	Aug	Sep									
		A. interioris	New Guinea eel	1st	0.0	–		0.7		7.1	3.9	4.0									
		A. bengalensis bengalensis	Indian Bengal eel	2nd	0.3	–		1.4		1.3		1.6									
		A. marmorata	Indo-Pacific eel	2nd	0.9	–	31.6			25.0	18.8	18.2									
		A. bicolor bicolor	Indian bicolor eel	4th	98.8	–	68.4	97.8	100.0	66.6	77.4	76.3									

Source: Aoyama et al. 2015; Shirotori et al. 2016; Yoshinaga et al. 2014, unpubl. Data

December 2011, but fewer in other months. On Sulawesi Island, *A. marmorata* has dominated for almost the entire year, and recruitment of *A. bicolor pacifica* was quite limited. However, even though the survey period was limited, *A. bicolor bicolor* was dominant on Java Island facing the Indian Ocean (Pelabuhan Ratu). Therefore, compared to the other 3 localities, a stable fishery catch of *A. bicolor* glass eels is possible. Glass eels collected at this site have been farmed and processed in Indonesia and have been sold in major Japanese supermarkets since 2013. However, since *A. bicolor bicolor* is widely distributed along the Indian Ocean coast (see Chap. 1), it is expected to be extremely difficult to implement, requiring coordinated monitoring among many countries that share resources of this species.

Around the mid-2010s, many eels caught in Southeast Asia were introduced into East Asian countries as a substitute species for Japanese eels. As mentioned earlier, >5 species of freshwater eels are distributed in the tropics and the distribution center of the genus, whereas Japanese eels are rare. However, tropical eels are not easily distinguished from Japanese eels, except for scientific observations, leading some eels imported from Southeast Asia to be incorrectly sold as Japanese eels. Although the number of eels brought from Southeast Asia is currently decreasing, this chapter introduces some cases of disguised species, and provides information on the challenges surrounding eels.

21.3.1 Case 1

End of March 2012: A non-specialized middleman sold 2 kg of glass eels of the Japanese eel imported from the Philippines to a trader in the Japanese central fish market, *Tsukiji Shijo*. The price was ~1/5 of the current price, and was accompanied by a DNA certificate stating: "This is to certify that the MARINE PRODUCT tested and presented in this laboratory is: A. JAPONICA 98%". However, the authors confirmed by genetic analysis that these were actually *A. marmorata* and some other species, but not *A. japonica*. The author could not find any information about the researcher who signed the certificate, and most of the certificate was plagiarized by an abstract of a paper reporting the genetic analysis of the Japanese eels, not the content of the expert opinion.

21.3.2 Case 2

At the end of January 2015, a wholesaler purchased 3 kg of glass eels, which were indicated as Japanese eels, from 2 men who the wholesaler had never met. The wholesaler later asked the author to assess the species because of his suspicion, probably because of a slightly different morphology. Genetic analysis revealed that most of the eels were *A. bicolor*, and their species composition suggested that they originated from the northern Philippines. The men who disguised the glass eels were later prosecuted and the case was settled with monetary compensation.

21.3.3 Case 3

In March 2015, elvers of Japanese eels purporting to be from Vietnam were posted on an Internet bulletin board (now closed), claiming to have a DNA certificate, and selling for about half of the normal price. The author assessed the species and found them to be European eels. The process is unclear, but the company that imported eels from Vietnam was an IT company, and an antisocial organization was also involved.

21.3.4 Case 4

These cases are known by the author, but are just the tip of the iceberg, as there were many suspected cases of deception in 2015. It has become common for a price hike to be reported in newspapers and TV news at the end of the demand season (January to March). Non-specialized middlemen who are engaged in the business of importing and exporting marine products other than eels usually obtain eels of unknown species from Southeast Asian countries and sell them at a lower price than usual. However, as mentioned above, there are almost no Japanese eels in Southeast Asia, which results in disguising this species. Some claimed to have been accompanied by expert certificates, but they were often poorly fabricated. We do not know how many of them were brought into the country under false pretenses. However, if eels were released into natural rivers, it is feared that they may have introduced parasites. The demand for tropical eel species is similar to that for temperate species in East Asian countries, and thus it is extremely difficult to collect specific species for aquaculture. The bycatch of glass eel includes some species that are found in threatened categories on the Red List, such as the Luzon eel (Vulnerable), and attention must be paid to the conservation of such species.

21.4 Changes in the Species Used for Inexpensive Eel Products

In Japan, freshwater eels are consumed in a variety of ways, from high-end to inexpensive restaurants, and as precooked products sold in supermarkets. Freshwater eels were once a luxury food in Japan; however, in the 1990s, European eels farmed in China were imported and sold in large quantities, causing prices to drop dramatically and availability to increase, thereby making them a common household food.

Domestic traders were so exhausted by the price decline that they demanded the establishment of safeguards. In contrast, in Japan, the import of freshwater eels has declined since 2010 due to the detection of banned chemicals, the revelation of counterfeit products, and the ban on exports of the European eel from the European Union. The current annual consumption is ~4–50,000 tons, and ~1/2 of the eel products sold in Japan are domestically produced, with the remainder coming from China.

In Japanese supermarkets, "domestic" and "Chinese" eel products (*kabayaki* in Japanese) are sold, but there is no obligation to label the species. To understand the actual situation of eel species in circulation, we have been conducting genetic species identification on the *kabayaki* provided by supermarkets and relatively cheap restaurants since 2011. Although the number of samples examined is limited and it is not possible to grasp the entire distribution process, we have identified some factors such as the change in species after 10 years of survey.

In 2011, the first year of our survey, most of the identified samples were Japanese eels. However, the species of eel served at a conveyor-belt *sushi* restaurant changed from the European eel in July to the Japanese eel in October, even though it was served as an ingredient in the same product. They may have been using cheap European eels when the purchase price was high and Japanese eels when the price had settled. As for the *kabayaki* sold in supermarkets, the relatively cheap ones from China were usually European eels, and more recently, American eels have been observed.

Both the Japanese and European eels were served at a beef-bowl restaurant in 2013. In 2014, 3 species were detected in several eel bowls that were purchased from the same store. In this restaurant chain, 3 pieces of *kabayaki* are processed in China and vacuum-packed, then heated and served (we observed this by looking into the kitchen). Since there were 4 pieces of the European eel and 2 pieces of the American eel, the 2 species were already together when they were processed in China. In 2014, the European eel was used at other beef-bowl restaurants, lunch boxes, and conveyor-belt *sushi* restaurants. However, it is interesting to note that the number of restaurants serving the European eel has decreased since 2015. One reason for this is that the international trade of European eels has reduced as range states develop management and policy measures in line with the CITES listing, and also because of negative campaigns by conservation groups. In 2015, only one conveyor-belt *sushi* restaurant used European and American eels, as the staff in charge of the restaurant did not fully understand the CITES listing of the European eel, and switched to Japanese eel the following year.

In 2017, European eels were still sold in some supermarkets, but the distribution of European eels was almost nonexistent compared to previous years. However, European eels are still farmed in China, and some beef-bowl chain restaurants began to handle them again after 2018. It is unclear whether these eels are illegally exported from the EU or caught in North Africa, where they can be legally exported. Indeed, the European eel was found in almost half of the commercial products sold in Hong Kong supermarkets between 2017 and 2018 (Richards et al. 2020). In any case, in the consumption of eels, which depends on natural resources, the reality of international trade is constantly changing; thus, it is important to keep an eye on the actual situation of the species in circulation.

References

Aoyama J, Watanabe S, Miyai T, Sasai S, Nishida M, Tsukamoto K (2000) The European eel, *Anguilla anguilla* (L.), in Japanese waters. Dana 12:1–5. https://www.aqua.dtu.dk/-/media/institutter/aqua/publikationer/dana/dana_vol_12_pp_1_5.pdf. Accessed 8 June 2023

Aoyama J, Yoshinaga T, Shinoda A, Shirotori F, Yambot AV, Han Y-S (2015) Seasonal changes in species composition of glass eels of the genus *Anguilla* (Teleostei: Anguillidae) recruiting to the Cagayan River, Luzon Island, The Philippines. Pacific Sci 69:263–270. https://doi.org/10.2984/69.2.8

Arai K, Itakura H, Yoneta A, Yoshinaga T, Shirotori F, Kaifu K, Kimura S (2017) Discovering the dominance of the non-native European eel in the upper reaches of the Tone River system, Japan. Fish Sci 83:735–742. https://doi.org/10.1007/s12562-017-1107-z

Han Y-S, Yu C-H, Yu H-T, Chang C-W, Liao I-C, Tzeng W-N (2002) The exotic American eel in Taiwan: ecological implications. J Fish Biol 60:1608–1612. https://doi.org/10.1111/j.1095-8649.2002.tb02454.x

Honga Y-K, Kimb J-E, Leeb J-H, Songa M-Y, Parka H-W, Lee W-O (2017) Occurrence of exotic eels in natural waters of South Korea. Anim Cells Syst 21:341–348. https://doi.org/10.1080/19768354.2017.1377108

Inoue H, Suzuki D, Kitano T, Kohno H (2021) Record of *Anguilla bicolor pacifica* from Iriomote Island, southern Japan. Jap J Ichthyol 68:29–34. https://doi.org/10.11369/jji.20-024; in Japanese with English abstract

Itoi S, Nakaya M, Kaneko G, Kondo H, Sezaki K, Watabe S (2005) Rapid identification of eels *Anguilla japonica* and *Anguilla anguilla* by polymerase chain reaction with single nucleotide polymorphism-based specific probes. Fish Sci 71:1356–1364. https://doi.org/10.1111/j.1444-2906.2005.01102.x

Kita T, Matsushige K, Endo S, Mochioka N, Tachihara K (2021) First Japanese records of *Anguilla luzonensis* (Osteichthyes: Anguilliformes: Anguillidae) glass eels from Okinawa-jima Island, Ryukyu Archipelago, Japan. Sp Div 26:31–36. https://doi.org/10.12782/specdiv.26.31

Koops H, Hartmann F (1989) *Anguillicola* infestations in Germany and German eel imports. J Appl Ichthyol 1:41–45. https://doi.org/10.1111/j.1439-0426.1989.tb00568.x

McConville J, Fringuelli E, Evans D, Savage P (2018) First examination of the Lough Neagh European eel (*Anguilla anguilla*) population for eel virus European, eel virus European X and Anguillid Herpesvirus-1 infection by employing novel molecular techniques. J Fish Dis 41:1783–1791. https://doi.org/10.1111/jfd.12885

Minegishi Y, Aoyama J, Tsukamoto K (2008) Multiple population structure of the giant mottled eel, *Anguilla marmorata*. Mol Ecol 17:3109–3122. https://doi.org/10.1111/j.1365-294X.2008.03822.x

Minegishi Y, Yoshinaga T, Aoyama J, Tsukamoto K (2009) Species identification of *Anguilla japonica* by real-time PCR based on a sequence detection system: a practical application to eggs and larvae. ICES J Mar Sci 66:1915–1918. https://doi.org/10.1093/icesjms/fsp158

Ogawa K (2006) Swimbladder namatode infection (Anguillicolosis). In: Hatai K, Ogawa K (eds) New atlas of fish diseases, 1st edn. Midori Shobo, Tokyo, p 87; in Japanese

Okamura A, Yamada Y, Mikawa N, Tanaka S, Oka H (2002) Exotic silver eels *Anguilla anguilla* in Japanese waters: seaward migration and environmental factors. Aquat Liv Res 15:335–341. https://doi.org/10.1016/S0990-7440(02)01190-7

Richards JL, Sheng V, Yi CW, Ying CL, Ting NS, Sadovy Y, Baker D (2020) Prevalence of critically endangered European eel (*Anguilla anguilla*) in Hong Kong supermarkets. Sci Adv 6: eaay0317. https://doi.org/10.1126/sciadv.aay0317

Sasai S, Aoyama J, Watanabe S, Kaneko T, Miller MJ, Tsukamoto K (2001) Occurrence of migrating silver eels *Anguilla japonica* in the East China Sea. Mar Ecol Prog Ser 212:305–310. https://doi.org/10.3354/meps212305

Shirotori F, Ishikawa T, Tanaka C, Aoyama J, Shinoda A, Yambot AV, Yoshinaga T (2016) Species composition of anguillid glass eels recruited at southern Mindanao Island, The Philippines. Fish Sci 82:915–922. https://doi.org/10.1007/s12562-016-1030-8

Tabeta O, Tanimoto T, Takai T (1976) Seasonal occurrence of anguillid elvers in Cagayan River, Luzon Island, The Philippines. Nippon Suisan Gakkaishi 42:424–426. https://doi.org/10.2331/suisan.42.421

Tanaka C, Shirotori F, Sato M, Ishikawa M, Shinoda A, Aoyama J, Yoshinaga T (2014) Genetic identification method for two subspecies of the Indonesian short-finned eel, *Anguilla bicolor*, using an allelic discrimination technique. Zool Stud 53:57. https://doi.org/10.1186/s40555-014-0057-8

Watanabe S, Aoyama J, Tsukamoto K (2004) Reexamination of Ege's (1939) use of taxonomic characters of the genus *Anguilla*. Bull Mar Sci 74:337–351. https://www.ingentaconnect.com/contentone/umrsmas/bullmar/2004/00000074/00000002/art00006. Accessed 1 June 2023

Watanabe S, Aoyama J, Tsukamoto K (2009) A new species of freshwater eel *Anguilla luzonensis* (Teleostei: Anguillidae) from Luzon Island of The Philippines. Fish Sci 75:387–392. https://doi.org/10.1007/s12562-009-0087-z

Yoshinaga T, Aoyama J, Shinoda A, Watanabe S, Azanza RV, Tsukamoto K (2014) Occurrence and biological characteristics of glass eels of the Japanese eel *Anguilla japonica* at the Cagayan River of Luzon Island, Philippines in 2009. Zool Stud 53:13. https://doi.org/10.1186/1810-522X-53-13

Yoshitake K, Yoshinaga T, Tanaka C, Mizusawa N, Reza MS, Tsujimoto A, Kobayashi T, Watabe S (2019) HaCeD-Seq: a novel method for reliable and easy estimation about the fish population using haplotype count from eDNA. Mar Biotechnol (NY) 21:813–820. https://doi.org/10.1007/s10126-019-09926-6

Zhang H, Mikawa N, Tamada Y, Horie N, Okamura A, Utoh T, Tanaka S, Motonobu T (1999) Foreign eel species in the natural waters of Japan detected by polymerase chain reaction of mitochondrial cytochrome *b* region. Fish Sci 65:684–686. https://doi.org/10.2331/fishsci.65.684

Chapter 22
River Improvement

Youichi Yasuda, Kouhei Yasuda, and Pietro Beretta Piccoli Marco

Contents

Freshwater eels generally use rivers during their growth period. However, while river improvement measures focusing on flood control and water utilization allow for safe evacuation during floods, they often do not contribute to their habitat during normal water conditions, and thus, result in a decrease in habitat density and habitat deterioration. To secure habitats for eels, it is important to maintain a balance between ecological preservation and disaster prevention. It is also necessary to conserve the habitats of various aquatic organisms other than eels to maintain the food chain. The application of consecutively assembled boulders in a monotonous river was devised to create an environment in which a diverse flow can be formed during normal water conditions, and a refuge space can be secured during floods. In this chapter, we introduce the results of the experiments on the flow condition at consecutively assembled boulders during a flood stage and emphasize the importance of approaches that improve the river environment for the fish that inhabit it, including freshwater eels.

Y. Yasuda (✉) · K. Yasuda · P. B. P. Marco
College of Science and Technology, Nihon University, Tokyo, Japan
e-mail: yasuda.youichi@nihon-u.ac.jp

K. Tsukamoto et al. (eds.), *Eel Science*, Fisheries Science Series,
https://doi.org/10.1007/978-981-99-5692-0_22

22.1 Balance Between Flood Control and Preservation of Aquatic Habitat through Consecutively Assembled Boulders

Most rivers in Japan have been renovated to ensure the safety of human life, lifelines, and economic viability in terms of flood damage. In addition, preservation of aquatic habitats in rivers has been designated as a third priority. There are calls to improve the river environment, but they lack concrete and effective implementation (Ministry of Land, Infrastructure, and Transport 2006). The most important aspect for improving the river environment is creating a river that is balanced between the flood and normal stages, and watershed flood control must be based on this balance (Yasuda 2021).

Reconstructed rivers have been flattened, and their floodplains are constrained, resulting in the loss of shallow pool structures in alluvial plains and affecting habitat conditions (Yasuda 2016). In natural rivers, the transport and sedimentation of boulders create diverse flows, and the imbrication of boulders may serve as a refuge environment during floods and preserve aquatic habitats during normal stages. Many researchers have focused on the transport, sedimentation, and stabilization of boulders (Church et al. 1998; Hassan and Reid 1990; Lamarre and Roy 2008; Papanicolaou et al. 2003; Strom et al. 2004, 2008; Wittenberg and Newson 2005) and have examined them in aquatic habitats and refuges environments (Biggs et al. 1997; Decker et al. 2013; Johansen et al. 2007; Lin et al. 2006). However, there are few studies on the structure of artificially assembled boulders (Yasuda and Fuchino 2021, 2022). Recent research has focused on improving flow regimes by installing consecutively assembled boulders (Yasuda 2021).

The loss of freshwater habitats due to human activities was previously studied by applying chronological Landsat images to 16 rivers in East Asia, including Japan, Korea, Taiwan, and China (Chen et al. 2014). They found that, on average, these 16 rivers lost 76.8% of their effective habitat area (Ae) between the 1970s and the 2010s. Itakura et al. (2015) investigated the relationship between catch reduction rates and revetment rates of shorelines in 18 rivers and 9 lakes. According to their investigation, higher reductions in eel catches might be caused by the installation of hydraulic structures (such as artificial banks) in rivers and lakes. Furthermore, Itakura et al. (2020) investigated the effect of artificial banks on aquatic community habitat by analyzing data obtained from 78 sites across 6 rivers in southern Japan, and found that the installation of a bank would negatively affect juvenile growth habitats for eels. They suggested that other types of habitat modifications or obstacles to the upstream migration of eels might also affect the fisheries catches of Japanese eels. Moreover, they reported that both the density of eels and the number of other diadromous species were significantly negatively correlated with distance from the river mouth and the drop height of hydraulic structures from the river mouth to each site.

Yokouchi et al. (2021) applied a generalized linear mixed model to 5 model river systems with different numbers, sizes, and positional patterns of small drop structures to analyze Japanese eels and their habitat characteristics. The results revealed that fish and benthic animals with lower swimming abilities, including the growth phase, might be an obstacle to the installation of small drop structures. Furthermore, the habitat distribution of Japanese eels at various growth stages might depend on the number of small drop structures and the upstream distance from the river mouth.

Recently, the installation of stone cages has attracted attention in an effort to improve the habitat of eels in fresh and brackish waters. This cage is a modified form of the traditional trap called *Ishikura* (see Chap. 18), which is made of piles of river stones to exploit the behavior of eels to hide in narrow spaces. Oto et al. (2022) examined the preference of Japanese eels for a floating stone void structure in *Ishikura* nets at different life history stages by varying the size of gravel in the nets. They used 3 different-sized stones (large 0.3 m in major axis; medium 0.2 m; small 0.1 m), and discovered the importance of establishing estuarine habitats with various-sized stones to allow many eels to survive in brackish water areas. However, questions remain regarding the effectiveness of *Ishikura* nets from the viewpoints of stability and cost to restore the habitat of freshwater eels (Kaifu and Wakiya 2019).

A balance between ecological preservation and disaster prevention is required for the eel habitat. It is desirable to preserve and restore the refuge area of Japanese eels by restoring the riverbed morphology and balance of erosion, transport, and sedimentation in watershed areas under river management of flood control and water supply. Furthermore, the habitat of various aquatic animals must be conserved for eels to grow within the food chain. Freshwater eels can act as indicators, umbrella and flagship species, and comprehensive surrogates for the conservation of freshwater biodiversity owing to their catadromous life history and global commercial and cultural importance (Itakura et al. 2020).

To create a variety of flows in artificially channelized rivers, as in the case of a meandering natural river, alternating gravel mounts are installed to ensure the river meanders during normal stages. As the flood discharge increases, the meandering flow changes to a straight flow, and the main flow rises to the water surface. During small and medium floods, the velocity field around alternate gravel mounts provides a refuge area for aquatic animals in stagnant areas downstream of mounts. It was also found that the mount shape can be protected by installing consecutively assembled boulders to ensure stability during flooding. In recent experiments (Beretta and Yasuda 2021, 2022), the spacing between mounts and the height of the mounts were varied to determine the optimum spacing and height based on the balance between the flood control and environmental aspects.

Regarding the velocity fields inside the assembled boulders, hydraulic confirmation of the mean velocity distribution, including the standard deviation, is required to determine the feasibility of aquatic habitats and refuge areas during normal and flood stages, respectively. However, little information is available regarding the velocity field. For aquatic habitats in normal stages, flow characteristics should be investigated using prototypes.

To verify the refuge space inside consecutively assembled boulders during floods, the authors experimentally investigated the velocity field inside assembled boulders with a bottom slope of 1/100. Experiments with a 1/10-scale model were conducted based on Froude similarity. In this case, consecutively assembled boulders were laterally inclined at a 1/10 slope and alternatively installed such that the interval between the assembled boulders was twice the width of the water passage. The transverse slope of the assembled boulders toward the center concentrated the main flow at the central part of the passage, and the undulation of the water surface at the sidewall was smaller than that at the center. Furthermore, the placement of assembled boulders along the sidewalls reduced the undulation of the water surface near the sidewalls. The main flow was located near the water surface and concentrated at the center of the water passage. The distribution and magnitude of the mean velocity and standard deviation inside the assembled boulders indicate that the installation of consecutively assembled boulders may provide a refuge space for aquatic animals during floods.

In a stream type B4 river (moderately entrenched, moderate gradient with 0.02–0.039 slopes, gravel river with infrequently spaced pools) (Rosgen and Silvey 1996) in Japan, the side protection wall for a road revetment curved along the river's geometrical shape, inhibiting the main flow in the normal and flood stages from moving. Protection blocks were installed near the sidewall, and shallow water flow was formed in normal stages. Consequently, because diverse flows are not formed, aquatic habitats are lost in this region. In December 2021, 3 continuous rows of assembled boulders were partially installed perpendicular to the wall on the side of the road revetment to improve the flow conditions considering the aquatic habitat. The installation interval was determined to minimize the effect of the backwater just below the assembled boulder, and was conducted at 2 locations on a trial basis. Long-sized boulders were selected ranging 0.4–1.2 m, and were delivered by a construction machine with power shovel. The structure of the assembled boulders was adjusted such that the top of the boulders was at almost the same level as that of water in the normal stage. Six months after installation, > 100 small fish were localized around the installed gravel, which was a vastly different and improved condition compared to that before installation.

This chapter presents experimental investigations and field verifications to demonstrate the potential of consecutively assembled boulders in a monotonous river to serve as a fishing reef for aquatic organisms during normal stages and as a refuge space inside assembled boulders during flood stages.

22.2 Experimental Setup

The experiment was conducted in a rectangular channel with a variable channel slope (length: 15 m, width: 0.80 m, and height: 0.60 m). The experimental conditions are listed in Table 22.1. A physical model was installed with a scale of Lr = 1/10 based on Froude's law of similarity. The 6 consecutively assembled boulders with diameters ranging from 0.06 m to 0.12 m were installed alternatively. The interval between the assembled boulders was 1.6 m and the lateral length of each group was

Table 22.1 Experimental conditions

Case	Q (m^3/s)	i (−)	Lr (−)	$h_{at6.6}$ (m)	h_c (m)	h_g (m)	L (m)	b (m)	b_s (m)
1	0.149	1/100	1/10	0.2899	0.141	0.154	1.60	0.40	0
2	0.149	1/100	1/10	0.2862	0.141	0.154	1.60	0.40	0.15

Q: discharge, i: channel slope, Lr: scale of model, $h_{at6.6}$: water depth at x = 6.6 m, h_c: critical depth, h_g: gate height at downstream end of channel, L: Interval between consecutive assembled boulders, b: lateral length of consecutive assembled boulders, and b_s: installation width of assembled boulders along the side

Fig. 22.1 Installation of consecutively assembled boulders in Case 2

0.4 m. The tops of the boulders near the sidewalls were laterally inclined such that they were ~ 0.045 m above the gravel bed, and the main flow was concentrated at the center and rose near the water surface. In this installation, it was assumed that the top edge of the assembled boulders was completely submerged during the small flood event. The gravel bed (~0.04 m in thickness) consisted of gravel with an average diameter of 0.016 m. To form a refuge space near the side during floods, assembled boulders were installed along the sidewall (Fig. 22.1). Furthermore, the protruded length was ~0.15 m from the sidewall.

The water surface profile, riverbed profile, and velocity were measured assuming a flood stage. The water surface and riverbed profiles were measured using a point gauge, and the velocity was measured using an electromagnetic velocity probe (sampling time and frequency of 30 ms and 50 ms, respectively) connected to an I-type detection unit (stream direction and span direction component detection) (KENEK, model VM-806H/VMT2–400-04P). The water level was adjusted using a sluice gate at the channel end to form an S1 curve downstream of the physical model.

22.3 Water Surface Profile and Water Level

According to experimental conditions shown in Table 22.1, the water surface profile is characterized as follows: In Case 1, undulation of the water surface occurred when passing over the first and second rows of consecutively assembled boulders.

However, the undulation of the wave surface was influenced by the consecutively assembled boulders installed at 6 locations, and became smaller downstream. In particular, this tendency increased near the sidewalls, unlike that in Case 2, except for the upstream region. Furthermore, the undulation of the water surface is formed in the center of the channel, and the degree of undulation increases from upstream to downstream, which might have been caused by the installation of the assembled boulders along the sidewalls, leading to mainstream concentration in the center of the channel. The hydraulic conditions in Cases 1 and 2 are identical, except for the installation of the assembled boulders along the sidewalls. In Case 2, this may have been caused by an increase in the kinematic energy at the center.

22.4 Velocity Fields

The mean velocity distributions of the x-directional component near the sidewall ($y/(B/2) = 0.95$) and the distributions of the standard deviations corresponding to the mean velocities in Case 2 are depicted in Figs. 22.2 and 22.3. Compared to the distribution of the mean velocity for Case 1, that in Case 2 was lower. This may have been caused by the installation of the assembled boulders along the sidewalls and the concentration of the main flow at the center. In addition, the mean velocity of the main flow at the center was larger than that in Case 1. The velocities inside the assembled boulders along the sidewalls were smaller than those in Case 1. The difference between Cases 1 and 2 was negligible in terms of the standard deviation inside the assembled boulders. These results indicate that the refuge space inside the assembled boulders can be easily maintained along the sidewalls during the flood stage. A similar experiment was conducted using freshwater eels that confirmed that the space inside the assembled boulders was helpful for eels (Yasuda 2021). Notably, the consecutively assembled boulders are stable even at the mainstream

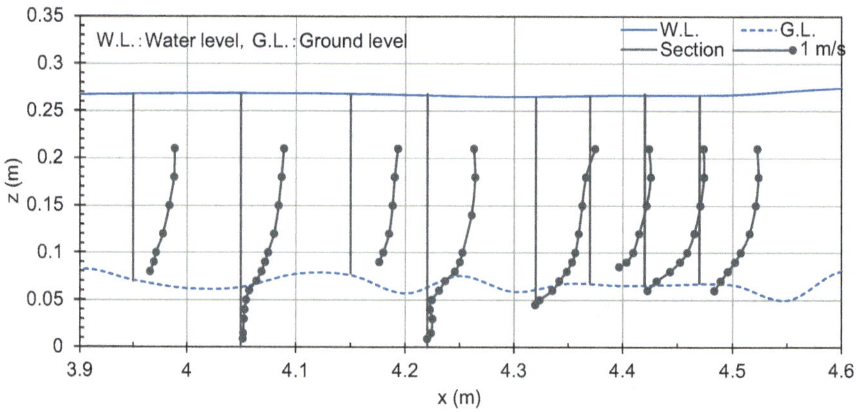

Fig. 22.2 Mean velocity distributions of x-direction component at $y/(B/2) = 0.95$ of Case 2

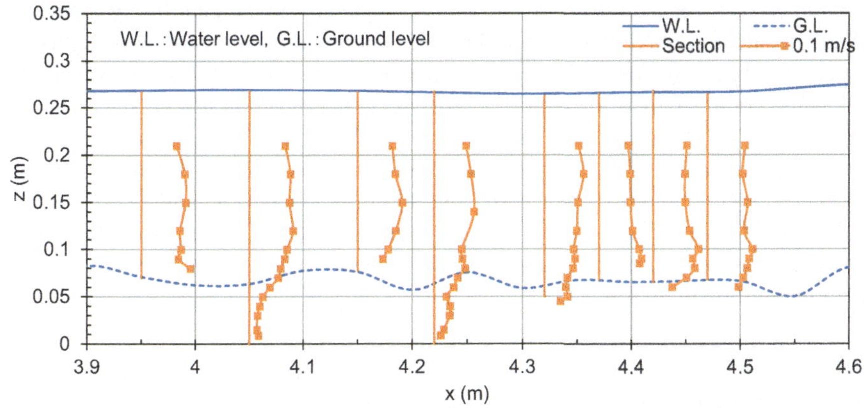

Fig. 22.3 Distributions of standard deviation x-direction component at y/(B/2) = 0.95 of Case 2

of the flood flow, as opposed to the loss of functionality indicated by *Ishikura* nets (Kaifu and Wakiya 2019). In addition, by placing them at locations where the flow is not stagnant, transported sand and gravel are eliminated because the flow near the top of boulders has a high velocity and turbulence during the flood stages. Furthermore, the spaces inside the boulders must be staggered to ensure flow through the gaps does not form in the spaces for freshwater eels and other aquatic organisms to be able to refuge. In addition, a combination of different gravel diameters must be used before and after the spaces.

22.5 Discussion for Experimental Results

Experimental results revealed that creating a refuge space inside the consecutively assembled boulders is effective in protecting aquatic animals during the flood stage. In this experiment, 1 discharge was applied as the maximum discharge that could flow through the experimental channel. Reducing the discharge and channel slope can result in smaller velocities and standard deviations inside the assembled boulders. From the perspective of the stability of the assembled boulders and the safety of the refuge space inside the assembled boulders, an experimental study of the maximum discharge may be useful for hydraulic design. If the discharge is represented as discharge per unit width, the value of 0.186 m^2/s in the physical model corresponds to 5.89 m^2/s in the prototype using Froude similarity. The most important parameter was the flow condition near the sidewall. These experimental results are applicable to a wide river, for which the lateral length of consecutively assembled boulders should increase. Furthermore, it is assumed that the top edge of the assembled boulders is fully submerged during a small flood event.

In general, the installation of large boulders in a river channel increases shape resistance during floods and increases water levels. In addition, it is important that the top edge of the assembled boulders be completely submerged during moderate flood events for maximum effect. In this experiment, alternating consecutively assembled boulders were installed in a flattened river channel, and the top of the assembled boulders was lowered in a cross-sectional direction toward the center of the channel. Consequently, the resistance was equivalent to the shape resistance of a natural river with a 1/100 slope (Hey 1979; Ferguson 2007; Rickenmann and Recking 2011). Notably, even with consecutively assembled boulders along the sidewalls, the water level did not increase during the flood event. Thus, the refuge space inside the assembled boulders along the sidewalls can be increased without obstructing flood control.

22.6 Field Verifications on Aquatic Habitat around Consecutive Assembled Boulders

Consecutively assembled boulders were partially installed in stream type B4 (Rosgen and Silvey 1996) with an 8 m width and 1/50 longitudinal average slope downstream of mountains in Japan. For field verification, in December 2021, 3 continuous rows of assembled boulders were partially installed perpendicular to the wall on the side of the road revetment to improve the flow conditions, while considering the aquatic habitat. The boulders were delivered using a construction machine with power shovel. The installation interval was conducted at 2 locations on a trial basis, and was determined to minimize the effect of the backwater just below the assembled boulder. Furthermore, the lateral length of the assembled boulders (ranging 0.4–1.2 m) was determined to cover the protection-block region. As this work is primarily conducted by fishing cooperatives, it is necessary to improve the environment while ensuring economic efficiency; therefore, it is essential to improve this technology. The structure of the assembled boulders was adjusted to ensure that the tops of the boulders had a similar water level at the normal stage (Fig. 22.4). Six months after installation, field inspection was conducted (on June 1, 2022). Inspection of the aquatic habitat conditions around the assembled boulders confirmed that the density of the habitat was >2.5 m^2. The species identified were Ayu *Plecoglossus altivelis*, Japanese dace *Tribolodon hakonensis*, dark chub *Nipponocypris temminckii*, and freshwater goby *Rhinogobius*. The results confirmed that the assembled boulders were stable and increased the flood width by eroding the inner sediments until the river formed a gentle slope on its inner curve. In the conventionally proposed *Ishikura* nets, because the gravel is encased in a net, if the net is installed at the main flow area during flood stages, it could easily slip between the net and riverbed, causing it to spill and lose its function as a habitat space. Furthermore, if the *Ishikura* nets are installed in the stagnation area (the inner side of the curved river meander where the flow path does not change during normal and flood stages), sediments including mud will accumulate in the flood flow (Kaifu and Wakiya 2019). In this case, the nets may also lose their function as a habitat.

Fig. 22.4 Flow condition around the three continuous rows of assembled boulders

Therefore, it was confirmed that the installation of consecutively assembled boulders is an effective river improvement method for both environmental and flood control. Freshwater eels and other aquatic animal habitats may be preserved inside consecutively assembled boulders unless a turbulent seepage flow forms.

22.7 Conclusions

The efficiency of installing consecutively assembled boulders for river improvement was determined through experimental investigations and field verification. It is possible to stabilize the assembled boulders and form a refuge space during floods. In addition, the concentration of the main flow at the center contributed to the control of the water level rise and amplification of the refuge space inside the assembled boulders along the sidewalls. The formation of diverse flows around the assembled boulders was possible during the normal stages.

To verify the refuge space inside consecutively assembled boulders during the flood stage, experiments with a 1/10 scale model were conducted based on Froude similarity. The experimental investigation revealed that the transverse slope of the assembled boulders toward the center concentrated the main flow at the center, and the placement of the assembled boulders along the sidewalls reduced the undulation of the water surface near the sidewalls and improved the velocity field inside the assembled boulders under a bottom slope of 1/100. The downstream water level was slightly lower when assembled boulders were installed along the sidewalls. The mean velocity inside the assembled boulders during flooding varied within ± 0.10 m/s (0.33 m/s in prototype), and the standard deviation was less than 0.05 m/s (0.16 m/s in prototype).

The utilization of assembled boulders produced a habitat density > 2.5 m^2 that was favorable to various aquatic species. The species identified, including both natural and released fish, were Ayu, Japanese dace, dark chubs, and freshwater gobies. The assembled boulders were stable, and the flood width was extended by eroding the inner sediments until the river had a gentle slope on the inner side of the curved river. Therefore, is it clear that the installation of consecutively assembled boulders is an effective river improvement method for both environmental and flood control.

References

Beretta PP, Yasuda Y (2021) Formation of refuge areas behind alternative gravel mounts for fishes during flood stages. Mod Environ Sci Eng 7:1021–1031. https://doi.org/10.15341/mese(2333-2581)/11.07.2021/001., http://academicstar.us/issueshow.asp?daid=3881. Accessed 8 June 2023

Beretta PP, Yasuda Y (2022) Experimental analysis on the formation of refuge areas for fishes behind alternative gravel mounts in channelized rivers during flood stages, Proceedings of the 39th IAHR World Congress (Granada, 2022)

Biggs BJF, Duncan MJ, Francoeur SN, Meyer WD (1997) Physical characterisation of microform bed cluster refugia in 12 headwater streams. NZ J Mar Freshw Res 31:413–422. https://doi.org/10.1080/00288330.1997.9516775

Chen J-Z, Huang SL, Han Y (2014) Impact of long-term habitat loss on the Japanese eel *Anguilla japonica*. Estuar Coast Shelf Sci 151:361–369. https://doi.org/10.1016/j.ecss.2014.06.004

Church M, Hassan MA, Wolcott JF (1998) Stabilizing self-organized structures in gravel-bed stream channels: field and experimental observations. Water Resour Res 34:3169–3179. https://doi.org/10.1029/98WR00484

Decker J, Swain E, Stith B, Langtimm C (2013) Assessing factors affecting the thermal properties of a passive thermal refuge using three-dimensional hydrodynamic flow and transport modelling. J Waterw Port Coast Ocean Eng 139:209–220. https://doi.org/10.1061/(ASCE)WW.1943-5460.0000165

Ferguson R (2007) Flow resistance equations for gravel and boulder bed streams. Water Resour Res 43:W05427. https://doi.org/10.1029/2006WR005422

Hassan MA, Reid I (1990) The influence of microform bed roughness elements on flow and sediment transport in gravel bed rivers. Earth Surf Process Landf 15:739–750. https://doi.org/10.1002/esp.3290150807

Hey RD (1979) Flow resistance in gravel-bed rivers. J Hydraul Div, Proc Am Soc Civ Eng 105:365–379. https://doi.org/10.1061/JYCEAJ.0005178

Itakura H, Kitagawa T, Miller MJ, Kimura S (2015) Declines in catches of Japanese eels in rivers and lakes across Japan: have river and lake modifications reduced fishery catches? Landsc Ecol Eng 11:147–160. https://doi.org/10.1007/s11355-014-0252-0

Itakura H, Wakiya R, Gollock M, Kaifu K (2020) Anguillid eels as a surrogate species for conservation of freshwater biodiversity in Japan. Sci Rep 10:8790. https://doi.org/10.1038/s41598-020-65883-4

Johansen JL, Fulton CJ, Bellwood DR (2007) Avoiding the flow: refuges expand the swimming potential of coral reef fishes. Coral Reefs 26:577–583. https://doi.org/10.1007/s00338-007-0217-y

Kaifu K, Wakiya R (2019) Conservation and restoration of growth habitat of Japanese eel, *Anguilla japonica*: does "Ishikura-kago" contribute population restoration? Ecol Civil Eng 22:109–115. https://doi.org/10.3825/ece.22.109; in Japanese

Lamarre H, Roy AG (2008) A field experiment on the development of sedimentary structures in a gravel-bed river. Earth 33:1064–1081. https://doi.org/10.1002/esp.1602

Lin JY, Chen YC, Tsao EH, Yang HC (2006) Restoration of Formosan landlocked salmon habitat as refuge during high flows. Doboku Gakkai Ronbunshuu B 62:320–329. https://doi.org/10.2208/jscejb.62.320

Ministry of Land, Infrastructure, and Transport (2006) Rivers in Japan, Preparation by Infrastructure Development Institute-Japan and Japan River-Association, The supervision of River Bureau. https://www.mlit.go.jp/river/basic_info/english/pdf/riversinjapan.pdf. 82 p. Accessed 1 Aug 2022

Oto Y, Sakanoue R, Hibino Y, Matsushige K, Uchida K, Mochioka N (2022) Prefered gap structure within stone piles of fishing gear by Japanese eel, *Anguilla japonica* at each life history stage: the search for an effective method to restore estuarine habitats. Nippon Suisan Gakkaishi 88: 152–161. https://doi.org/10.2331/suisan.21-00043; in Japanese with English abstract

Papanicolaou AN, Strom K, Schuyler A, Talebbeydokhti N (2003) The role of sediment specific gravity and availability on cluster evolution. Earth Surf Process Landf 28:69–86. https://doi.org/10.1002/esp.427

Rickenmann D, Recking A (2011) Evaluation of flow resistance in gravel-bed rivers through a large field data set. Water Resour Res 47:W07538. https://doi.org/10.1029/2010WR009793

Rosgen DL, Silvey HL (1996) Applied river morphology, 2nd edn. Wildland Hydrology, Pagosa Springs, Colorado, pp 5-72–5-75

Strom K, Papanicolaou AN, Evangelopoulos N, Odeh M (2004) Microforms in gravel bed rivers: formation, disintegration, and effects on bedload transport. J Hydraul Eng 130:554–567. https://doi.org/10.1061/(ASCE)0733-9429(2004)130:6(554)

Strom K, Lcolaou AN, Papanicolaou AN (2008) Morphological characterization of cluster microforms. Sedimentology 55:137–153. https://doi.org/10.1111/j.1365-3091.2007.00895.x

Wittenberg L, Newson MD (2005) Particle clusters in gravel-bed rivers: an experimental morphological approach to bed material transport and stability concepts. Earth Surf Process Landf 30: 1351–1368. https://doi.org/10.1002/esp.1184

Yasuda Y (2016) Contribution due to river engineering in preservation for diadromous aquatic animals. Aqua 38:387–396. (in Japanese)

Yasuda Y (2021) Improvement of flow condition in channelized river due to stacked boulders. IOP Conf: earth environ 626, the 2nd international conference on advances in civil and ecological engineering research October 20-23, 2020, online conference, pp 1–8

Yasuda Y, Fuchino N (2021) Hydraulics of roughness mild slope with stacked boulders in half trapezoidal section. J Modern Environ Sci Eng 7:1099–1108. https://doi.org/10.15341/mese(2333-2581)/12.07.2021/001. chrome-extension://efaidnbmnnnibpcajpcglclefindmkaj/http://academicstar.us/UploadFile/Picture/2022-5/2022513114521645.pdf. Accessed 8 June 2023

Yasuda Y, Fuchino N (2022) The efficacy of artificially assembled boulder installations in improving migration routes for aquatic animals. In: Ray RL, Panagoulia DG, Abeysingha NS (eds) River basin management - under a changing climate. IntechOpen, p 19. https://doi.org/10.5772/intechopen.105198

Yokouchi K, Itakura H, Wakiya R, Yoshinaga T, Mochioka N, Kimura S, Kaifu K (2021) Cumulative effects of low-height barriers on distributions of catadromous Japanese eels in Japan. Anim Conserv 25:137–149. https://doi.org/10.1111/acv.12725

Chapter 23
Conservation

Matthew Gollock and Hiromi Shiraishi

Contents

23.1 Introduction

Defining conservation is an on-going discussion and "actions that are intended to establish, improve or maintain good relations with nature" was recently proposed (Sandbrook 2015). This recognizes that conservation comes in many forms, and that management decisions, commercial activities, human habitation, and policies that do not have this primary purpose may have additional "conservation benefits". For example, fisheries management measures may benefit anguillids, whereas some actions labelled as "conservation" (e.g. restocking) still have uncertainties as to their benefits and limitations (ICES 2016). As such, there are many ways to categorize conservation mechanisms and interventions, but this chapter will primarily focus on species-focused activities as opposed to more holistic frameworks and measures. For example, while the Convention on Biological Diversity may indirectly benefit anguillids (Engler-Palma et al. 2013), it is very unlikely that there will be interventions under this framework that specifically focus on these species. Similarly, not all species-focused mechanisms will necessarily be solely focused on conservation, but they may have direct and/or indirect benefits for anguillids.

M. Gollock (✉)
Zoological Society of London, London, UK
e-mail: matthew.gollock@zsl.org

H. Shiraishi
Chuo University, Tokyo, Japan

K. Tsukamoto et al. (eds.), *Eel Science*, Fisheries Science Series,
https://doi.org/10.1007/978-981-99-5692-0_23

Ultimately, this review will not be exhaustive, but will focus on mechanisms in which the primary aim is to improve the status of anguillids.

In recent decades, it has been recognized that the status of a number of anguillids is of concern, and measures to address this are required (Righton et al. 2021). The suite of threats relating to the decline in anguillids includes barriers to upstream and downstream migration, hydropower and pumping station mortality, habitat loss and degradation, diseases and parasites, pollutants, unsustainable exploitation and trade, climate change, and predation. These vary depending on the species and location, and there may be more localized threats in addition to those listed above. Conservation of anguillids in the context of these threats has gained more attention as our understanding of the status of these species has improved. Arguably, this began after concerns were raised about the status of the European eel (*Anguilla anguilla*) in the late 1980s and early 1990s by the European Inland Fisheries Advisory Commission (EIFAC) and the International Council for the Exploration of the Sea (ICES), intergovernmental organizations that provide impartial advice on aquatic resources (ICES 1994). As a result, current conservation attention for the European eel is greater than that for the other anguillids. Indeed, to date, conservation measures have mostly been weighted towards temperate anguillids (i.e., *A. anguilla, A. australis, A. dieffenbachii, A. japonica,* and *A. rostrata*) in the Global North; however, many of these interventions will likely have applications for other species and locations (Righton et al. 2021). More recently, there have been studies in which interventions have been implemented with the conservation of tropical anguillids in mind (Shanmughan et al. 2022). Similarly, most conservation measures related to anguillids have been implemented in continental waters. This is due to the basic practicalities and our understanding of how threats impact the species; it is more realistic to mitigate the impacts of river barriers than changes in oceanic processes. It is also important to note that the prioritization of interventions is not always based on those that will maximize the benefit to anguillids and may also be influenced by factors such as economics and/or the will of interested parties.

23.2 Conservation Status

23.2.1 IUCN Red List of Threatened Species

There are a range of frameworks used to assess the conservation status of flora and fauna, but the International Union for Conservation of Nature (IUCN) Red List of Threatened Species is arguably the most well-known and utilized. This process relies on a set of categories and criteria with the aim of providing "…an explicit, objective framework for the classification of the broadest range of species according to their extinction risk" (IUCN 2012). The present framework aims to provide consistency across both taxa and users and facilitate an objective evaluation of the factors that affect extinction risk. This is extremely useful in the context of comparing the Red List status across species and prioritizing interventions; however, for those with

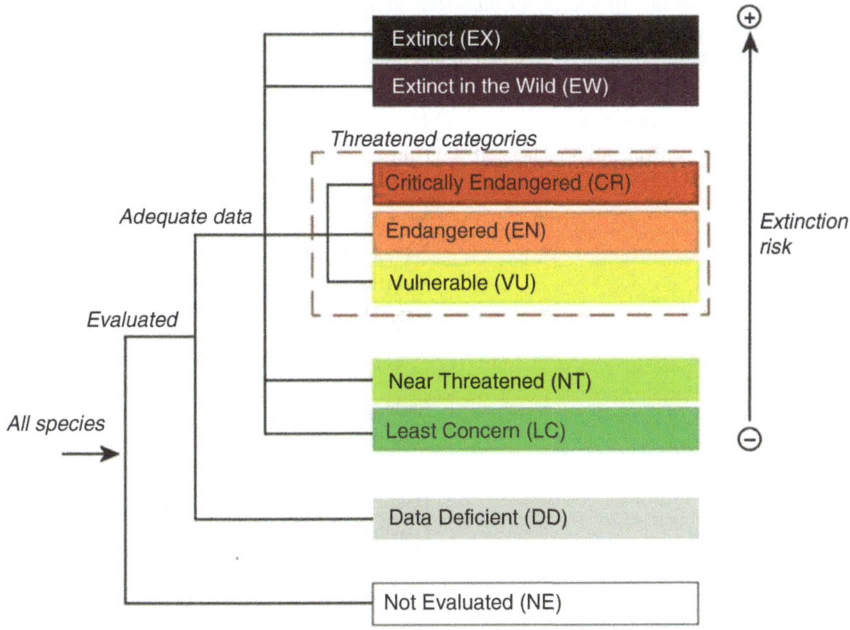

Fig. 23.1 IUCN Red List categories

complex life histories, such as anguillids, it can be challenging to apply the categories and criteria (Jacoby et al. 2015). Despite any limitations of the Red List, there is enormous value in its application, and it is globally utilized as a tool to prioritize conservation interventions by governments, NGOs, and researchers. It also helps prioritize conservation and research efforts within taxa; for example, there is arguably a greater need for measures relating to *A. anguilla* (Critically Endangered [CR]), compared to *A. reinhardtii* (Least Concern [LC]).

The IUCN Categories can be seen in Fig. 23.1. These are assigned using specific criteria that relate to changes in metrics, such as population size and geographic range, over a period of three generations or 10 years, whichever is longer, in the context of threats and conservation measures. Ultimately, the assigned category '...should be both precautionary and credible'. In the context of anguillids, temperate species are more likely to be assessed in a threatened category than tropical species (Table 23.1), which may well reflect the situation on the ground. However, temperate species are exploited in far greater numbers and are more likely to have long-term data series. As such, there is a dichotomy as to whether they are indeed at greater risk, or there is not the information to determine whether other species are similarly under threat. Most species have not changed category between assessments, and those that did were a result of better understanding these species, as opposed to a measurable change in status (Table 23.1).

Consequently, in addition to the 6 species of anguillids in the Threatened categories, there is considerable concern for the four listed as "Data Deficient".

Table 23.1 List of the 16 species of anguillids and the IUCN Red List Categories. Prior to 2013–2014, species were assessed *ad hoc*. Australasian species were not assessed in 2013–2014 and all Red List assessments were updated in 2019–2021 except *Anguilla rostrata* which will be published in 2023

Species	2008–2012	2013–2014	2019–2021
A. anguilla	Critically Endangered	Critically Endangered	Critically Endangered
A. australis			Near Threatened
A. bengalensis		Near Threatened	Near Threatened
A. bicolor	Least Concern	Near Threatened	Near Threatened
A. borneensis		Vulnerable	Vulnerable
A. celebesensis		Near Threatened	Data Deficient
A. dieffenbachii			Endangered
A. interioris		Data Deficient	Data Deficient
A. japonica		Endangered	Endangered
A. luzonensis		Near Threatened	Vulnerable
A. marmorata	Least Concern	Least Concern	Least Concern
A. megastoma		Data Deficient	Data Deficient
A. mossambica		Least Concern	Near Threatened
A. obscura		Data Deficient	Data Deficient
A. reinhardtii			Least Concern
A. rostrata		Endangered	

This category is applied when there is extremely limited information relating to the life history, ecology, population status, and/or potential threats. Eels are most frequently assessed under Criterion A (population decline), and assessments are most robust when they can draw on long-term datasets. Therefore, the establishment of monitoring programs for anguillids can be the most fundamental conservation measure.

23.2.2 IUCN Green Status of Species

While the Red List is a valuable tool for assessing a species' status in the context of threats and population abundance, it lacks quantitative metrics for evaluating the effectiveness of conservation measures and their linkage to recovery. As a result, in 2012 the IUCN began the development of the Green Status of Species to evaluate a species' progress towards recovery (Grace et al. 2021). This assessment involves evaluating a species' "recovery status," which can then be used to determine the historical and future impact of conservation measures.

Following testing of the Green Status across a range of taxa, with varied life histories, biomes, and Red List categories, it was observed that "species recovery is conceptually different from extinction risk and reinforces the utility of the IUCN Green Status of Species to more fully understand species conservation status" (Grace et al. 2021). For instance, the Green Status can identify counterfactual scenarios where the Red List status of a species remains unaltered in the future despite the implementation of interventions; however, without these measures, it would have continued to decline – this is known as "conservation dependence". Additionally, the Green Status assessment can help address issues relating to a shifting baseline. For example, data for both the European and American eels indicate that abundance is at a very low level compared to historic baselines. If this situation persists for an extended time, the 3-generation period over which the assessment is made will move away from the decline, towards the plateau, and as a result, the Red List status may eventually improve. Although this would appropriately reflect a reduction in extinction risk, it would fail to account for the historical depletion of the species.

At present, there are no published studies including Green Status assessments of anguillids, as this process is in its early stages. However, the assessment has been tested on European and Japanese eels (Gollock et al. unpublished data). Similar to the Red List, there are challenges to applying the criteria to anguillids, due to their life history, particularly the scale of some species' geographical range and panmictic breeding strategy. Nevertheless, the information garnered from Green Status assessments will undoubtedly complement that of the Red List to inform conservation interventions. Furthermore, these frameworks are extremely useful in determining the effectiveness of conservation measures at the population level.

23.3 Conservation Action

Mechanisms for identifying and/or implementing conservation measures that benefit anguillids can occur at multiple spatial scales. These can be global, regional, national, and sub-national and may all contribute to improving the status of anguillid species within and between range States, provided they are appropriately co-ordinated. However, it has been observed that this coordination varies among species and range States (Righton et al. 2021).

23.3.1 Global Frameworks

Currently, anguillid species are listed in 2 global conservation treaties. The Convention on International Trade in Endangered Species of Wild Fauna and Flora (CITES) is an agreement between governments that aims "to ensure that international trade in specimens of wild animals and plants does not threaten the survival of the species". While the focus is on international trade, there is a clear linkage to national exploitation. Species can be listed in 3 Appendices depending on the degree of protection needed. Trade in specimens of Appendix I species is permitted only under exceptional circumstances. Appendix II includes species in which "trade must be controlled in order to avoid utilization incompatible with their survival". The primary mechanism for trade in Appendix II is a determination known as a non-detriment finding (NDF), which sets parameters for which exports from a country will be considered sustainable. Appendix III "contains species that are protected in at least 1 country, which has asked other CITES Parties for assistance in controlling the trade". CITES is a legally binding convention and therefore national implementation laws are required to ensure that trade in protected species is legal, sustainable, and traceable.

Global eel production is mainly from farming, which accounted for 98% in 2019, with a significant proportion being concentrated in East Asia (CITES 2022). However, closed-cycle captive breeding is not yet commercially viable, and as a result, there is a global demand for juvenile eels as seed for farming. At the 14th meeting of the Conference of the Parties to CITES in 2007, the European eel was listed in Appendix II due to concerns relating to the species status, and the role unsustainable exploitation and associated trade played in this. This listing came into force in 2009, and in 2010, the European Union (EU) ceased import and export outside of the Member States as it was decided a NDF could not be produced due to the concerning conservation status of the species (CITES 2022). There is still a demand for European eel in key import and consumer markets in East Asia and this has been met through the opening and/or expansion of harvest and export in non-EU European eel range States. Additionally, there has been diversification and/or expansion of supply through increased exploitation and export of other species, particularly *A. rostrata* and *A. bicolor* (CITES 2022). Available data also suggest that the annual demand for glass eels is driven, at least in part, by the availability of Japanese eels (Fig. 23.2). There is evidence of significant illegal exploitation and trade of the European eel and other anguillids. As such, while only 1 species of anguillid is presently listed in CITES, it has been recognized that it is pragmatic to examine the genus as a whole, as there are clear linkages in the exploitation and trade of several species (CITES 2022). At the 74th CITES Standing Committee meeting in 2022, the Dominican Republic announced its intent to list *A. rostrata* in Appendix III, but no further information was available at the time of writing.

The inclusion of the European eel in CITES Appendix II highlights the complexities of global eel trade, but a number of interventions have proven beneficial to the species. There has been an increase in collaboration across range, transit and import countries to address both illegal fisheries and export of the European eel. For instance, Europol launched Operation Lake in 2015, a partnership among EU range States that led to a rise in seizures and arrests related to the illegal trade of the European eel. The sixth phase of Europol's Operation Lake VI (November 2021 to June 2022) involved law enforcement authorities from 24 countries and led to the arrest of 49 individuals and the confiscation of 1255 kg of glass eels. Outside of the EU, Canada reported having encountered incidents of illegal exports and/or transit of European eels, with investigations on-going (CITES 2022). While not directly linked to the CITES-listed European eel, similar efforts have been implemented for other species, such as Operation Broken Glass, which was established by the U.S. Fish and Wildlife Service to combat the trafficking of *A. rostrata* (Gollock et al. 2018). Additionally, scientific advances have supported the implementation of

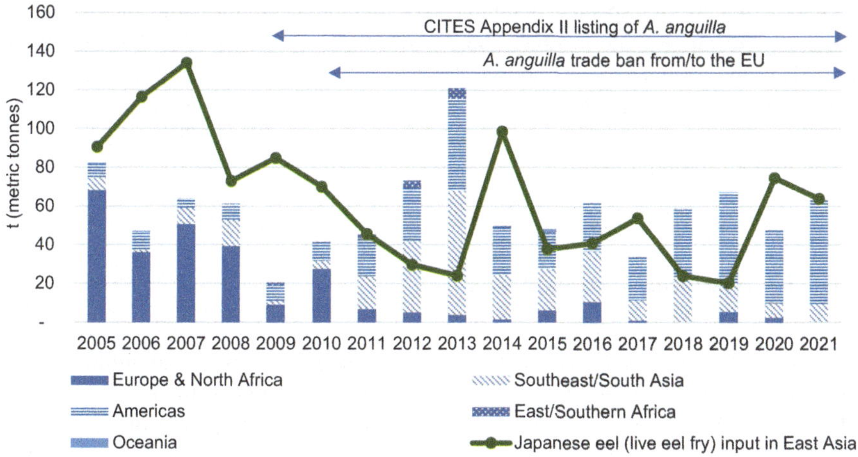

Fig. 23.2 Reported imports of live eel fry for farming (all sizes) into East Asia and *A. japonica* glass eel input, 2005–2020, by weight (t). Source: East Asian Customs (reported imports), Joint Statement, Annex 1 of the Joint Press Release, and Anon (2021b) (*A. japonica* live eel fry input). Note: **Europe and North Africa** (likely to be *A. anguilla*): Algeria, Belgium, Denmark, Egypt, France, Germany, Greece, Ireland, Italy, Morocco, the Netherlands, Romania, Spain, Tunisia, and the UK; **Americas** (likely to be *A. rostrata*): Canada, Cuba, Dominican Republic, Haiti, and the USA; **Southeast Asia** (likely to be and other tropical Anguilla species): Bangladesh, Timor Leste, Indonesia, Malaysia, Philippines, Singapore, Thailand, and Vietnam; **East/Southern Africa** (likely to be *A. mossambica* and other tropical species): Madagascar, Mauritius and South Africa; **Oceania** (likely to be *A. australis*): Australia

CITES, e.g. the use of rapid molecular screening to identify species and expedite enforcement decisions (Cardeñosa et al. 2019).

The Convention on the Conservation of Migratory Species of Wild Animals (CMS) provides a global platform that facilitates international collaboration of range States for the benefit of migratory animals and their habitats. Similar to CITES, CMS signatories identify species for listing on Appendices that provide guidance for the relevant interventions. Migratory species that are considered "endangered" are included in Appendix I of the Convention, and CMS Parties enforce strict protection measures relating to their habitats, migration routes, and associated threats. Species that "need or would significantly benefit from international co-operation," often through collaborative agreements, are listed in Appendix II. These agreements range from legally binding treaties to less formal instruments that are tailored to the needs of the species, range States, and conservation measures to be implemented.

In 2014, the European eel was listed in CMS Appendix II. The proposal suggested several "co-operative actions", such as convening knowledge sharing workshops to discuss coordinated conservation and management activities; strengthening collaboration between range States with contiguous coastlines and/or transboundary river basins; and drafting an agreement to recognize the importance of the

Sargasso Sea as the species potential breeding area. At the time of writing, 3 European eel range State meetings held under the auspices of CMS between 2016 and 2018, and a single species action plan was being drafted. CMS has the potential to be a valuable mechanism for the conservation of anguillids, as it can address the entire species range and all aspects of species recovery, unlike CITES, which focuses solely on international trade.

23.3.2 Regional Frameworks

Arguably, the most advanced regional mechanism focused on anguillids is EC Regulation 1100/2007 (European Council 2007), also known as the Eel Regulation. This came into force in 2007 to 'establish measures for the recovery of the stock of European eel.' The primary action from the Regulation was that all EU Member States with *A. anguilla* populations should develop Eel Management Plans (EMPs) to address declines of the species and associated threats (see Sect. 23.3.3 National Measures). EMPs were to include measures that addressed threats to the eel and/or potentially increase populations – these included reducing commercial and recreational fisheries, improving passage around barriers, implementing restocking and other eel movements, temporary hydropower cessation, habitat improvement, and/or reducing predation. It was left to the Member States to develop EMPs and associated measures to address issues encountered at the sub-national level. However, the overall aim of EMPs was "to reduce anthropogenic mortalities so as to permit with high probability the escapement to the sea of at least 40% of the silver eel biomass relative to the best estimate of escapement that would have existed if no anthropogenic influences had impacted the stock" (European Council 2007).

To date, 19 Member States have developed EMPs that cover almost 90 Eel Management Units (EMUs). There have now been 3 evaluations of progress in implementing EMPs by ICES, and an independent evaluation of the Regulation itself. Overall, the Regulation was considered to be an important mechanism to address sources of mortalities including fisheries and other anthropogenic activities, and ultimately improve the status of the European eel. However, there were significant challenges to implementation, and recovery of the species needed to be considered on a decadal scale (European Commission, Directorate-General for Maritime Affairs and Fisheries 2019). Indeed, it was stated that "the direct environmental benefits of the EMP actions (e.g. eel stock recovery through reduced fishing mortality and increased spawner escapement) and the indirect environmental benefits (reconnected coastal, estuarine and riverine ecosystems) are long-term in nature, still nascent and yet to be fully quantified in terms of their environmental benefits" (European Commission, Directorate-General for Maritime Affairs and Fisheries 2019). In assessing the EMPs and associated measures, it was clear that there was variation across the Member States, and gaps in the data meant determining their effectiveness was challenging (ICES 2022a). Ultimately, the 40% escapement target is presently being achieved in only 10 of 53 reporting EMUs, but these are not consistently calculated, which makes comparison between values complicated (ICES 2022a). In the evaluation of both the Regulation and EMPs, it has been proposed that including mortality metrics could be more useful for directly assessing the effectiveness of implemented measures.

An overlapping initiative of the EU Regulation is the General Fisheries Commission for the Mediterranean (GFCM) – an organization focused on conservation and sustainable use of marine resources – Recommendation GFCM/42/2018/1 on a multiannual management plan for European eel in the Mediterranean. The primary aim of the Recommendation was "to establish a multiannual management plan for the fisheries catching European eel in the Mediterranean Sea in line with the precautionary approach to provide and maintain high long-term yields and to guarantee a low risk of stock collapse". The GFCM membership includes both EU and non-EU States and as such it is complementary to the implementation of EC Regulation 1100/2007. The most recent meeting focusing on this Recommendation (WGMEASURES-EEL), held in February 2022, appraised the GFCM research programme on European eel, and the effectiveness of existing and proposed management measures. The research programme had collected and analyzed data from across the Mediterranean range to inform implementation of future measures and recognized "...the importance of being able to assess and monitor the effectiveness of implemented management measures in terms of reductions in mortality..." (GFCM 2022).

Aligned with the Regulation and associated EMPs, and the GFCM Recommendation, is the annual Advice from ICES, informed by the meeting and report of the EIFAAC/ICES/GFCM Working Group on Eel (WGEEL). The EU, the UK and others specifically request the Advice in the context of the status of the European eel to inform how management measures are implemented. This Advice is based on analysis of data submitted by range States; at present, the most robust metric is an index of juvenile recruitment (Fig. 23.3), though yellow and silver series are being developed. The most recent Advice in 2022 stated that recruitment was, depending on the location, 0.5–9.7% (provisional) of the 1960–1979 baseline (ICES 2022b). This resulted in the Advice stating "when the precautionary approach is applied, there should be zero catches in all habitats in 2023. This applies to both recreational and commercial catches and includes catches of glass eels for restocking and aquaculture". It was also stated that "...based on ecosystem-based management considerations..." that "...all non-fisheries related anthropogenic mortalities should be zero..." and "...the quantity and quality of eel habitats should be restored; this includes restoring connectivity and the physical, chemical, and biological properties of the habitats...".

This is obviously important in the context of fisheries measures being implemented as part of EMPs and the scale of interventions for other forms of anthropogenic mortalities. The ICES data series serve as valuable resources in identifying conservation interventions across the entire species range, including filling knowledge and monitoring gaps.

Apart from the measures relating to the European eel, a number of initiatives exist aiming to improve the status of other anguillids. The non-binding "Joint Statement"

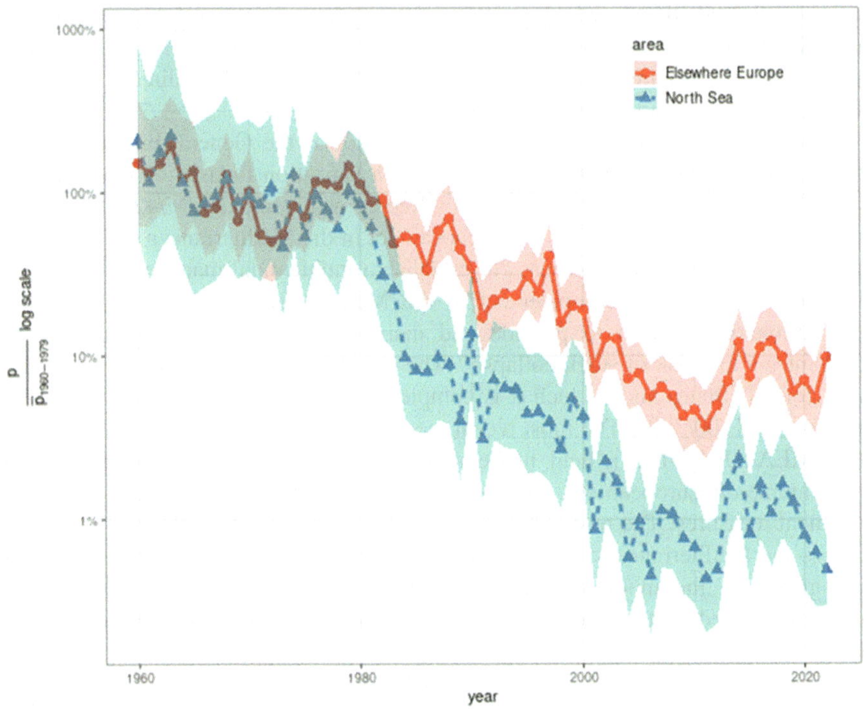

Fig. 23.3 WGEEL glass eel recruitment index for the continental North Sea and Elsewhere Europe series with 95% confidence intervals updated to 2022 (ICES 2022b). The index was estimated using a GLM (glass eel ~ area: year + site) fitted on 57 time-series comprising either pure glass eel or a mixture of glass eels and yellow eels. Note the logarithmic scale on the y-axis. Number of series Elsewhere Europe = 31, North Sea = 26

relating to "International Cooperation for Conservation and Management of Japanese Eel Stock and Other Relevant Eel Species" was developed in 2014 by China, Japan, the Republic of Korea, and Taiwan (Anon 2014). The statement focuses on exploitation, aquaculture input of juveniles, and trade (CITES 2022). While input limits for juvenile eel (*A. japonica* and other *Anguilla* spp.) into farms have been established and agreed upon annually, they have not changed since 2014, at the time of writing (CITES 2022). In April 2022, the first scientific meeting on the Japanese eel was held to share knowledge and experiences and provide advice on conservation and management measures for the species (see Chap. 19).

Regarding the American eel (*A. rostrata*), there are currently no trans-boundary mechanisms in place. However, in 2018, a workshop was held in the Dominican Republic to convene range States and a report was produced including recommendations as how co-operation could progress. A follow-up to this workshop was held in 2021 amongst harvesting range States with a view to developing a roadmap to improve coordination and collaboration. Further, a new ICES Working Group on American Eel (WGAMEEL) was convened for the first time in 2022 (ICES 2023).

WGAMEEL initially includes only Canada and the U.S. with the intention to involve more countries if the WG becomes established in the long term.

In Southeast Asia, the Southeast Asian Fisheries Development Center (SEAFDEC) established a project entitled "Enhancing sustainable utilization and management scheme of tropical anguillid resources in Southeast Asia," to strengthen the management framework related to these species in Cambodia, Indonesia, Myanmar, the Philippines, Thailand, and Vietnam from 2017 to 2019. The aims were to carry out baseline surveys on national utilization, assess species abundance and genetic structure, improve culture techniques, and develop policy recommendations (Southeast Asian Fisheries Development Center 2019). It was determined that *A. bicolor* and *A. marmorata* were the primary catch species, but the available data could not accurately assess any changes in abundance. Regardless, a number of policy and management recommendations were proposed relating to developing catch monitoring data collection systems, carrying out stock assessments, and implementing measures to ensure sustainable use.

23.3.3 National Measures

As previously stated, many national measures to improve the status of anguillids may occur as a result of global and/or regional frameworks. However, there are also complementary efforts that occur within range States outside of these mechanisms as a result of stakeholder efforts.

There are a range of measures delivered at the national level for the European eel across its range under the auspices of EC Regulation 1100/2007. The 2022 evaluation of EMP measures highlights the variation in those implemented within Member States (ICES 2022a). However, in all countries, measures related to commercial fisheries were by far the most common (~33% of all reported) (ICES 2022a). Indeed, in recent years, the EU has begun implementing closed periods for eel fishing in marine and adjacent brackish waters. In 2022, these were for 3 consecutive months, but at the time of writing, the EU had proposed to extend the closure from 3 consecutive months to a more flexible 6-month period to "*cover the main migration period of* both glass eel and silver eel". The next most common measures related to hydropower and obstacles (17% of all reported) which are recognized as causing both mortality and reduced habitat availability. These measures have synergies with those implemented for the European eel in the GFCM countries.

Outside of these devolved regional frameworks, there are a number of examples in which national-level organizations, both government and non-government, are implementing actions for the benefit of anguillids. As previously stated, the establishment of abundance monitoring is a hugely valuable activity, as it allows data to be used to inform the prioritization of measures. In recent years, eel monitoring programmes for citizen scientists have been established; this is a valuable outreach method that can also be used to fill data gaps (Pecorelli et al. 2019). In the Philippines, community-based protected areas have been established using anguillids as a flagship species, and fisheries-independent data collection within these is shared with government agencies (DEFRA 2021). Indeed, eels have been proposed as flagship species for aquatic systems in Japan (Itakura et al. 2020).

Regarding fisheries and trade, the prohibition of catching silver eels has been introduced in almost all prefectures in Japan where they naturally occur, in order to maximize spawner escapement (CITES 2022). The Philippines and Indonesia have also developed national policy that state minimum size metrics of anguillids for export (Gollock et al. 2018). In addition to commercial fisheries, there are a number of indigenous groups that have ceremonial and spiritual use for anguillids and are amending these in response to the status of certain species. In New Zealand, Māori people have implemented management practices, called "rāhui," which involve imposing fishing bans and seeding areas with small eels (Jellyman 2012). In 2008, indigenous peoples assembled in Ottawa, Canada, expressed concerns about the decline of the American eel, and prepared an Aboriginal People's American Eel Resolution (Gollock et al. 2018).

A number of countries are aiming to address the impacts of hydropower and pumping stations and barriers to migration. In New Zealand, for example, assisted upstream elver passage and downstream "trap and transport" of silver eels over hydropower structures has been occurring for decades (Boubée et al. 2008). These methods are also frequently used across European and American eel range States, and in many cases are now a requirement of power production and/or establishment of any new structures. Dam removal and installation of eel ladders are common in the US, with over 1200 structures having been removed since 1999 (American Rivers 2019). Japan has also made efforts to ease barriers using a variety of methods in recent years (Ministry of Environment 2017).

23.4 Conclusions

Eel conservation occurs in many forms, be it the collection of monitoring data, easing barriers to migration, reducing anthropogenic mortalities or implementing policy that benefits these species. These efforts will have varying levels of effectiveness, but considering the complexity of the eel life history and how potential threats might impact them cumulatively and synergistically, they will undoubtedly have a greater impact when co-ordinated at domestic and international levels. In reality, there are many cases where this does not occur, and efforts may be skewed to where there is greater resource availability and/or stakeholder will (Righton et al. 2021). For example, in the case of the European eel, despite increasing efforts within Mediterranean range States, our understanding of the species and associated interventions are weighted to the northern portion of the range (ICES 2022b). Ultimately, regional mechanisms to conserve anguillids are limited and presently absent in South Asia, East Africa, and Oceania.

Consequently, there are significant gaps in conservation capacity, both within and between species ranges, and there are undoubtedly opportunities to share learnings on successes and challenges. Increased collaboration between range States will catalyze knowledge sharing, and in a number of cases, trans-boundary measures could be hugely beneficial for freshwater biodiversity more broadly. More fundamentally, there is a need for long-term monitoring of anguillid populations, and also assessments of conservation interventions to determine their effectiveness. Long-term datasets are available for only a few species and for limited parts of the range. As such, it is essential that monitoring is established to address these knowledge gaps to better inform conservation, management and policy.

References

American Rivers (2019) American rivers dam removal database. figshare. Dataset. https://doi.org/10.6084/m9.figshare.5234068.v10

Anon (2014) Joint statement of the Bureau of Fisheries of People's Republic of China, the Fisheries Agency of Japan, the Ministry of Oceans and Fisheries of the Republic of Korea and the Fisheries Agency of Chinese Taipei on international cooperation for conservation and management of Japanese eel stock and other relevant eel species. https://www.jfa.maff.go.jp/j/saibai/pdf/140917jointstatement.pdf

Boubée J, Jellyman D, Sinclair C (2008) Eel protection measures within the Manapouri hydro-electric power scheme, South Island, New Zealand. Hydrobiologia 609:71–82. https://doi.org/10.1007/s10750-008-9400-6

Cardeñosa D, Gollock MJ, Chapman DD (2019) Development and application of a novel real-time polymerase chain reaction assay to detect illegal trade of the European eel (*Anguilla anguilla*). Conserv Sci Pract 1:e39. https://doi.org/10.1111/csp2.39

CITES (2022) Status of use and trade of anguillid eels. https://cites.org/sites/default/files/eng/com/sc/74/E-SC74-64-01.pdf

DEFRA (2021) Final report - sustainable community-based stewardship of freshwater resources in the Northern Philippines. https://www.darwininitiative.org.uk/documents/DAR24016/25935/24-016%20FR%20-%20Edited.pdf

Engler-Palma C, VanderZwaag DL, Apostle R, Castonguay M, Dodson JJ, Feltes E, Norchi C, White R (2013) Sustaining American eels: a slippery species for science and governance. J Int Wildlife Law Policy 16:128–169. https://doi.org/10.1080/13880292.2013.805060

European Commission, Directorate-General for Maritime Affairs and Fisheries (2019) Evaluation of the eel regulation: final report, Publications Office. https://data.europa.eu/doi/10.2771/679816

European Council (2007) Council regulation (EC) no 1100/2007 of September 18, 2007 establishing measures for the recovery of the stock of European eel. Off J Eur Union 248:17–23

GFCM (2022) Working group on the management of European eel (WGMEASURES-EEL). GFCM, Rome

Gollock M, Shiraishi H, Carrizo S, Crook V, Levy E (2018) Status of non-CITES listed anguillid eels. https://cites.org/sites/default/files/eng/com/ac/30/E-AC30-18-01-A2.pdf

Grace MK, Akçakaya HR, Bennett EL, Brooks TM, Heath A, Hedges S et al (2021) Testing a global standard for quantifying species recovery and assessing conservation impact. Conserv Biol 35:1833–1849. https://doi.org/10.1111/cobi.13756

ICES (1994) Report of the joint ICES/EIFAC working group on eel (WGEEL). ICES Expert Group reports. https://doi.org/10.17895/ices.pub.19262471.v1

ICES (2016) Report of the workshop on eel stocking (WKSTOCKEEL). ICES, Toomebridge. https://doi.org/10.17895/ices.pub.8436

ICES (2022a) Workshop for the technical evaluation of EU member states' progress reports for submission in 2021 (WKEMP3). ICES Sci Rep 4:41. https://doi.org/10.17895/ices.pub.19768585

ICES (2022b) European eel (*Anguilla anguilla*) throughout its natural range. In Report of the ICES Advisory Committee, 2022. ICES Advice 2022, ele.2737.nea, https://doi.org/10.17895/ices.advice.19772374

ICES (2023) Working Group on American Eel (WGAMEEL; outputs from 2022 meeting). ICES Scientific Reports. https://doi.org/10.17895/ices.pub.21762842

Itakura H, Wakiya R, Gollock M, Kaifu K (2020) Anguillid eels as a surrogate species for conservation of freshwater biodiversity in Japan. Sci Rep 10:8790. https://doi.org/10.1038/s41598-020-65883-4

IUCN (2012) IUCN red list categories and criteria: version 3.1. Second edition. IUCN, Gland, Switzerland and Cambridge, UK

Jacoby DM, Casselman JM, Crook V, DeLucia MB, Ahn H, Kaifu K, Kurwie T, Sasal P, Silfvergrip AMC, Smith KG, Uchida K, Walker AM, Gollock MJ (2015) Synergistic patterns of threat and the challenges facing global anguillid eel conservation. Glob Ecol Conserv 4:321–333. https://doi.org/10.1016/j.gecco.2015.07.009

Jellyman DJ (2012) The status of longfin eels in New Zealand - an overview of stocks and harvest. National Institute of Water and Atmospheric Research, Christchurch

Ministry of Environment (2017) Concept of habitat conservation for Japanese eels. https://www.env.go.jp/content/900491527.pdf; in Japanese

Pecorelli J, Macphie K, Hebditch C, Clifton-Dey D, Thornhill I, Debney A (2019) Using citizen science to improve the conservation of the European Eel (*Anguilla anguilla*) in the Thames River Basin District. Freshw Sci 38:281–291. https://doi.org/10.1086/703398

Righton D, Piper A, Aarestrup K, Amilhat E, Belpaire C, Casselman J, Castonguay M, Díaz E, Dörner H, Faliex E, Feunteun E, Fukuda N, Hanel R, Hanzen C, Jellyman C, Kaifu K, McCarthy K, Miller MJ, Pratt T, Sasal P, Schabetsberger R, Shiraishi H, Simon G, Sjöberg N, Steele K, Tsukamoto K, Walker A, Westerberg H, Yokouchi K, Gollock M (2021) Important questions to progress science and sustainable management of anguillid eels. Fish Fish 22:762–788. https://doi.org/10.1111/faf.12549

Sandbrook C (2015) What is conservation? Oryx 49:565–566. https://doi.org/10.1017/S0030605315000952

Shanmughan A, Dahanukar N, Harrison A, Pinder AC, Ranjeet K, Raghavan R (2022) Demographics and exploitation of two near threatened freshwater eels, *Anguilla bengalensis* and *Anguilla bicolor*, in small-scale subsistence fisheries and implications for conservation. Aquat Conserv Mar Freshw Ecosyst 32:269–281. https://doi.org/10.1002/aqc.3765

Southeast Asian Fisheries Development Center (2019) Enhancing sustainable utilization and management scheme of tropical Anguillid eel resources in Southeast Asia. SEAFDEC, Bangkok

Correction to: Reproduction

Shigeho Ijiri, Hajime Matsubara, Moemi Horiuchi, and Yukinori Kazeto

Correction to:
Chapter 12 in: K. Tsukamoto et al. (eds.),
***Eel Science*, Fisheries Science Series,**
https://doi.org/10.1007/978-981-99-5692-0_12

The original version of this chapter was inadvertently published with an error. Figure 12.6 was published with incorrect labels. The figure has been replaced with this correction.

The updated version of this chapter can be found at
https://doi.org/10.1007/978-981-99-5692-0_12

K. Tsukamoto et al. (eds.), *Eel Science*, Fisheries Science Series,
https://doi.org/10.1007/978-981-99-5692-0_24

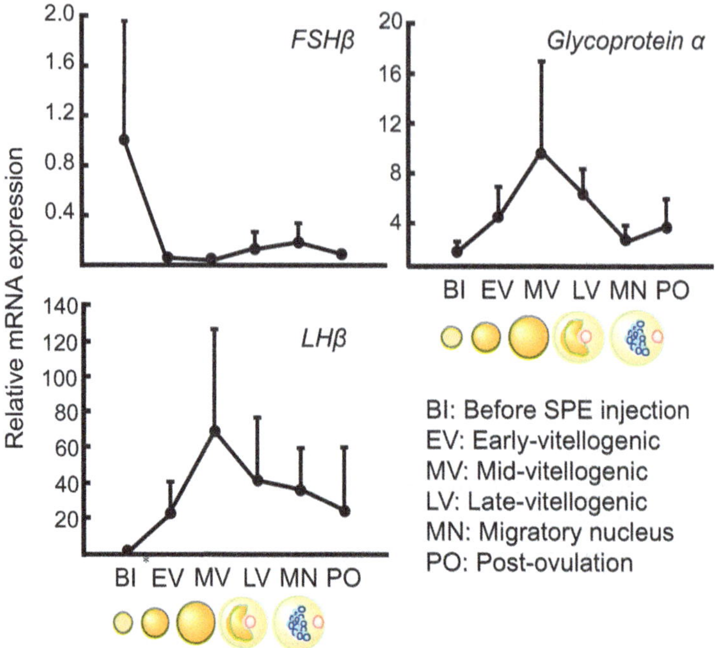

Fig. 12.6 Changes in mRNA levels of FSHβ, LHβ, and glycoprotein α in pituitaries during induced sexual maturation in the eel

SPRINGER NATURE

GPSR Compliance

The European Union's (EU) General Product Safety Regulation (GPSR) is a set of rules that requires consumer products to be safe and our obligations to ensure this.

If you have any concerns about our products, you can contact us on ProductSafety@springernature.com

In case Publisher is established outside the EU, the EU authorized representative is:

Springer Nature Customer Service Center GmbH
Europaplatz 3
69115 Heidelberg, Germany

The manufacturer's authorised representative in the EU is Springer Nature Customer Service Centre GmbH, Europaplatz 3, 69115 Heidelberg, Germany. If you have any concerns regarding our products, please contact ProductSafety@springernature.com

Printed and bound by CPI Group (UK) Ltd, Croydon, CR0 4YY
30/09/2026
02225322-0011